D0773549

DEADLY CULTURES

Mark Wheelis, Lajos Rózsa,
and Malcolm Dando, Editors

Deadly Cultures

Biological Weapons since 1945

HARVARD UNIVERSITY PRESS
Cambridge, Massachusetts, and London, England
2006

Library of Congress Cataloging-in-Publication Data

Deadly cultures : biological weapons since 1945 / Mark Wheelis, Lajos Rózsa,
   and Malcolm Dando, editors.
      p. cm.
   Includes bibliographical references and index.
   ISBN 0-674-01699-8 (cloth : alk. paper)
   1. Biological weapons—Testing.   2. Biological weapons—Research.
I. Wheelis, Mark.   II. Rózsa, Lajos, 1961–   III. Dando, Malcolm.
UG447.8.D43 2005
358'.3882'09—dc22   2005050225

Contents

Preface

This book originated when two of us (Malcolm Dando and Mark Wheelis) were at a meeting of the Pugwash Conferences on Science and World Affairs in Geneva, whose topic was strengthening the biological disarmament regime. We lamented the fact that there was no recent scholarly history, based on primary sources, of offensive biological weapons (BW) programs. How, we wondered, could effective policy be generated if there was widespread ignorance of BW developments in recent decades? There were good recent sources for BW history up to 1945, and older sources for the period until the early 1970s, but the critical period since then—which saw the negotiation of the Biological Weapons Convention, the discovery of the illegal Soviet, Iraqi, and South African BW programs, the arduous and ultimately unsuccessful attempts to negotiate an inspection regime for BW, and resurgent fears of bioterrorism—was discussed only in popular books based on secondary or incomplete sources. We resolved at that time to fill this obvious gap.

Shortly thereafter, at a NATO Science Program Advanced Study Institute in Budapest, we met Lajos Rózsa, and the three of us applied for a grant from the same NATO program to fund meetings of the authors selected to write the chapters of the book, and to provide some support for the archival research that would be necessary. That grant (977940) allowed us to hold two sets of meetings: one at the beginning of the project, when authors presented outlines of their chapters; and one near the end, when first drafts were available. One of the authors, Nicholas Sims, did not participate in these NATO-funded meetings and did not accept any NATO research funds, for reasons of conscience.

In addition to the authors, two senior scholars in the field—Julian

Perry Robinson and Graham Pearson—agreed to comment on the drafts, and both have been actively involved in advising authors upon each chapter and in collaborating with the editors in writing the final chapter. The contributions of these two scholars have greatly strengthened the entire book; any remaining weaknesses persist despite their efforts.

An additional grant from the Carnegie Corporation of New York (D03044) provided financial assistance for the authors' research, the authors' final meeting, and the editorial work. We are grateful to both NATO and Carnegie for their support, which has been critical to producing a book this complex.

We are also most grateful to all those who have reviewed and commented on draft chapters; their contributions have greatly improved the quality of the book. In addition to Robinson and Pearson, all authors read all chapters and made constructive comments (even when they may have disagreed with particular points). Martin Hugh-Jones read the entire manuscript and made very helpful comments, as did two anonymous reviewers selected by Harvard University Press. Ann Hawthorne's editing improved the writing significantly and helped to unify the disparate chapters. Catherine Rhodes provided organizational support that assisted the editors greatly. Thanks to Claudia Graham, of UC Davis Mediaworks, who drew all the figures. Michael Fisher, Executive Editor for Science and Medicine (now Editor in Chief) at Harvard University Press, has been actively involved in the project from early on, and we are grateful for his support and advice.

We hope that this book will assist in the crucial public discussions and state-level policy decisions that must take place in coming years if the renewed development and potential use of biological weapons are to be prevented.

Abbreviations

ACDA	Arms Control and Disarmament Agency
ADAC	Agricultural Defence Advisory Committee
AHG	Ad Hoc Group of Governmental Experts
ARMET	Armement et Etudes of the EMA
ASF	African swine fever
BG	*Bacillus globigii*
Biopreparat	Main Directorate for Biological Preparations
BRAB	Biological Research Advisory Board
BW	biological weapons; in quoted material may stand for biological warfare
BWC	Biological Weapons Convention
BWS	Biological Warfare Sub-Committee
BZ	3-quinuclidiny benzilate
CASDN	Comité d'Action Scientifique de la Défense Nationale
CB	chemical and biological
CBM	confidence-building measure
CBR	chemical, biological, and radiological
CBW	chemical and biological weapons; in quoted material may stand for chemical and biological warfare
CCD	Conference of the Committee on Disarmament
CD	Conference on Disarmament
CDAB	Chemical Defence Advisory Board
CDC	Centers for Disease Control and Prevention
CDEE	Chemical Defence Experimental Establishment
CEB	Centre d'Etudes du Bouchet
CEECB	Commission des Etudes et Expérimentations Chimiques et Bactériologiques
CIA	Central Intelligence Agency
CIAS	Commandement Interarmées des Armes Spéciales
CIEECB	Commission Interarmées d'Etudes et d'Expérimentations Chimique et Bactériologique
CINBC	Comité Interarmées du NBC

CmlC	US Army Chemical Corps
CRSSA	Centre de Recherches du Service de Santé des Armées
CW	chemical weapons; in quoted material may stand for chemical warfare
CWC	Chemical Weapons Convention
CWS	Chemical Warfare Service
DEA	Department of External Affairs
DERA	Defence Evaluation and Research Agency
DIA	Defense Intelligence Agency
DMA	Délégation Ministérielle à l'Armement
DND	Department of National Defence
DOD	Department of Defense
DRB	Defence Research Board
DRC	Defence Research Committee
DREO	Defence Research Establishment Ottawa/Shirley's Bay
DRES	Defence Research Establishment Suffield
DRPC	Defence Research Policy Committee
DRPS	DRPC Staff
EMA	Etat-Major des Armées
ENDC	Eighteen Nation Committee on Disarmament
FBI	Federal Bureau of Investigation
FFCD	full, final, and complete declarations
FMD	foot and mouth disease
FOIA	Freedom of Information Act
4P	prophylactic, protective, or other peaceful purposes
GABA	gamma-aminobutyric acid
GDR	German Democratic Republic
GML	General Medical Laboratory
ICRC	International Committee of the Red Cross
ISC	International Scientific Commission
ISG	Iraq Survey Group
ISSBW	Inter-Services Sub-Committee on Biological Warfare
JCS	Joint Chiefs of Staff
KGB	Committee for State Security
LAC	large area concept
LMRV	Laboratoire Militaire de Recherches Vétérinaires
LSD	lysergic acid diethylamide
LTBT	Treaty Banning Nuclear Weapon Tests in the Atmosphere, in Outer Space and Under Water (Partial, or Limited, Test-Ban Treaty)
MOD	Ministry of Defense
MRD	Microbiological Research Department
MRE	Microbiological Research Establishment

MSE	Al Muthanna State Establishment
NARA	National Archives Research Administration, College Park, Md.
NATO	North Atlantic Treaty Organization
NBC	nuclear, biological, and chemical
ND	Newcastle disease
NGO	nongovernmental organization
NPO	military-scientific production facility
NPT	Treaty on the Non-Proliferation of Nuclear Weapons (Non-Proliferation Treaty)
NSC	National Security Council
POW	prisoner of war
PRO	Public Records Office, London
R&D	research and development
RRL	Roodeplaat Research Laboratories
SADF	South African Defence Force
SALT	Strategic Arms Limitation Treaty
SBVA	Service Biologique et Vétérinaire des Armées
SGA	Special Group Augmented
SGTEB	Sous Groupe de Travail et d'Etudes Biologiques
SHAD	Shipboard Hazard and Defense
SHAPE	NATO Supreme Headquarters Allied Powers Europe
SHAT	Service Historique de l'Armée de Terre
SIPRI	Stockholm International Peace Research Institute
SOD	Special Operations Division of the US Army Biological Laboratory at Fort Detrick
STA	Service Technique de l'Armée
THC	tetrahydrocannabinol
TRC	Technical Research Center; or Truth and Reconciliation Commission
UN	United Nations
UNMOVIC	United Nations Monitoring, Verification and Inspection Commission
UNSCOM	United Nations Special Commission
USGPO	US Government Printing Office
VEE	Venezuelan equine encephalomyelitis (sometimes encephalitis)
WMD	weapons of mass destruction
WP	Warsaw Pact Treaty

DEADLY CULTURES

CHAPTER 1

Historical Context
and Overview

MARK WHEELIS,
LAJOS RÓZSA,
AND MALCOLM DANDO

The threat of biological weapons (BW) use by either states or terrorists has never attracted so much public attention as in the past five years, when heads of state and other governmental leaders have spoken repeatedly about the dangers from weapons of mass destruction and the threat of bioterrorism. BW proliferation among states has been a concern since before the end of the Cold War, and concerns about BW in terrorist hands became a prominent issue in the early 1990s. The US anthrax letter attacks of 2001 seemed to validate these concerns, and intelligence suggesting interest in BW among international terrorist organizations has exacerbated them, especially in the US. Currently the threat of BW attracts keen international attention at the highest levels.

These contemporary concerns relate largely to the threat of BW acquisition and use by rogue states or by terrorists. However, the BW threat has much deeper roots, and it has changed markedly over the past 60 years. During most of the Cold War period, major global powers invested substantial resources to develop a strategic BW capability aimed at the military forces, civilian populations, or agricultural resources of their adversaries. Indeed, early in this period BW were considered to rival nuclear weapons in strategic importance.

Despite the shifting view of the nature of the BW threat, it has been evident for over 60 years that biological agents can be used to cause mass casualties and large-scale economic damage. However, BW are not well understood, and there has been little historical analysis—and hence little

appreciation—of the various ways in which such weapons have been regarded over those decades. The following chapters attempt to fill this gap,
by providing a concise and accurate history of offensive BW programs
since 1945, and by drawing possible lessons from that history with regard
to strengthening the long-standing total prohibition of BW. Whenever
possible we use primary sources to provide an accurate account of why
some countries sought biological weapons and why some abandoned
such programs. Because primary sources are not uniformly available for
all countries and periods, however, there is inevitable variation in the
depth of coverage.

Despite these limitations, there is an immense amount of material

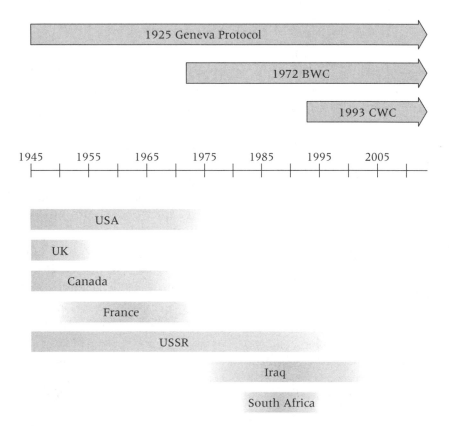

Figure 1.1 Major arms control and disarmament treaties limiting BW (top)
and known offensive BW programs (bottom) since 1945.

available. Multivolume works could easily be written on several of our topics. Consequently, our aims of brevity and comprehensiveness are necessarily often in conflict. Each chapter cites reliable secondary sources, when available, to which the interested reader is referred for additional material.

Our reliance on verifiable primary sources has another consequence: we cannot treat in any detail events in the very recent past for which documentation is not yet available. For instance, the failed attempt to negotiate a verification protocol for the Biological Weapons Convention (BWC), the extensive reorganization of the US government following the 2001 terrorist attacks, and the nature of current US biodefense programs are all topics for which much of the primary-source documentation remains classified. Until these documents are declassified, these important topics will not be amenable to rigorous historical treatment and must receive only passing comment here.

Biological Weapons before 1945

The history of biological weapons before 1945 has been addressed in several monographs. Most important are *The Problem of Chemical and Biological Warfare*, produced in six volumes by the Stockholm International Peace Research Institute (SIPRI), which covers the period until approximately 1970; and *Biological and Toxin Weapons Research, Development and Use from the Middle Ages to 1945*, edited by Erhard Geissler and John Ellis van Courtland Moon.[1] Our volume is conceived as a sequel to the latter, but it owes much to the earlier SIPRI series as well.

Geissler and Moon's volume covers BW from the Middle Ages to World War I, BW use in World War I, BW programs from the 1920s to 1945 in a number of individual states, and BW use in World War II. The resulting picture shows that after the "Golden Age" of bacteriology at the end of the nineteenth century, when scientists first unraveled the causes of infectious diseases, the military applications of this knowledge intrigued several countries, some of which initiated offensive programs. These began with efforts by Germany in World War I (and by France on a much more limited scale) to attack military draft animals covertly with the diseases anthrax and glanders. Following the war, there was widespread speculation in the press and in military circles that the next major conflict would involve extensive use of chemical weapons (CW) and BW. Because of these fears, many countries started to develop BW as a deterrent and for retaliation, if not for first use. These included Canada, France, Germany, Hungary, Italy, Japan, the Soviet Union, the UK, and the US. Thus most belligerents entered World War II with at least exploratory BW programs, and most (with the notable exception of Germany) increased their activities significantly during the war.

Despite these efforts, no European or North American country other than the UK mass-produced a usable BW during the war, and in the case of the UK, the weapon was unsophisticated—five million cattle cakes of linseed meal laced with anthrax spores. The intent was to disseminate the cattle cakes into German fields through the flare chutes of bombers, and thereby to cripple German domestic animal production, in retaliation if the Germans used such unconventional weapons against the Allies. The Germans did not use unconventional weapons, the cattle cakes remained unused, and the stockpile was destroyed after the war. Neither CW nor BW were used in combat in the European theater.

In Asia, however, the situation was different. Japan made extensive use of BW (and CW) against China, targeting both troops and civilians. The methods were unsophisticated, involving the release of live plague-infected rats or fleas and the contamination of wells or foodstuffs with agents of intestinal disease. Despite the unsophisticated methods, tens or hundreds of thousands of Chinese are thought to have been killed by Japanese BW, including thousands used as human guinea pigs for infectious-disease experiments.

In addition to marking the start of a major BW arms race, the period before 1945 saw the first major international effort to ban biological warfare. The Geneva Protocol, signed in 1925 (entered into force in 1928), banned the use of both BW and CW in warfare. This prohibition is regarded as having entered the realm of customary law, binding on all states regardless of whether they are parties to the Protocol.

Biological Weapons after 1945

This book begins in 1945, picking up the narratives where they left off in Geissler and Moon. The world was in a very fluid state then. A global war had just been fought, killing millions, leaving millions homeless, and wrecking the economies of most nations of the world. Recovery would require years. Nuclear weapons had been used for the first time, with devastating effectiveness, and the major powers were pursuing their acquisition. The United Nations was being formed, with grand hopes of international organizations that might make war obsolete. The emerging Cold War would soon transform much of international relations into a hall of mirrors where nothing was as it appeared. The entire conceptual fabric of arms limitation, deterrence, retaliation, and war fighting was changing rapidly.

In this context, many of the victorious powers emerged from the war with active BW development programs. Although these programs had not produced usable military weapons (apart from the UK cattle cakes), they had suggested sufficient promise to make continued pursuit of BW attractive. Chapters 2 through 6 examine the programs in these countries: the US, the UK, Canada, the Soviet Union, and France, whose BW program was terminated by the German occupation at the beginning of the war but resumed shortly after the war's end. The offensive BW programs in these countries are traced from the flux of the post–World War II period until program termination. Major themes are the tripartite cooperative arrangements among the US, UK, and Canada; the changing role of BW vis-à-vis nuclear weapons; and the reasons for continuation, and then termination, of the programs.

All the principal BW powers of the immediate postwar period eventually discontinued their programs. The UK ended its program in the 1950s with a quiet and gradual shift from offensive to defensive work. The US very publicly and unexpectedly renounced offensive BW in 1969. Canada, which had never had an independent offensive BW program, but which had close collaborative arrangements with the US and UK programs, pledged in 1969 to discontinue cooperative offensive research and development. Russia, which inherited the Soviet Union's offensive program, apparently ended it in the early 1990s, although concerns remain about residual activities in military microbiology facilities that are still closed to outsiders.

Several states pursued an offensive BW capability beginning well after World War II. Two of these are discussed here (Chapters 8 and 9): Iraq and South Africa. Iraq's program mostly pursued a battlefield military capability, whereas South Africa's was designed to develop weapons for assassination and covert use. Both programs ended in the 1990s—Iraq's with defeat in the Gulf Wars, and South Africa's voluntarily, when the government changed and apartheid ended.

South Africa and Iraq are probably not the only countries to have begun offensive BW programs since 1945. Intelligence agencies of several countries have identified a number of possible proliferators of BW. We do not include chapters on these allegations, because the documentary materials upon which they are based are classified, and very little can be said about them. In the final chapter we will return to the issue of suspected proliferation and its implications for BW control.

Although quite a few countries had offensive BW programs after 1945, there are no confirmed instances in which they used these weapons. However, there are allegations of covert state use of BW, and several instances of terrorist use, or attempted use. Chapters 13 and 14 address these instances or allegations of BW use against humans.

Several of these allegations concern agricultural targets, highlighting the fact that biological weapons include agents that attack crops and domestic animals, as well as agents that affect humans. Our chapters on the national programs of offensive BW development make it clear that, as before 1945, most of them have included antiagricultural efforts. Chapters 10 and 11 address the development and use of anticrop and antianimal BW.

At the end of the 60-year period covered here, BW are conceived in much more nuanced terms than at the beginning. In 1945 the landscape of BW included two major types of agent: infectious agents (viruses, bacteria, and fungi) that multiply within the host body to cause disease; and the protein toxins that are excreted from many bacteria and can cause acute toxemia at low doses. Agents were described as either lethal or incapacitating, depending on the proportion of victims that died and on their intended use. BW were clearly distinguished from CW, which were low-molecular-weight synthetic chemicals. Now, however, BW are recognized as part of a spectrum of agents that includes traditional CW agents like mustard and chlorine at one end, and traditional infectious agents like anthrax and plague at the other. In between are protein toxins; small-molecule toxins of biological origin like some snake and insect venoms; bioregulators and hormones—natural mediators of human, animal, and plant physiology that can be highly toxic at very small doses;

and synthetic chemical analogs of all of these. Many of these midspectrum agents, particularly the bioregulators and their analogs, offer potential as effective incapacitating weapons and behavior control agents. As Chapter 12 makes clear, there has been continuing military interest in midspectrum incapacitants.

The last half of the twentieth century saw not only BW development but also active efforts to ban BW. The Geneva Protocol had already banned their use in war, but it did nothing to prevent the development of BW and the accumulation of major stockpiles held for retaliation and for deterrent value. In the late 1960s momentum built to achieve a robust ban that would encompass development and possession, to prevent countries from entering hostilities equipped with large stockpiles of BW, lest circumstances tempt them to use these prohibited materials. The Biological Weapons Convention was the result of these efforts. It bans the development, production, stockpiling, and transfer of both biological and toxin agents not intended for prophylactic, protective, or other peaceful purposes. It also bans munitions to disseminate biological agents for hostile purposes. The BWC conceives agents broadly, encompassing antihuman, anticrop, and antianimal infectious agents, toxins, and midspectrum agents. Some consider that the BWC also extends to biological agents intended for antimateriel purposes, a belief that is reflected in, for example, US legislation implementing the BWC.[2]

The BWC is clear, and it is robust enough to encompass new agents that biotechnology is making possible. However, it has very weak provisions for resolving suspicions of noncompliance, and considerable diplomatic effort has been devoted to developing a new international agreement that would add an inspection regime to the BWC prohibitions. So far these efforts have proved unsuccessful, and it is unclear if the political climate will be conducive to resumption anytime in the near future. Chapters 15 and 16 cover the political and legal perspectives on the development of the biological disarmament regime.

Central Issues

In considering BW development programs since 1945, we address three central issues: (1) Why have states continued or begun programs for acquiring BW? (2) Why have states terminated BW programs? and (3) How have states demonstrated to other states that they have truly terminated their BW programs? Imbuing these issues are four recurrent themes:

- Changing perceptions of biological weapons, and of their utility or disutility
- The limitations of intelligence
- The shifting balance between secrecy and transparency, suspicion and confidence
- The influence of treaties and of international technological collaboration upon national BW programs

Our final chapter summarizes the history of BW since 1945 and offers some lessons for the future. We make no claim that these lessons are complete or comprehensive. Because our history is incomplete, drawing as it does upon incomplete source material (especially for contemporary events), the lessons we draw are also necessarily limited. Nevertheless they possess value, and should greatly enrich current policy debates. Although awareness of history does not guarantee that we will avoid repeating mistakes, ignorance virtually assures repetition.

We live in a time in which the basic knowledge needed to develop BW is more widely available than ever before, perhaps even to the point of allowing individuals or terrorist organizations to construct effective BW. The importance of preventing biological warfare, terrorism, or criminality is more acute now than when the two major powers of the Cold War had lavish offensive BW programs. The problem is serious, and lessons from history are urgently needed.

CHAPTER 2

The US Biological
Weapons Program

JOHN ELLIS VAN COURTLAND MOON

For more than two decades after 1945, the US pursued a biological weapons (BW) program. Believing that the USSR was developing a BW capability, the US felt it had no choice but to prepare to wage biological warfare if it became necessary or militarily desirable to do so. The BW effort was bolstered by arguments from deterrence theory and by the conviction that the US must be prepared to retaliate. The Cold War conditioned the US to see the USSR as a formidable and determined enemy whose evil intentions would lead to development of every possible weapon to secure world domination. This perception was strengthened by the culture of secrecy. BW advocates argued that it was a flexible weapon system, relatively cheap to develop and maintain and easy to hide. However, despite efforts to build a major program, achievements in preparedness were not impressive. The reasons are several: the weapon was not a top priority of either the political or military establishment, the organizational structure was diffuse, the program lacked the focus and direction necessary for success, and it was driven by the flawed logic of retaliation in kind.

The program also had to contend with the general revulsion against BW, reflected in the 1925 Geneva Protocol and reinforced in 1946, when the United Nations identified as especially heinous "atomic and all other major weapons adaptable now and in the future to mass destruction."[1] In 1969 the secretary general of the UN unhesitatingly classified chemical and biological weapons (CBW) as means of mass destruction.[2]

In 1945, however, BW were viewed differently. On 24 October, George W. Merck, chairman of the US Biological Warfare Committee, submitted his final report to Secretary of War Robert P. Patterson, summarizing the World War II biological warfare program. In his report Merck concluded that the program was essential to national security.[3] Five months later the Military Intelligence Division of the US Army echoed Merck's opinion.[4]

Twenty-four years later, in a complete reversal, on 25 November 1969, the US unilaterally renounced its offensive biological warfare program. President Nixon was openly contemptuous of germ warfare, dismissing it as useless.[5] After the president had declared that the US would also renounce toxins, National Security Advisor Henry Kissinger was asked by a reporter whether such unilateral renunciation would weaken deterrence. He replied: "We also believe that we have other weapons of retaliation, including chemical and nuclear weapons, which we could use if toxins were used against us."[6]

In 1972 the US joined other major powers in signing the subsequently negotiated Biological Weapons Convention (BWC). In 1975 the US ratified both the Geneva Protocol and the BWC. A sea change had taken place in American policy.

American post–World War II BW policy thus falls into two phases: an offensive one from 1945 to 1969, and a defensive one from 1969 to the present. This chapter analyzes both phases, addressing questions of policy, intelligence, organization, facilities, research and development (R&D), testing, envisioned use, and preparedness.

Offensive Phase: 1945–1969

Policy

Policy is not an integrated whole, free from internal contradictions. It falls into four categories: legal, declaratory, agreed, and implemented. Legal policy consists of treaties and laws limiting what a government can do. Declaratory policy consists of public declarations on the course a nation chooses to pursue—for example, "no first use" of CBW. Agreed policy is the result of discussions within a government. The subsequent agreement is often kept secret, leaving the government free either to ignore or to forget it. Finally, there is implemented policy: how a nation actually behaves. For example, although the US made some exceptions to the Geneva Protocol, it adhered to the Protocol's "no first use" prohibition with regard to lethal CBW. Implemented policy is determined by capability and circumstance: what a government can do and what it wants to do. Ultimately, policy is action.

Despite Merck's warning about the dangers of ignoring the potential of BW, the development of these weapons was largely neglected in the late 1940s. But the question remained: How important would BW prove in any future war? In 1947 the acting Army chief of staff posed that question in evaluating a series of future developments in weapon systems: "the probability of biological warfare being extremely effective cannot be ruled out . . . it may hold a considerable future potential."[7]

An obstacle to any offensive BW program lay in the moral qualms about continuing to pursue a program that had all the markings of black science. How would scientists reconcile such an enterprise with their moral convictions? But many scientists, even those who detested war, had no problem. Theodore Rosebury, a distinguished biologist, resolved his moral dilemma by invoking social responsibility and patriotism.[8]

During this period advocates of BW preparedness defined what they saw as its advantages, establishing a theoretical base for its development. The case for BW was often argued simultaneously with the case for chemical and radiological warfare. Five arguments shaped the case. First, the USSR, with its large conventional forces, could overrun Europe and a large part of the Near and Middle East within six months. Second, toxic weapons could be used to counter Soviet conventional military superiority. Third, these weapon systems would not destroy property, and thereby freed the victor from the task of urban reconstruction. Fourth, they would be useful in obtaining surprise over the enemy. Fifth, they could have considerable psychological impact.[9]

Among these systems, BW were seen as superior.[10] They would shorten a war by threatening a "significant proportion of a populace," thereby leading to a negotiated settlement.[11] However, before there could be any consideration of policy, the value of BW as a military option had to be examined.

On 16 March 1948 Secretary of Defense James Forrestal informed President Truman that the Joint Research and Development Board recommended a "review of national policy on biological warfare." Priority would go to its military value. Once this estimate was completed, broad policy questions would "be in order."[12]

Given the inherent difficulties of the task, it is not surprising that American CBW policy remained unchanged throughout the 1940s and early 1950s. The issue was reviewed by the Joint Chiefs of Staff (JCS), who recommended that the policy of retaliation be continued as "an interim measure . . . subjected to review after detailed operational evaluations of chemical warfare, biological warfare, and radiological warfare have been made."[13]

The interim policy was challenged five days after the outbreak of the Korean War. On 30 June 1950 the Stevenson Committee, chaired by Earl P. Stevenson, urged major emphasis on BW preparedness along with a change in chemical, biological, and radiological (CBR) policy, a recommendation rejected by Secretary of Defense George Marshall in October 1950.[14]

During the Korean War (1950–1953) the Communists alleged that the US was using CBW against North Korea and Communist China. American officials feared that the Communist powers were charging the US with germ warfare to justify their intended use of BW. The war brought urgency to the need for speeding up the CBW program and attention to the possible need to revise policy, but no revision was achieved during the Truman administration.[15] In 1956, however, during the Eisenhower administration, a policy change was enunciated by the National Security Council (NSC): "To the extent that the military effectiveness of the armed forces will be enhanced by their use, the United States will be prepared to use chemical and bacteriological weapons in general war. The decision as to their use will be made by the President." Time and circumstance permitting, the US would consult its allies before initiating CBW.[16]

An earlier discussion in the NSC revealed the extent of the shift. Dr. Arthur S. Flemming, director of the Office of Defense Mobilization, raised the issue:

> Dr. Flemming said he wished to raise the question with respect to paragraph 12, regarding chemical, bacteriological, and radiological weapons in general war. He asked whether he was correct in believing that our previous policy had been that we would have recourse to such weapons only in retaliation against their use by an enemy. Did the language of paragraph 12 thus amount to a change in policy in respecting the use of such weapons? The President commented that the chief purpose of para-

graph 12 was to encourage research and development in these weapons fields. Mr. Anderson [Dillon Anderson, special assistant to President Eisenhower for national security affairs] added that previously our policy respecting the use of these weapons called for their use only in retaliation. Accordingly Dr. Flemming's surmise was correct, and the present paragraph 12 constituted a change in our policy.[17]

Final authority rested in the hands of the president. Successive presidents could continue the "no first use" commitment, but they were not bound to do so. The Roosevelt declaratory policy of "no first use" regarding chemical weapons (CW) had been publicly announced. This new agreed policy was not.

Despite the change, President Eisenhower was personally committed to "no first use." At the 412th meeting of the NSC, he stated that "what this government had always done with respect to these weapons was first of all to make sure that we had sufficient chemical and biological weapons to retaliate if the enemy used it on us."[18]

Although the new doctrine had given him additional authority, Eisenhower was reluctant to initiate any offensive CBW operations. But the NSC discussed qualifying adherence to "no first use" so as to give US field forces flexibility in the use of CBW agents. As revealed in the minutes of the 435th meeting of the NSC, chaired by Eisenhower, a distinction was drawn between the use of lethal and nonlethal CBW agents.

At this meeting, Gordon Gray, special assistant to the president for national security affairs, referred to the budget director's judgment "that we were spending too much money on chemical and biological weapons if we did not intend to use them and too little money if we did intend to use them." Dr. Herbert F. York of the Department of Defense (DOD) expounded on the value of incapacitating agents, and General Lyman Lemnitzer explained the operational use of CBW, especially incapacitants. This discussion led the president to caution that the use of incapacitants would allow the enemy to charge the US with waging germ warfare.[19] Biological nonlethal weapons were not yet available, although it was expected that they would soon be developed.[20]

However, the argument that the use of nonlethal agents in war was legitimate was not persuasive either domestically or internationally. Despite a public relations offensive by the US Army Chemical Corps (CmlC) to persuade the public and Congress of the humanity of CBW, few were convinced. The later use of nonlethal agents in Vietnam intensified opposition to the distinction strategy.[21] Kennedy's CBW policy largely resembled Eisenhower's,[22] although there was also a move to abolish BW.[23]

In 1961 a proposal, "Joint Declaration on Disarmament," met with opposition from the military chiefs, who were not comfortable with the provisions dealing with the elimination of CBR weapons. The JCS argued that "the US is, and should remain, free to use nuclear weapons if placed in a position of individual or collective self-defense. The proposal should not be made even for CBR weapons because it would open the door to inclusion of all 'weapons of mass destruction' including nuclear weapons."[24] On 11 August 1961 the revised "Joint Declaration of Disarmament" was released "for approval by the General Assembly of the United Nations."[25]

Another initiative during the Kennedy administration was a proposal to pursue BW disarmament separately from general disarmament. In 1963 Secretary of State Dean Rusk pushed a program to ban BW. The logic for such an unconditional ban was based partly on the lag in BW preparedness.[26] William C. Foster, director of the Arms Control and Disarmament Agency (ACDA), wanted to study the proposal. McGeorge Bundy, the president's special assistant, agreed that an ACDA study reassessing the utility of CBW should precede any action.[27]

Despite its pursuit of disarmament, the Kennedy administration implemented a policy that complicated later attempts to abolish CBW: the use of herbicides and riot-control agents in Vietnam. On 30 November 1961, President Kennedy approved the use of herbicides in operations against the Vietcong. While food crops were sprayed by South Vietnamese units, defoliants were used by American air units to destroy Vietcong food supplies and expose key enemy routes. This program was begun with the realization that it could provoke serious protests.[28] Riot-control agents, deemed nonlethal, were used for various tactical situations, including the flushing out of enemy troops from caves and fortifications.

The protests were not long in coming. The Kennedy and later the Johnson administrations constantly argued that because they were not using lethal agents or poisons, they were not violating international law or the Geneva Protocol.[29]

Intelligence

Throughout the offensive planning period, US intelligence speculated about the Soviet threat. Some concern was expressed about an overt attack, but the main BW threat to the US was seen as sabotage, since the Soviet bombing range in the late 1940s was limited.[30]

On 5 October 1948, the Committee on Biological Warfare reported on the covert BW threat. Sounding an alarm that was to reverberate throughout the Cold War and into the "war on terror," it stressed that "the United States is particularly vulnerable to this type of attack." Small amounts of biological agents could be used to kill or incapacitate "a significant portion of the human population within selected target areas." The food supply of a nation could be directly attacked. "Stamps, envelopes, money and cosmetics" might be used to disseminate biological agents. Ventilating systems and water supplies could be contaminated.[31] The USSR was cited as the nation most capable of using BW against the US. But would it? The intelligence estimates were always carefully hedged.[32]

What then did the US really know about the supporting Soviet infrastructure for the launching of BW? Two months after the end of World War II, the Joint Intelligence Committee (JIC) had drawn up a list of twenty potential targets in the USSR, citing cities that contained "facilities for scientific research and development." The JIC admitted that its intelligence was limited regarding "the locations and functions of the leading scientific research and development laboratories." Gorki, Kuibyshev, Sverdlovsk, and Kazan were identified as engaged in bacteriological research. Moscow and Leningrad were listed as "primary centers for scientific research and development" though not specifically tied to BW.[33]

If the USSR was pursuing an active BW program, what agents was it cultivating? Citing German and Japanese sources, the Joint Intelligence Staff (JIS) named "cholera, dysentery and anthrax" but then confessed that it had no further "factual information." The degree of ignorance regarding Soviet capabilities is revealed by the report's final speculations: "With intensive effort, the USSR should be capable of being one of the most advanced nations of the world in the field of BW. It is believed that the USSR would require only a few years preparation (no more than five) to wage open large-scale biological warfare, and it is conceivable that the USSR may be prepared to do so at the present time."[34] Some of the capabilities assigned to Soviet subversive BW strained credulity.[35]

The outbreak of the Korean War heightened American anxieties. A 1951 CIA estimate opined: "At present, the Soviets are capable of producing a variety of agents in sufficient quantities for sabotage or small-scale employment. By 1952 at the latest, the Soviets probably will be capable of mass production of BW agents for large-scale employment. The Soviets would most likely develop and produce for employment against the United States one or more of the BW agents listed in A." Appendix A consisted of a list of antipersonnel, antianimal, and anticrop agents with their attendant diseases in what may well be a classic example of mirror imaging.[36]

Another 1951 estimate described Soviet BW sabotage operations in terms that eerily foreshadow the current fears of terrorist operations: that saboteurs would enter undetected through the porous borders of the US and would establish sleeper cells, eluding detection until they were prepared to strike at personnel, animals, or crops.[37]

By the end of the Eisenhower administration, more attention was focused on the danger of large-scale open attacks on the US. In a 1960 report the CIA estimated that the advantages of a Soviet clandestine BW attack would diminish as the USSR developed its intercontinental ballistic missile (ICBM) capability. The Soviets would be more likely to use BW "as a subsidiary operation in conjunction with a deliberate Soviet initiation of general war."[38] Heightening the anxiety of American policymakers was the belief that the USSR had a vastly superior stockpile of CBW. As Dr. York stated at the 435th meeting of the NSC, the US stockpile was "one-fourth that of the USSR."[39]

From 1960 on, the ICBM revolution had a profound effect on evaluations of how the danger might come. Increasing attention was paid to Soviet ICBM capability. Now nuclear weapons trumped CBW, rendering them redundant for purposes of strategic attack.[40] Uncertainty predominated. Lack of firm intelligence reinforced the belief that the USSR was better prepared than the US and that the ruthless Soviets, if it were to their advantage, would use CBW in war. During the Kennedy-Johnson era, bafflement still prevailed. For the JCS, determining enemy intentions was shaped by available Soviet doctrinal literature. In a general war, the Soviets would use every weapon at their disposal.[41] While the CIA noted that the Soviets might be able to load their ICBMs with CBW warheads, it was skeptical that they would follow that course: "The USSR could adapt BW and CW munitions to its long range bombers and missiles . . . The technical and especially the operational problems involved would be severe, however."[42]

The most complete available CIA evaluation of the Soviet BW program was issued on 21 April 1961. The report detailed the USSR's interest in antipersonnel and antianimal diseases. Past Soviet research had centered on four antipersonnel bacterial agents: plague, anthrax, brucellosis, and tularemia. Now the CIA detected a shift of emphasis to viral and rickettsial diseases. Interest in toxins, such as botulinum toxin and exotoxins, and on antilivestock diseases, such as foot and mouth disease and rinderpest, was noted. The agency reported that the USSR was emphasizing aerosol technology and the development of arthropod vectors.

How serious was the Soviet BW threat? The CIA was cautious, stating that "there are still no firm indications that biological weapons have been standardized, produced and distributed" to military units. How well prepared was the Soviet Union against BW attack? The report gave a mixed judgment. Civil defense was "extensive" and "well organized." Military units were well trained and effectively prepared against BW. But a massive BW attack, especially against civilian targets, would probably overcome all defenses.[43]

To a great extent, then, even this detailed report was steeped in uncertainty. In 1964 the CIA judged the possibility of Soviet use of BW agents against the US to be low for at least five years.[44] Uncertainty was also evident in a 1967 evaluation.[45]

Organization

The first decade of the Cold War marked a frantic effort to rearm. But priorities had to be defined. How important was a particular weapon system in relation to another? The shadow of the nuclear sword fell over all programs.

The NSC was the supreme coordinating body of the federal government. Its core membership consisted of the president, the vice president, the secretaries of defense and state, army, navy, and air force, and the chairman of the National Security Resources Board. Theoretically, it supervised all aspects of BW policy, intelligence, and preparedness.

The DOD had the main responsibility for the defense of the US. Under its aegis were several advisory groups: the Research and Development Board and its think tank, the Committee on Biological Warfare. According to historian Dorothy L. Miller, "The board . . . integrated all military research and development . . . It kept the Joint Chiefs of Staff informed on the capabilities of BW so that they could determine the position BW should occupy in the national defense effort."[46] In its reports, "it consistently advocated a strong research and development program in BW," thereby seeking to influence the development of policy.[47]

In 1953 and 1954 the Office of the Secretary of Defense was reorganized. An assistant secretary of defense (R&D) now became the chief advisor to the secretary of defense on CBW R&D.[48] Previously, on 7 September 1949, it had been decided that day-to-day responsibility for CBW planning belonged to the Chemical Corps.[49] From the start of the postwar period, the CmlC drew up the budget. The Office of the Chief Chemical Officer was a command and staff organization. The CmlC had several divisions that dealt with CBW, among them the Research and Engineering Division, and the Special Operations Division (SOD) at Camp Detrick (renamed Fort Detrick in 1956). The SOD, activated on 17 March 1949, was "to perform research and development in the field of covert or sabotage operations."[50] It carried out tests with BW simulants throughout the country.

Project proposals originated in the CmlC Technical Committee, defined as "a committee established and maintained by the chief of a developing agency to effect coordination among the developing and using agencies during research, test, type classification and procurement activities."[51]

Outside the Army, several other organizations and commands played roles in the program: the Office of the Surgeon General, the Navy, the Air Force, the Public Health Service, the Department of Agriculture, the Department of the Interior, the National Academy of Sciences, and the Central Intelligence Agency.

The Office of the Surgeon General, whose Biological Branch was activated in 1950, was largely responsible for biological warfare defense.[52] The Navy had played a role in BW preparedness during World War II. Its laboratories in California had worked on *Yersinia pestis*. It provided funds for BW R&D and assigned some of its personnel to Camp Detrick.[53] The Air Force shared responsibility on BW R&D with the CmlC.[54]

The CIA's Technical Support Staff specialized in gadgetry and carried out a number of questionable experiments on unaware human targets. Moreover, it had a close relationship with Camp Detrick. As its director, William Colby, explained to the Church Committee in 1975:

> CIA's association with Fort Detrick involved the Special Operations Division (SOD) of that facility. This division was responsible for developing special applications for biological warfare agents and toxins. Its principal customer was the U.S. Army. Its concern was with the development of both suitable agents and delivery mechanisms for use in paramilitary situations. It performed "certain research and development" in the laboratory facilities of the Special Operations Division of the Army Biological Laboratory at Fort Detrick.[55]

Too many organizations were involved in US biological warfare planning. This dispersal of responsibility forms a sharp contrast to the Manhattan Project. The need to centralize CBW coordination within the CmlC was recognized by the Eisenhower administration. In 1955 and 1957 the CmlC's authority over CBW R&D coordination was strengthened.[56]

Facilities

Several facilities were dedicated to biological warfare preparedness after World War II: Camp (later Fort) Detrick, Maryland; Edgewood Arsenal, Maryland; the X-201 Plant, Pine Bluff, Arkansas; Dugway Proving Ground, Utah; Plum Island, New York; and the Chemical Corps School, Army Chemical Center (which later became Fort McClellan), Alabama.

Camp Detrick, established in 1943, had from 3,000 to 4,000 personnel during the final phase of World War II. By March 1946 it was down to between 300 and 400 employees.[57] Miller describes it as consisting of "research laboratories, pilot plants, and chamber test facilities. It also conducted agent and agent-simulant tests."[58] Its main testing site was the 8 Ball Bomb Test Building, completed in 1950 at a cost of $715,468.[59] Edgewood Arsenal performed a multitude of R&D tasks during the postwar period.

A crucial need for the biological warfare program was a BW production plant. Between late 1950 and mid-June 1954, the X-201 Plant was built in Pine Bluff, Arkansas. According to Miller, "This plant tested production processes, produced limited quantities of agents and filled munitions, and developed data for storing, handling, and performing surveillance of biological weapons."[60] Although it was normally on operational standby, it "could go into full scale production in 72 hours."[61] It had a secure location and extensive buildings.

After 1949 the Dugway Proving Ground became the major testing site for the CmlC. Plum Island in Long Island Sound became the center for the investigation and production of antianimal agents banned from the continental US.[62] The fifth major biological warfare facility, the Chemical Corps School, Army Chemical Center, was a center for BW training and indoctrination.[63]

Research and Development

The importance of BW R&D was stressed in a 1948 report of the Committee on Biological Warfare to the Research and Development Board of the

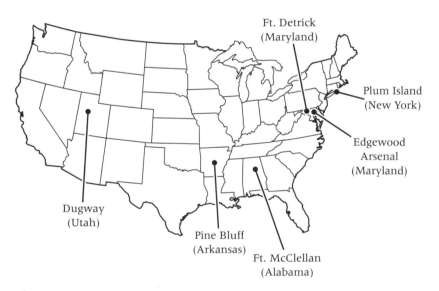

Figure 2.1 Major sites involved in the US BW program.

DOD. The committee firmly believed that BW attack could be deterred only by being prepared to use BW and by announcing to any potential enemy US determination to retaliate in kind.[64] The JCS and the secretary of defense agreed that work on BW preparedness should be intensified.[65] Any agent chosen for military use would have to pass through seven procedures: screening through laboratory work, pilot plant testing, field testing, assessment, standardization, weaponization, production.

CmlC funding between 1946 and 1954 rose steadily but not dramatically until the Korean War. After the war it declined slightly.[66] Historian Henry Stubblefield noted that "the greater part of appropriations for Chemical Corps construction and improvement went into long-range programs for research and development and industrial facilities."[67]

In April 1952 the Office of the Chemical Corps listed 271 projects, among them CBR proposals, for its formidable FY 1953 program. The BW programs had a considerable range, including development of weapons, testing of potential agents, the development of dissemination techniques, warheads for missiles, decontamination procedures, detection devices, and protective equipment.[68] Similarly long lists were drawn up by the CmlC's Technical Committee throughout the US offensive program: a cornucopia of R&D proposals were floated. Many of them got no further than the laboratory.[69]

Numerous biological agents were screened in the Camp Detrick laboratories. Only a few became agents of major interest. The criteria for antipersonnel agents were infectivity, virulence, persistence, wide but concentrated dispersion, and stability. Initially the US program sought major killing power. The antipersonnel, antianimal, and antiplant agents favored in 1949 are listed in table 2.1. Other agents studied during the offensive period of the US program are listed in table 2.2. Some of these (for instance, VEE) were eventually standardized; most were not.

What uses did the planners project for different biological agents? In February 1951 the Committee on Biological Weapons envisioned a number of operational uses for selected priority agents:

> If, for example, the objective were to kill many people, botulinum toxin might be used, whereas Bacterium tularense [*Francisella tularensis*] would be more suitable if protracted disability . . . were desired. Likewise, spores of *Bacillus anthracis* could best be employed to produce per-

Table 2.1   Agents and diseases of greatest interest in 1949

Antipersonnel
Bacteria
   Brucellosis: *Brucella suis:* after 1949, available as an agent
   Tularemia: *Francisella tularensis*
   Melioidosis: *Burkholderia pseudomallei*
   Q fever: *Coxiella burnetii*
   Glanders: *Burkholderia mallei*
   Plague: *Yersina pestis:* "insufficiently stable"
   Anthrax: *Bacillus anthracis*
   Psittacosis: *Chlamydia psittaci*

Toxins
   Botulism: Toxins of *Clostridium botulinum*

Antianimal
Viruses
   Foot and mouth disease
   Rinderpest
   Epizootic viral diarrhea of cattle
   Virus III disease of swine (probably African swine fever)
   Fowl plague
   Newcastle disease
   Fowl malaria

Bacteria
   Anthrax

Antiplant
   Plant inhibitors and plant pathogens
   Wheat blight
   Rusts
   Defoliants
   Inhibitors for top application or soil and water application

sistent contamination of terrain . . . Should it be desired to try to start a spreading epidemic, Pasturella [*Yersinia*] pestis would be the agent of choice, but if one preferred to exclude the possibility of spread from man to man, *Brucella suis* or *Brucella melitensis* might be chosen.[70]

One of the major problems in waging BW was dissemination.[71] There was difficulty in wedding agents to munitions. In 1949 the munitions ca-

Table 2.2   Other agents and diseases of interest during the offensive phase
of the US program

Antipersonnel
Viruses
   Smallpox virus (variola)
   Venezuelan equine encephalomyelitis
   Yellow fever
   Rift Valley fever

Fungi
   Coccidiomycosis: *Coccidiodes immitis*

Toxins
   Ricin
   Shellfish toxin
   Staphylococcal enterotoxin Type B

Anticrop
   Black stem rust of cereals: *Puccinia graminis*
   Rice blast: *Piricularia oryzae*

---

pable of disseminating biological agents were the 4-pound bomb, the British experimental bomb B/E1, the M114, and the E61 bomb. Later developments were the M33 and the 500-pound bomb. All of them were found unsatisfactory in one respect or another. Later the Air Force sought to replace the M114 bomb with the E61 bomblet, which would carry and dispense *Bacillus anthracis*. Its advantages were its compact size, its carrying capacity, and an extensive dispersal range. However, Miller notes, "some questioned the advisability of putting so much emphasis on this development. In the nuclear weapon the Air Force already had a devastating strategic weapon. Their argument was: why not go after a controlled debilitating weapon?"[72]

In February 1953, becoming impatient, the JCS "directed the military services to develop and completely test by 1 July 1954 a lethal munition agent combination that could withstand a wide range of meteorological conditions." The agent of choice was *Yersinia pestis;* but since its development was lagging, the JCS suggested *Bacillus anthracis* in liquid suspension. It would be placed in E61 bomblets placed in the E113 500-pound bomb.[73]

For antiagricultural operations, the favored munition was the feather bomb, a modified pamphlet bomb that dispersed feathers dusted with agent.[74] During the Eisenhower administration, one project sought to wed CBW agents to the Snark missile. The Air Force sought BW warheads for its missiles. A drone project was inaugurated.[75] At the CmlC, however, doubt continued that "the combination of agent and munition could be achieved in the near future and action was postponed."[76]

Additional expenditure was devoted to vectors for dissemination, especially the breeding of the *Aedes aegypti* mosquito, designed to spread yellow fever. Other projects reportedly included mosquitoes to spread malaria and dengue fever; fleas to carry plague, cholera and dysentery; and ticks to spread tularemia.[77]

Following up on the policy shifts of the Eisenhower administration, the Kennedy administration placed its emphasis on spraying rather than bombs and on incapacitating rather than lethal weapons.[78] By then a number of BW-capable munitions were available: bomblets, spray tanks, mines, portable generators, cluster bombs, and the Sergeant rocket.

Throughout this period, American BW R&D benefited greatly from a tripartite arrangement among the US, the UK, and Canada. Australia joined the group in 1964. Joint meetings were held, and personnel from the three countries participated in numerous sea and land field trials. Generally, the tripartite division of labor established during World War II was continued: the UK focused on basic research, while field trials were conducted in the US at Dugway and in Canada at Suffield. Production was centered largely in the US.[79]

Testing

Throughout the offensive phase of the US BW program, testing was fairly constant, taking the form of city simulant tests, large area land tests, and sea tests. Designed to measure the dispersion possibilities of BW agents, simulant tests were carried out against buildings and in the New York subway. Field tests were also staged at Detrick and Dugway. At first only animal subjects were used; but finally, breaking a long-respected taboo, test experiments used human volunteers.

In the decade following the end of World War II, BW tests were held at the Pentagon (1949); the Norfolk, Hampton, and Newport News area (1950); the San Francisco Bay area (1950); Minneapolis (1953); St. Louis

(1953); Camp Detrick (1950s); and Dugway (1950s).[80] The experiments were clothed in secrecy. The inhabitants of San Francisco, Minneapolis, and St. Louis were unaware of what was being done.

Most of the Detrick trials were carried out in the 8 Ball facility. The tests used hot agents like *Francisella tularensis,* the causative agent for tularemia, on monkeys, goats, sheep, mice, rabbits, and guinea pigs. More than 2,000 rhesus monkeys were exposed in this testing chamber.[81] At Dugway, tests of the M33 bomb, charged with *Brucella suis,* were designed to evaluate the highly sought agent-munition combination. The Air Proving Ground Command delivered a tepid verdict, comparing the simulated BW attack so unfavorably with a nuclear attack that any military leader or official would wonder: Was it worth it?[82]

During the Truman administration and the first year of the Eisenhower administration, field tests, designed for defensive equipment and offensive weapons, were carried out at several installations besides Dugway.[83] Later in the Eisenhower administration, BW testing accelerated. A major program carried out at Dugway Proving Ground was the 1954 St. Jo project, designed to test an agent-weapon combination, bomb clusters loaded with anthrax, against an unprotected population. It was a success: a large number of animals were infected, and the aerosolized agent spread "up to 40 miles downwind."[84] Challenged by a wide variety of environmental conditions, the agent-weapon combination worked well when delivered by aircraft. The CmlC concluded that "although development of the agent component of the 'St Jo' program is still in progress, acceptance and identification of the munition proper is considered a timely step in establishing DOD BW readiness as required."[85]

Operation Whitecoat, the first testing project to use human volunteers, was carried out from 1955 through 1973 at Dugway and at Detrick. Responsibility was shared between the CmlC and the Office of the Surgeon General. The aim of the program was to assess human vulnerability to BW aerosolized agents. Its justification was that no "direct experimental evidence" was available regarding the vulnerability of field troops to a BW attack.[86] Historian Ed Regis reveals that during this period 2,200 U.S. Army Seventh-day Adventists were exposed to a variety of diseases: "Q fever, tularemia, sandfly fever, typhoid fever, Eastern, Western, and Venezuelan equine encephalitis, Rocky Mountain spotted fever, and Rift Valley fever." These experiments resulted in no deaths, relapses, permanent sicknesses, or incapacitations. In the process, "The experimenters learned that larger doses of the Q fever agent shortened the incubation period, that previous vaccination prevented the disease, and that the infection was highly responsive to oxytetracycline."[87] The results of the BW spray tests satisfied the CmlC by demonstrating that spraying was the superior means of disseminating BW agents.[88]

Human testing was not confined to the Whitecoat trials. A 1996 report by the Chemical Weapons Exposure Study Task Force gives the most complete listing of human testing throughout the offensive phase of the program. Subjects were exposed to *Bacillus subtilis, Francisella tularensis,* ricin, botulinum toxin, Venezuelan equine encephalitis, Coe virus, rhinovirus, *Mycoplasma pneumonia, Coxiella burnetii, Brucella* species, *Bacillus anthracis,* smallpox virus, influenza virus, staphylococcal enterotoxin, and *Rickettsia rickettsii,* the agent of Rocky Mountain spotted fever. Several test sites were utilized, from Detrick to the Ohio State Penitentiary. The subjects were exposed to aerosol inhalation, airborne particles, syringe injections, and inoculations. Although the 1966 report listed injuries, it did not report any fatalities. Presumably, when lethal agents were used, adequate protection was provided.[89]

The objective of another series of trials, Operation Large Area Coverage (1957–1958), was to test the theory that large air masses would carry biological agents over considerable distances (see Chapter 3). According to the CmlC, "These tests proved the feasibility of covering large areas of a country with BW agents" sprayed from airplanes.[90]

BW testing accelerated during the Kennedy and Johnson administrations. The Kennedy administration launched Project 112, which included both land and sea tests. The ocean and coastal tests were codenamed Operation SHAD (Shipboard Hazard and Defense). In 1961 Secretary of Defense Robert McNamara ordered a review of America's military capabilities. Among the 150 projects initiated, Project 112 was aimed at "research, testing, and development for chemical and biological weapons." The purpose of SHAD was "to identify US warships' vulnerabilities to attacks with chemical or biological weapons and to develop procedures to respond to such attacks while maintaining a war-fighting capability." Tests were held off the coast of Hawaii and California, in the open spaces of the Pacific, in the Marshall Islands, in Alaska, in the Panama Canal Zone, in Florida, and in the Great Plains region of Canada. Agents were released from air, from ships, and from shore. The following agents were tested: *Francisella tularensis, Serratia marcescens, Escherichia coli, Bacillus globii,* staphylococcal enterotoxin Type B, *Puccinia graminis* var. *tritici* (stem rust of wheat), and simulants. Agents and simulants were usually discharged in aerosol form from disseminators, bomblets, or spray tanks. Monkeys were used as test subjects. The tests were used to determine the effectiveness of these agents in various conditions, including the effectiveness of "selected protective devices in preventing penetration of a naval ship by a biological aerosol," the impact of "meteorological conditions on weapon system performance over the open sea," the penetrability of jungle vegetation by biological agents, "the penetration of an arctic inversion by a biological aerosol cloud," "the feasibility of an offshore release of *Aedes aegypti* mosquito as a vector for infectious diseases," "the feasibility of a biological attack against an island complex," and the decay rates of BW agents under certain conditions.[91]

Ironically, given the rather cavalier approach to testing simulants in urban areas, a good deal of sensitivity was shown in the Pacific Ocean operations. Concern regarding ecological effects and the possible spread of BW agents via seabirds led to a series of protocols. Safeguards were established and requirements set by the Kennedy administration.[92]

Land-based dissemination tests continued. In May 1965, SOD covert agents carried out two localized tests with an anthrax simulant, dried *Bacillus globigii,* disseminated by spray generators in specially built briefcases. One test was carried out at the Greyhound bus terminal in Washington, D.C.; the other at the north terminal of the National Airport. Air samples taken subsequently revealed that if lethal agents had been used, the effects would have been deadly.[93]

From 7 through 10 June 1966 tests were carried out in the New York City subway system by SOD agents.[94] Neither the New York Transit Authority nor the New York Police Department was informed.[95] The agent used was *Bacillus subtilis* var. *niger,* which was disseminated by dropping germ-filled light bulbs from platforms between cars of the express trains along the tracks of three lines.[96] The report on the New York tests showed that the Army was clearly satisfied with the results: "Dropping an agent device onto the subway roadbed from a rapidly moving train proved an easy and effective method for the covert contamination of portions of subway lines . . . Test results show that a large portion of the working population of New York City would be exposed to disease if one or more pathogenic agents were disseminated covertly in several subway lines at a period of peak traffic."[97]

Envisioned Use

The key question addressed by testing was: What military uses would be served by the development of BW? In 1950 the Air Force presented a plan for possible use of BW in a future general war that was divided into a number of phases: In Phase I the Strategic Air Command would strike enemy targets with biological and atomic weapons. These combined strikes would be continued throughout Phase II. From Phase III onward the BW strikes would be "determined by the rate of increase in available aircraft, availability of suitable targets, and the exigencies of the situation."[98]

A 1952 study by the Weapons Systems Evaluation Group (WSEG) analyzed attacks against military targets and attacks against cities. Military targets were considered unpromising, since the enemy would probably be prepared. Cities, however, were judged to be promising targets; their populations were more vulnerable than soldiers.[99]

How would the USSR use CBW against the US? An April 1955 report to the National Security Council once more emphasized sabotage as a likely means of attack.[100] Almost a year later, the NSC's estimate of the situation envisioned several ways in which the Soviet Union might use CBW "as secondary means of attack" in general nuclear war. The report concluded by weighing the advantages to each side in a "total nuclear war" in which unconventional weapons were used: "Initiation of BW may be more to the advantage of NATO than to the USSR initially; however, to whose ultimate net advantage it could be is problematical." If the war were limited to Western Europe and remained nonnuclear, "Use of chemical and biological weapons appears to offer NATO a means of destroying large enemy forces without resorting to the use of nuclears; however these weapons are no substitute for conventional forces."[101] If a limited war took place outside Europe, "Chemical and biological [sic] could contribute extremely effective fire power, in addition to that of conventional weapons against masses of Soviet Bloc manpower . . . In a 'brush fire' war . . . incapacitating (non-lethal) chemical and biological weapons offer a means of US assistance without the undesirable mass destruction or excessive loss of life."[102]

Three months later, on 7 July 1960, an interdepartmental committee issued an extensive study of "United States and Allied Capabilities for Limited Military Operations to 1 July 1962." It featured several scenarios: the waging of limited war over Berlin, in Laos, Iran, the offshore islands in the Formosa Strait, and Korea. In each case, consideration was given to whether nuclear, chemical, or biological weapons should be used. The study drew dual conclusions regarding the use of BW: "From a military point of view, it would *not* be advantageous for U.S. and allied forces to initiate the use of lethal CW/BW agents, principally because current programs provide only a limited capability and because our allies lack protective equipment and training." However, "U.S. employment of non-lethal CW/BW agents would, under certain circumstances, enhance the capabilities of U.S. and allied forces." Incapacitating CBW agents were seen as potentially useful in two cases: Laos and Korea. The study carefully weighed the political and military advantages and disadvantages of using CBW in these theaters.[103]

Despite the attention paid to the possible use of "non-lethal agents," the option of indiscriminate use of BW was hardly dead. It was reflected in some of the Kennedy administration's military studies. In August 1962 the Office of the Deputy Chief of Staff for Military Operations noted that several factors had to be borne in mind in evaluating the advantages of large-area spray attacks: since there would be "no property destruction," target accuracy was "not required"; the attack would be "independent of local weather conditions"; it could be launched "as a follow-up to a nuclear attack assuring that we prevail"; it would be far less expensive than other weapon uses; and it would complicate the enemy's defense problems.[104]

The 1962 *Field Manual on Chemical and Biological Weapons Employment* argued that one advantage of BW was that discriminating target identification was not necessary. BW could cover large target areas, penetrate buildings and defensive fortifications, interdict communication lines, and kill or incapacitate personnel in cities, factories, ports, rail centers, oil fields, and missile sites even when only limited intelligence was available.[105]

The question of how biological agents would be used was largely treated in a speculative and hypothetical manner. There was, however, one detailed Pentagon plan that integrated the use of BW into an invasion of Cuba: the Marshall Plan.[106] No comparable plans were drawn against the USSR or Communist China.

Preparedness

Throughout the five years following World War II, a number of reviews assessed the CBR capability of the United States, emphasizing the vulnerability of the US to BW attack and the need to build a retaliatory capability and to conduct a vigorous CBW R&D program. These reports had a common theme: the US was not prepared, offensively or defensively, for CBW attack. The lack of a reliable retaliatory weapon was decried.

The Stevenson Report had the greatest impact on the US CBW program. On 30 June 1950 it urged a major emphasis on BW preparedness and concluded that "the United States should be prepared to defend itself against biological warfare and to wage biological warfare offensively."[107]

From 1945 to 1950, BW preparedness was complicated by a number of overall problems. As Miller observes, "Many authorities were agreed that the efficiency of biological weapons would never be known short of actual use in large-scale military operations and that until that time any evaluation could only be an 'educated guess.'"[108]

During the period 1945–1950, the policy of retaliation and the emphasis on the development of conventional and nuclear weapons discouraged attention to more esoteric and untried weapons. "As a result, funds were limited, and research and development did not progress to the point where BW could be incorporated in firm military plans."[109]

Moreover, deployment overseas posed serious logistical problems: given that such munitions were highly vulnerable to temperature conditions, how could the viability of the agents be preserved during both transport and storage abroad, to assure their effectiveness if needed?[110]

As a result of such problems, the development of BW remained largely stalled "in the research and development phase."[111] Coordination among the services lagged. The CmlC found it difficult to implement the preparedness directives of the DOD. Private industry was unenthusiastic about participating in the CBW program, leaving the CmlC to design and construct its own facilities.[112] The Air Force, which presumably would launch any major BW offensive, also lagged in reaching its goals; as late as 1947, "No quantities of BW weapons were in stock."[113]

On 19 December 1949 the JCS had completed their "Long-Range Plans for War with the USSR," codenamed Dropshot. In its estimate of the relative combat strength of the USSR and NATO in biological warfare, the JCS assessment of predictable US progress was cautiously optimistic while remarking that it took "up to 18 months to attain quality production" of BW.[114] Years of neglect could not be swiftly overcome. Despite the challenge of the Korean War, despite new efforts devoted to BW planning and construction, reliable test data were still lacking, and operational planning was deficient. There seemed to be no hope of achieving preparedness in the near future.[115]

When would the US be realistically prepared for biological warfare? That question had troubled successive secretaries of the army and defense during the Truman administration. In 1952 Secretary of the Army Frank Pace was pessimistic in his report to the secretary of defense: "If the signal to 'retaliate' were given tomorrow or within the next year, the United States could make little more than a token effort in CW and BW."[116]

By 1953 only two BW munitions were available: the M33 antipersonnel bomb and the M115 anticrop bomb. But the M33 suffered from "critical deficiencies." And there were deeper problems revolving around targeting and the impact of these weapons if they were used: "Particularly curious was the fact that the Air Force had been unable to come up with lucrative targets for biological munitions."[117] This slow progress led to skepticism regarding the value of BW.[118] There were persistent problems with training, with management, with technical knowledge on how to wed agent to munition, with coordination between the CmlC and the Air Force.[119]

Reports on preparedness differed significantly by agency. The CmlC touted its achievements and trumpeted its imminent successes. In a report delivered in July 1953 it listed its progress during fiscal year 1953: standardization of two anticrop agents, bomb development, and advances in anticrop spray devices. It was most optimistic in predicting forthcoming achievements.[120] The Joint Strategic Plans Committee (JSPC) did not agree. Asked to evaluate CBW readiness in light of the goals set by the Stevenson Report, it delivered a devastating verdict in

August 1953, stressing the failure in biological warfare preparedness: "The desired degree of success . . . in research and development in BW may never be realized."[121]

CBW preparedness continued to lag in the early years of the Eisenhower administration. During these years, the Office of the Special Assistant for Science and Technology expressed dissatisfaction with the CBW program. On 8 May 1958 Herbert Scoville Jr., a member of the BW-CW Panel, reacted with exasperation at the lack of progress, excoriating the CmlC.[122]

On 30 July 1958 G. B. Kistiakowsky Jr., of the White House Office of Science and Technology, complained to Dr. J. R. Killian Jr., the president's special assistant for science and technology, regarding BW preparedness. He concluded: "The Chemical Corps has no directives authorizing it to develop weapons and operational doctrines; only to do R&D in BW. It is very clear that what is now available could be substantially improved by an effort directed at a specific weapons system."[123] Almost a year later, another devastating report was sent to Killian on management problems within the DOD, including those affecting CBW: "A special management problem is raised by single Service support of programs that are clearly of tri-Service interest . . . Examples of this are found in the BW/CW program of the Army."[124]

By the end of the Eisenhower administration, preparedness still lagged. The report of the Atomic, Chemical, Biological, and Radiological Division was harsh: "The US does not now have and has not programmed for procurement by 1965 a CW/BW capability adequate either to contribute effectively to deterring the Soviet Bloc from initiating chemical and biological warfare, nor to retaliate effectively if the enemy does initiate."[125]

In the Kennedy years, additional effort was dedicated to the BW program as the new administration committed itself to flexible response, seeking options besides nuclear weapons with which to react to Soviet aggression. In funding, the CmlC reached its highest point since fiscal year 1957, although it still had to compete with many other rearmament programs.[126]

The key to improving American preparedness lay in testing. Project 112 centered on CBW. In his mission statement to the JCS, Secretary of Defense McNamara directed them to "consider all possible applications, including use as an alternative to nuclear weapons. Prepare a plan for the

Table 2.3 Standardized weapons

Standardized and/or weaponized: antipersonnel
  Bacillus anthracis: lethal
  Francisella tularensis: lethal
  Brucella suis: incapacitating
  Coxiella burnetii: incapacitating
  Yellow fever virus: lethal
  Venezuelan equine encephalitis virus: incapacitating
  Botulinum toxin: lethal
  Staphylococcal enterotoxin Type B: incapacitating
  Saxitoxin: lethal

Standardized and/or weaponized: antiplant
  Puccinia graminis var. *tritici:* stem rust of wheat
  Piricularia oryzae: rice blast disease

development of an adequate biological and chemical deterrent capability, to include cost estimates, and appraisal of domestic and international political consequences."[127]

The working committee, convened by the JCS, recommended that "the nation enter on both a short-term crash program in biological agent manufacture and munitions production and a five-year program of research, testing, and development of new chemical and biological weapons systems."[128] During this final period of the offensive phase, the BW program received the highest amount of support in its history. Detrick personnel were confident that they were making progress. The recent tests had proved highly satisfactory: BW agents could be effectively disseminated. Nevertheless the stockpiles of biological agents and weapons were still not impressive. And the uncertainty factor remained: How would these agents function on the battlefield?

Defensive Phase: 1969–2003

Policy

Shortly after the Nixon administration took office, Secretary of Defense Melvin Laird wrote to the president's national security advisor,

Henry Kissinger, asserting that there was a need to review the CBW program:

> I am increasingly concerned about the structure of our chemical and biological warfare programs, our national policy relating to such programs, and our public posture vis a vis chemical and biological warfare activities . . .
>
> It would seem reasonable to have the subject brought before the National Security Council at an early date. I request the necessary studies and reviews be initiated immediately, to facilitate early consideration by the NSC.[129]

On 28 May 1969, following Nixon's approval, Kissinger directed the NSC to carry out the CBW review.[130] There were four options:

1. Retain a Full Capability Including Both Lethal and Incapacitating Biological Weapons.
2. Retain a Capability for Incapacitating Weapons Only.
3. Research and Development Program Only, but for both Offensive and Defensive Purposes.
4. Research and Development Program for Defensive Purposes Only and to Protect against Technological surprise.[131]

On 10 November the Interdepartmental Political-Military Group (IPMG) submitted its report. After defining the differences between biological and chemical warfare, it declared that the American BW capability was limited:

> No large inventory of dry (powdered) anti-personnel lethal or incapacitating biological agents is maintained and only eight aircraft spray disseminators are in the inventory. No missile delivery capabilities are currently maintained for delivery of biological agents, although a bomblet-containing warhead for the SERGEANT missile has been standardized, but not produced in quantity. Small quantities of both lethal and incapacitating biological agents are maintained in special warfare devices.

The IPMG report then analyzed the basic CBW issues.[132] The report pitted military flexibility, the availability of options, and the need for preparedness against unforeseen threats, against the political advantage domestically and internationally in renouncing BW. It also posed the advantages of a clear "no first use" policy against the ambiguities inherent in making distinctions among agents.

On 25 November 1969 President Richard Nixon startled both the BW community and the world by announcing that the US was unilaterally renouncing its biological warfare program:

> Biological weapons have massive, unpredictable and potentially uncontrollable consequences. They may produce global epidemics and impair the health of future generations. I have therefore decided that: The United States shall renounce the use of lethal biological agents and weapons, and all other methods of biological warfare. The United States will confine its biological research to defensive measures such as immunization and safety measures.

Nixon presented his renunciation of BW "as an initiative toward peace. Mankind already carries in its own hands too many of the seeds of its own destruction. By the examples we set today, we hope to contribute to an atmosphere of peace and understanding between nations and among men."

The president also announced that he would submit the 1925 Geneva Protocol prohibiting the use in war of CBW to the US Senate for its advice and consent. He asserted that the US had "long supported the principles and objectives of this Protocol. We take this step toward formal ratification to reinforce our continuing advocacy of international control of these weapons."[133]

What was most startling about the assertion above was that it totally ignored the 1956 agreed change of policy regarding CBW. It was as though the US had never abandoned its "no first use" commitment. On 22 January 1975, in signing the instrument of ratification, President Gerald Ford reasserted the Nixon claim.[134]

Why, at this moment, after the US had successfully tested these weapons in the Pacific and other ocean areas, did it suddenly renounce them? How important was the banning of BW to the administration? Was it merely a convenient and popular thing to do?[135] What influenced this decision? Was it the "Skull Valley" incident (13 March 1968) in Utah, in which an accidental release of VX at Dugway may have caused the death of 3,000 sheep off the base? Was it the mounting unpopularity of the war in Vietnam, in which "nonlethal weapons" were used? Was it the questioning from Congress and the articulate opposition of a large portion of the scientific and international community regarding the legality of this use of "nonlethal weapons"? Would sacrificing BW be a small price to pay in order to retain a CW capability? Did the lukewarm attitude of the military toward BW play a role? Despite reservations regarding BW, the JCS argued fairly tenaciously for retaining the option of using them in retaliation. Nevertheless, Secretary of Defense Laird insisted on total abolition. Nixon agreed. It is puzzling that in his memoirs the former president did not mention this praiseworthy achievement.[136]

Other factors influenced American policymakers: the US renunciation would encourage the adoption of a universal ban and, hopefully, persuade the Soviets to follow suit. The fear of proliferation had grown: BW could prove attractive to poorer countries that could not afford to match the great powers by developing nuclear weapons. Among some policymakers, there were moral considerations: biological weapons were abhorrent. And finally, what the US was giving up by destroying its stockpile was not substantial in comparison with its chemical and nuclear reserves. Its destruction would not deprive the US of a well-developed weapon system.

The decision to abandon toxins went through the same process as the decision to end the BW program. On 21 January 1970 the Interdepartmental Political-Military Group delivered its report which began by defining toxins as "chemical substances."[137] On 14 February 1970, after a furious intragovernment debate over whether toxins were chemicals or biologicals, President Nixon followed up the renunciation of BW with a renunciation of toxin weapons.

In 1975 the US ratified the BWC and, after a 50-year delay, the Geneva Protocol—following an extended debate with Senate leaders over whether the Protocol banned the use of herbicides and riot control agents in war. In August 1969 the Department of the Army terminated all production of BW. The order to destroy the weapons followed. It was swiftly accomplished.[138]

A cache of deadly weapons, however, escaped attention. The CIA retained several agents in its refrigerating units, including cobra venom and saxitoxin. Although the cache was subsequently discovered and destroyed, the organization expressed no regret at having held it.[139]

The reaction of the BW community to the end of the offensive program

Table 2.4   US BW Stockpile in 1970

Listed by Defense Secretary Laird in July 1970

Antipersonnel agents in liquid suspension
  Lethal: none
  Incapacitating
    Venezuelan equine encephalomyelitis   4,991 gallons
      (VEE) virus
    Coxiella burnetii (Q fever)   5,098 gallons
      Total   10,089 gallons

Antipersonnel dried agents
  Lethal
    F. tularensis (rabbit fever)   804 pounds
    B. anthracis (anthrax)   220 pounds
  Incapacitating
    VEE virus   334 pounds
      Total   1,358 pounds

Filled munitions
  M1   4,450 (toxin and simulant)
  M2   71,696 (toxin, biological, and simulant)
  M4   21,150 (biological and simulant)
  M5   90 (simulant only)
  M32   168 (biological and simulant)

Anticrop biological agents
  Wheat rust   158,684 pounds
  Rice blast   1,865 pounds
    Total   160,549 pounds

Listed by DOD in September 1970

Liquid agents stored as frozen pellets (bulk)
  Incapacitating
    VEE   4,991 gallons
    Q fever   5,098 gallons
      Total   10,089 gallons

Table 2.4   (continued)

Dry bulk agents	
Incapacitating	
VEE	334 pounds
Lethal	
Tularemia	804 pounds
Anthrax	220 pounds
Total	1,358 pounds
Bulk toxins	
Clostridium botulinum toxin	13 pounds
Staphylococcus aureus enterotoxin	71 pounds
Total	84 pounds
Agents in filled munitions	
Anthrax	167.5 pounds of agent
Tularemia	569.7 pounds
Toxins (botulinum toxin saxitoxin)	10.9 pounds
Total	748.1 pounds
Anticrop agents	
Rocky Mountain Arsenal	
Wheat stem rust	
Stockpile	127,078 pounds
Nonstockpile	26,385 pounds
Total	153,463 pounds
Beale Air Force Base	
Wheat stem rust	
Stockpile (unprocessed field harvest)	5,157 pounds
Nonstockpile	34 pounds
Total	5,191 pounds
Fort Detrick	
Rice blast	1,865 pounds
Agent/munition combinations[a]	
M2 botulinum toxin	
M4 tularemia	
M32 tularemia	

Table 2.4   (continued)

Unused hardware[b]	
MI	5,315
M2	14,046
M4	34,568
M5	2,604
M32	348

Sources: Memorandum for the President from the Secretary of Defense, Melvin Laird, Subject: National Security Decision Memoranda 35 and 44, 6 July 1970, doc. 22 of Biowar electronic compilation of the National Security Archives, George Washington University; DOD, Environmental Impact Statement for Disposal of Biological Agents and Weapons, 17 September 1970, table A, enclosure 10, Folder: Chemical, Biological Warfare (Toxins, etc.), vol. 3, box 311, National Security Council Subject Files, Richard M. Nixon Presidential Materials Staff, NARA; Fort Detrick, "Demilitarization Plan for Biological Stockpile: Summary Report," Frederick, Md., 28 January 1970, NSC H-Files, Policy Papers, NSDM-35 (4 of 4), box H-213, Nixon Papers, NARA.
a. Although the report does not say so, it is apparent that M1 munitions were filled with botulinum toxin, saxitoxin, or a mixture of the two. It would also appear that some M2 munitions were filled with anthrax spores.
b. There is no documentary record of the stockpiles of Air Force, Navy, or Marine Corps spray tanks; Army warheads and submunitions for the Sergeant missile; or any non–spray tank delivery for the Air Force's stockpile of anticrop agent. Presumably these were destroyed also. I am grateful to Gregory Koblentz for providing this information.

was predictable: incredulity mixed with anger. They believed in their weapon, in what they were doing out of consciousness of a current danger, out of professional pride and patriotism.

During the subsequent decades the US stuck publicly to its renunciation of BW. However, allegations that the Communists had used CBW in Southeast Asia and Afghanistan from 1978 to the early 1990s (see Chapter 13) led to questions about whether the US should adhere to the BWC. On 4 January 1982 a National Security Directive presented two options: "(a) Taking the issue to the United Nations Security Council; and (b) As an ultimate step, withdrawing from the Biological Weapons Convention."[140]

Intelligence

At the start of the Nixon administration, the CIA concluded that there had been a shift of emphasis by the Soviets from defensive to offensive

BW R&D.[141] No sooner was the BWC in place than accusations of Soviet cheating began to circulate in American government circles. Between 1975 and 1992 a number of startling revelations unveiled an extensive Soviet BW program. Instead of complying with the prohibitions in the treaty, the USSR had systematically violated it.

On 23 August 1976 the Defense Intelligence Agency (DIA) evaluated the charges that the USSR was violating the BWC.[142] In 1980 the US charged that an anthrax outbreak at Sverdlovsk was the result of an accident at an illegal BW facility (since proven correct; see Chapter 13). The 1980s had also brought a number of serious charges that the USSR and its allies were violating the Geneva Protocol and the BWC by using lethal chemical agents and trichothecene mycotoxins throughout Southeast Asia and Afghanistan (also discussed in Chapter 13).[143]

In September 1983 the CIA published a study titled "Implications of Soviet Use of Chemical and Toxin Weapons for US Security Interests." Claiming that the increased Soviet chemical and biological warfare activity posed a threat to the West, the CIA asserted a by-now-common theme: the Soviets were ahead of NATO in CBW, and intelligence in this field was hampered by its low priority.[144]

From 1984 to the close of his administration, President Ronald Reagan, in his noncompliance reports to Congress, charged that the Soviets were violating the BWC and the Geneva Protocol.[145] Despite considerable skepticism from the scientific community, successive American administrations have never retreated from these charges. One reflection is in order. Is it credible to argue that the Soviets would have transferred advanced weapons to their allies in Southeast Asia?

Although the use of trichothecene mycotoxins remains controversial, it is now an inescapable fact that the USSR and its successor Russia pursued an aggressive BW program.[146] Confronted with American intelligence evidence a few months after the collapse of the Soviet Union, President Boris Yeltsin admitted "that the USSR had systematically violated the Biological and Toxin Weapons Convention . . . In September 1992, Moscow declared that the offensive biological weapons program had been terminated . . . Yeltsin offered to allow U.S. and British inspectors into the facilities concerned, a deal that came to be known as the trilateral agreement."[147] But the success of this process (spanning 1993 and 1994) was limited.

Because all trilateral inspections ceased after 1994 and because four Russian military BW facilities remain closed to outside inspection, skepticism persists regarding the Russian Federation's abandonment of its BW program. Is this resistance due to pressure from a hard-core military establishment unwilling to let former enemies into top-secret facilities, or are continued violations of the BWC under way?

The major postwar concern, however, has centered on CBW proliferation and its relationship to terrorism. These fears were greatly heightened by the discovery of the extensive Iraqi CBW program at the end of the 1991 Gulf War and by the 11 September 2001 terrorist attacks in New York City and Washington and the October anthrax letters. However, the subsequent failure to locate CBW in occupied Iraq at the close of the 2003 war, despite confident CIA assertions of the existence of such stockpiles, is a sobering reminder regarding the limitations of intelligence.[148]

Former BW Facilities

Starting in 1971, the former US BW facilities were quickly converted. Detrick became the Frederick Cancer Research Center and home to a number of military organizations: the field operating agency of the Surgeon General of the Army Medical Intelligence and Information Agency, the Naval Medical Logistics Command, and the US Army Medical Research Institute of Infectious Diseases, all of which conducted only defensive BW research or conventional medical research.[149]

The conversion of Pine Bluff has an ironic ring. According to a draft of the announcement on its conversion, it was to become the National Center for Toxicological Research, after the cleanup of its BW, to be taken over by the Food and Drug Administration in fiscal year 1973: "The Center will examine the biological effects of a number of chemical substances which are found in man's surroundings, such as pesticides, food additives, and therapeutic drugs."[150]

Dugway Proving Ground became the exclusive facility for outdoor testing. Edgewood became the site for detection research. But despite the general divestment, the US military kept control over all the facilities except for the area at Fort Detrick yielded to cancer research. In 2003 Plum Island was turned over to the Department of Homeland Security.

Preparedness

Today US preparedness is restricted by treaty and law to defense.

Given recent events, two questions persist regarding the current US preparedness against a surprise CBW attack. First, to what extent can defensive preparedness be separated from offensive preparedness, and how do we guard against the crossing of that thin line? Second, given that active defense against a CBW attack depends upon forewarning, how confident are we that our intelligence agencies can detect an attack before it comes or detect the agent after it comes?

Recently, moreover, skepticism has arisen as to whether the US honors the distinction between offensive and defensive research. The discovery that BW work was being carried out at the Nevada Test Site has raised questions among BW scholars and some journalists as to whether this R&D crosses the thin line between offensive and defensive research.[151] That concern has been raised more sharply in the aftermath of 9/11.

The flurry of activity that followed the shock of 11 September 2001 is too recent to evaluate, especially since much of the documentation is classified. The extensive governmental reorganization that accompanied the creation of the Department of Homeland Security is an ongoing political process as various constituencies and organizations jockey for position and power. Similarly, an extensive range of costly biodefense measures have been initiated whose effectiveness cannot yet be determined. Some of these measures have raised serious concerns in arms control, disarmament, and scientific quarters. These concerns have centered upon the issues of security versus transparency, and the maintenance of a distinction between defensive and offensive BW work.

How can one know if one's government is pursuing only defensive research? Some secrecy is essential to keeping advanced CBR capabilities out of the hands of terrorists or violent states. But the danger of the culture of secrecy is that it can subvert safeguards protecting this frontier. Hence the paradox: the more dangerous the weapons, the greater the need for secrecy; but the more dangerous the weapons, the greater the need for transparency if national policy is to secure the boundary between offense and defense.

It is impossible to separate completely offensive and defensive research. To know how the enemy can use BW, one must understand the agents and tactics he may use. A new project undertaken by the Homeland Security Department, the National Biodefense Analysis and Countermeasures program, challenges the prohibitions of the treaty. Under the aegis of this organization, genetic research, development, and testing and even the modeling of production processes could be undertaken. Although such activities would challenge the prohibitions in the BWC, a Homeland Security official has justified these pursuits as defensive in intent. But regardless of intent, the proposed activities would in effect give the US a modern offensive BW capability. And is there any current need? The likelihood that terrorists could create genetically engineered germs is highly improbable. Intent can become a Trojan horse opening the door to offensive capability. Finally, if the US undertakes these dubious projects, it will promote proliferation and will be suspected of noncompliance by other parties, especially those habitually suspicious of American intentions. The US may once more be sowing dragon's teeth.[152]

Conclusion

Why did the American BW program make so little progress in the post–World War II period? The answer lies in the policy, the intelligence, the organization, war plans and doctrine, and the degree of preparedness pursued throughout the offensive phase.

Policy

The Nixon policy decisions on biological and toxin weapons marked the end of the offensive period in biological warfare planning. It also marked the close of a prolonged debate between those who wanted to develop and use BW as though they were ordinary weapons of war and those who thought that they were reprehensible weapons banned under customary international law. In 1956, when the "no first use" policy was abandoned, the former won the policy battle. However, they subsequently lost the war. The pivotal issue was the president's will. None of the presidents from Eisenhower to Nixon had any itch to use CBW lethal agents.

In the post–World War II period, the nuclear sword overshadowed all other military preparedness efforts. Reliance on nuclear weapons for deterrence led to the relatively low priority of CBW. The unparalleled use of force in strategic bombing during World War II and the dropping of the atomic bomb were the climax of a movement that equated more force with more power. Subsequently there was a gradual movement away from the utilization of maximum force. The use of BW always carried the danger of escalation to nuclear weapons no matter how hard military officials might try to separate them. Was it worth the risk to keep such an option?

Despite the 1956 change of policy, repugnance against these weapons persisted. Also there was a certain degree of confusion. Some of the chief architects seem to have occasionally forgotten that the agreed policy had changed. Others moved away from lethal to nonlethal BW. However, the moral concerns regarding the use of any CBW agents (lethal or nonlethal) were reinforced by their use in Vietnam. Overoptimism regarding the potentialities of BW in the 1940s and 1950s turned to disillusion in the late 1960s. Experience tested the promise dreamed of in 1945; these weapons were not worth the fiscal, moral, and public relations costs entailed in developing them.

The Intelligence Dilemma

It was virtually impossible to give an accurate reading of who had what or how much they had. The pronouncements of the intelligence community remained as ambiguous as those of the Delphic Oracle. That ambiguity tended to promote misreadings. The effect of this intelligence blindness was that the US judged the intentions of its enemies by the character of the regime: evil governments will do evil things; therefore, the US must be prepared to retaliate in kind because it must be ready to meet every possible contingency so as to punish the evildoers with their own poison.

How detectable were US BW tests to the USSR? How detectable would tests by either the US or the USSR be to one another today? During the late 1940s and most of the 1950s, the intelligence community lacked the reconnaissance means the US commands today: U-2s and reconnaissance satellites. But even with those means, the US has found that accurate intelligence is difficult to collect and assess, especially regarding BW capabilities. The controversy over Iraqi WMD emphasizes the intelligence problem. Intelligence tends to gather around belief, which in turn attracts supporting information. We end as we began in our consideration of the effectiveness of intelligence on BW: in speculation, ambiguity, and uncertainty.

Organization

The CBW advocates never had the kind of leverage they needed to develop an adequate BW program. The CmlC remained a division of the Army. This arrangement hindered the emergence of priority for the program that would have allowed it to fulfill expectations. Moreover, the advocates of BW tended to see their program in isolation, not as a component within the entire weapons program. Finally, CBW was never fully integrated into either military doctrine or forces. Military men generally did not feel comfortable with these weapons.

War Plans and Doctrine

What does a nation do with CBW if it builds up its capabilities? The uncertain assessments of the value of BW and the often-inconclusive intelligence estimates of enemy preparedness complicated the making of war plans that integrated CBW into doctrine and forces. Gauging the accuracy and the effect of BW was more problematic than estimating the destructive effects of nuclear weapons. From the discussions regarding their use, it is obvious that they were seen more as a strategic than a tactical weapon system. Moreover, no war has been won by CBW. Claims that these weapons could tilt the balance in a future conflict were hypothetical constructs.

Preparedness

Because BW agents had not been used in military operations, uncertainty and unpredictability haunted all preparedness efforts. Certainly, progress was made in CBW preparedness between 1945 and 1969. What kept the preparedness program going was organizational dynamics: the CmlC kept pressing it. Moreover, during the Truman and Eisenhower administrations, fiscal constraints curtailed spending on arms. There was a momentary spurt of increased funding during the Korean War, but CBW programs had to compete with other military programs. Although a high point was reached in the Kennedy-Johnson era, the progress was never felt to be sufficient. Complaints regarding US BW unpreparedness continued. Because no firm priority was assigned to the CBW program, attention to it tended to fluctuate, often driven by intelligence perceptions of what the "enemy" was doing.

Why were the results of the BW program so meager, measured by its advocates' predictions? There are several reasons. First, the use of BW would summon unpredictable consequences, making them an undesirable means of waging war. Second, the program was never given the priority it needed to produce a fully developed weapon system. Third, the moral revulsion against the use of BW could not be fully exorcized. Fourth, BW were redundant at best when compared with other WMD. Fifth, the rationale for developing BW capability was unpersuasive to any nation armed with nuclear weapons. The BW effort was based on an Achilles' heel theory: any weakness in a nation's armor could give the enemy a fatal advantage. But military strength depends upon overall command of superior means: munitions availability, production facilities, means of delivery, training, organization, morale, mobilization and deployment ability. The overall balance of force is what counts in war, not single weapon systems.

Having looked at the offensive BW program within a historical context, we are faced repeatedly with the question that haunts all arms programs: Should we build up a military capability against a hypothetical threat that may never materialize? Was the offensive BW program necessary? Within the logic of the program itself and contemporary fears about the intentions of other countries, yes. Indeed, the effort was imperative. But within a broader context of international security, perhaps not. Defensive BW preparations are, of course, necessary, and it is impossible to separate offensive from defensive preparedness completely. But we should reflect on the sense of pursuing what amounted to an ultra kill capability. Contemporary scenarios envisioned the use of BW as an accompaniment or follow-up to nuclear attacks; the population of a country struck by nuclear weapons would be highly vulnerable to infectious diseases. But what purpose would be served? Mere slaughter is not strategy.

CHAPTER 3

The UK Biological Weapons Program

BRIAN BALMER

British involvement with biological weapons (BW) research and policy shares with its international counterparts an atmosphere of exceptional secrecy. During the Cold War, radical changes in outlook and practice occurred behind firmly closed doors as Britain moved from its aim of developing a biological bomb to a defensive stance that emphasized the need to protect against a potential biological attack. This chapter draws on archival material in the Public Record Office, Kew, to provide an overview of this transition from the early Cold War, when the Chiefs of Staff gave equal priority to biological warfare and atomic warfare, through the shift to a defensive outlook in the 1950s, the emergence of large-scale outdoor trials in the 1960s and 1970s, and finally the "civilianizing" of much of the BW research program. The resurgence of an offensive outlook with regard to biochemical agents (midspectrum agents) in the 1960s is treated in Chapter 12.

Existing accounts of British BW focus variously on research at Porton Down,[1] international collaboration,[2] science policy and scientific advisers,[3] and the ethics of scientists involved in the program.[4] This chapter inevitably covers some of the same ground, concentrating on the period until the early 1970s, when the public record closes. It supplements previous accounts with new material that allows us to locate BW more fully in relation to the strategic thinking of the Chiefs of Staff, especially in relation to atomic warfare. It also provides more, recently declassified, details of the significant series of sea trials between 1949 and 1952. And, although the details of negotiations around the Biological Weapons Convention (BWC) have been left to another chapter (see Chapter 15), additional declassified material from the 1960s has revealed details of important debates around policy and the future of the research program.

The Early Program

In Britain, concerns about BW predated research. The threat from a biological warfare attack, together with earlier well-publicized claims that German spies had undertaken biological warfare trials on the London Underground, had prompted the British government to establish an expert advisory committee in 1936.[5] Apart from requesting some research on decontamination methods, however, the British authorities did not sanction a formal research program on BW until after the outbreak of war. In late 1940 a team of scientists headed by Paul Fildes, a renowned bacteriologist, arrived at the new Biology Department Porton in Wiltshire to commence covert research. They established close links with the Canadian and US biological warfare efforts, and their research soon resulted in trials of prototype weapons.[6] Over the course of the war, Fildes' team developed a stockpile of anthrax-contaminated cattle feed cakes, to be used as an antilivestock weapon in the event of needing to retaliate in kind against a German biological warfare attack. As their main aim, they also tested an antipersonnel anthrax bomb, codenamed the N-bomb. Toward the end of the war an order for a consignment of anthrax bombs was placed with the US, but delays and then the ending of war prevented the filling of the order. Possibly of more importance than these achievements, the scientists had demonstrated that workable BW could be made. As David Henderson, Fildes' deputy, wrote shortly after the war, "the overall achievement in five years of experimental study has been to raise biological warfare from the status of the improbable, where the difficulties involved were believed by many (without experimental evidence) to be insurmountable, to the level of a subject demanding close and continuous study."[7]

Policy, Strategy, and Advice: "BW Is of the Highest Priority"

This close and continuous research continued into the postwar period. In September 1945 the Chiefs of Staff urged the Cabinet Defence Committee to permit the BW research program to continue: "in the interests of National Defence, work on BW research should continue in peace time . . . As an island we are an ideal target for attack by BW methods as our attacker need have no fear that diseases which may spread would recoil upon himself or upon his allies."[8]

After also being informed that the US was continuing its biological warfare research, in October 1945 the Defence Committee issued a directive that "approved the continuation of BW research in peace-time."[9] Offensive research, according to a statement of biological warfare policy written five years later, was "implicit" in this general directive. Indeed, a

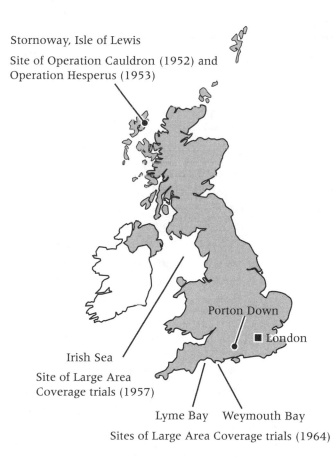

Figure 3.1 Major sites involved in the UK BW program, and sites of open-air tests in the British Isles.

year after the directive was issued, the Chiefs of Staff had decided on a broad program of research maintaining the same policy rationale: "at the time it was emphasised that for the successful development of defensive measures it was essential to study the offensive field."[10]

Following the highly secretive and largely word-of-mouth arrangements of the war, a far more formalized web of advisory committees was quickly established in the immediate postwar period to oversee biological warfare research and policy. The three key committees were:

- Biological Research Advisory Board (BRAB). Independent scientists providing technical advice to Porton Down and also passing on technical advice and information to other committees.
- Chiefs of Staff Biological Warfare Sub-Committee (BWS). Until April 1947, the Inter-Services Sub-Committee on Biological Warfare (ISSBW). Discussed and formulated offensive and defensive policy on biological warfare.
- Defence Research Policy Committee (DRPC). Scientists and Services representatives responsible for balancing priorities across the board in defense research.[11]

Within these new decision-making and advisory structures, links were soon made between biological, chemical, and nuclear weapons. In 1946 the Cabinet Defence Committee of the Atlee government had secretly committed Britain to independent acquisition of a nuclear bomb, a move intended to maintain Britain's status as a leading global player. Yet, as increasing US reluctance to share nuclear secrets culminated in outright refusal, embodied in the August 1946 McMahon Act, it is unsurprising that the Chiefs of Staff also considered a serious future for both chemical and biological warfare.

In this respect, BW policy was soon incorporated within more general defense policy that recognized deterrence as a key element. In July 1947 the Chiefs of Staff agreed on "a cardinal principle of policy to be prepared to use weapons of mass destruction. The knowledge of this preparedness is the best deterrent to war in peace-time."[12] WMD were specified as nuclear, biological, and chemical weapons, and BW enjoyed equal priority with nuclear weapons. A month earlier the DRPC had recommended that "research on chemical and biological weapons should be given priority effectively equal to that given to the study of atomic [weapons]."[13]

In a later policy statement the committee added that BW "may eventually prove comparable with the atomic bomb."[14] Six months earlier the ISSBW had reported: "We have been advised by the Deputy Chiefs of Staff Committee that research on BW is of the highest priority and in the same category of research as atomic energy and research."[15]

Driven by this policy, the DRPC recommended an increase in funding that would "make it possible to bring into service by 1957 biological weapons comparable in strategic effect with the atomic bomb, and defensive measures against them."[16] The committee soon formulated this recommendation as a threefold set of objectives for R&D:

- A biological weapon for strategic use
- A means of storage of selected agents coupled with a small quality plant, or, alternatively, the means of large-scale production
- Defensive measures, including detection, mass immunization, personnel protection, and medical treatment[17]

The Air Staff had formulated a request for an antipersonnel, strategic biological weapon as early as November 1946, in parallel with a request for a strategic chemical weapon. The BW requirement, labeled OR/1006, was meant to take high priority, and the bomb was "intended for strategic use against industrial targets and should contain the most effective biological agent for the incapacitation of workers."[18] The thinking behind this request was elaborated shortly afterward in an Air Staff assessment of biological warfare.[19] Here the potential use of biological bombing was expanded beyond industrial targets to include attacks on isolated strongholds and, with less enthusiasm, to waging economic or tactical warfare. The assessment noted that strategic attacks against industrial strongholds or other targets deep behind enemy lines were particularly suited to the Air Force. This use of biological warfare was similar to that envisaged by the Air Staff for Britain's nascent nuclear bomb.

Although the assessment was somewhat vague about the agents that would fill the bomb, they were relatively clear about the bomb itself and its effects. It would be a 1,000-pound cluster bomb, a parent bomb that contained smaller, "child" bombs, each housing a pathogenic load. The parent bomb would break up at a predetermined height, and the cluster of child bombs would be released to cover as wide an area as possible. The authors also provided an estimate of the damage that could be wrought by this type of bomb: "It has been estimated that approximately 200 tons of this type of bomb would be required to cover an industrial city of 100 square miles in order to produce a 50 per cent risk of death. This figure is based on the assumption that each individual weapon produces a 50 per cent risk of death over a square mile of city."[20]

Advisors on the BRAB initially rejected this specification for an anti-personnel bomb. These scientists, including Paul Fildes, who had left Porton to return to civilian work, were skeptical that a workable bomb could be produced by the target date, which the initial request had set as 1951. Describing the achievements of the N-bomb and cattle cake as "war-time expedients," they argued that the postwar program should be stepped up if it was to make any significant progress.

By October 1947 the requirement for a biological bomb had been reformulated. Both the chemical and biological bombs from the previous requests had been rolled into a single requirement, OR/1065, for a strategic toxic weapon. Without now specifying a chemical or biological agent, the weapon was still intended to cause "widespread incapacitation of the workers" and produce "maximum adverse effects on morale."[21] The Air Staff wanted either a single weapon that could take different fillings or two different weapons. A weapon developed with alternative fillings would, they hoped, be able to induce temporary incapacitation or produce a high percentage of deaths or permanent disability. The target was specifically intended to be personnel in industrial areas but would also include pockets of resistance and lines of communication.[22] Alternatively, two different weapons could be produced, with priority going to an incapacitating weapon. A lower-priority, lethal weapon would target places such as nuclear plants, armament centers, and research establishments in order to kill skilled personnel.

"A Complementary Weapon"

With OR/1065 in place, and with equal priority accorded to nuclear and biological warfare research, the Chiefs of Staff began to discuss the role of the putative bioweapon as a complement to nuclear weapons. Attempts were made within the Air Ministry to predict the effects of the two weapons and to make some sort of comparison between them. The architects of these exercises acknowledged, however, that there existed "no definable measure whereby strategic effects may be gauged or compared."[23] Additionally, a telling draft contribution to the Chiefs of Staff 1949 Report on Biological Warfare put the complementarity philosophy in clear terms:

> We assume that, as a party to the Geneva Protocol of 1925, this country will not initiate biological warfare but will resort to its use only as a retaliatory measure. However, lest over-optimistic results should be expected from retaliation on a massive scale, it should be borne in mind that the large-scale use of biological warfare from the air as a weapon of war has not been tried nor will it have been possible to test experimentally its effect on men. In these circumstances it would be unwise to expect decisive results from an all-out offensive with such a totally untried weapon. It should therefore be regarded, not as a competitor with the other types of weapon in our armoury, but as a complementary weapon to be used when its peculiar characteristics can be fully exploited.[24]

The first Soviet atomic bomb trial in September 1949, which was received rather more calmly in Britain than in the US,[25] did not appear, at least initially, to shake these military priorities. Not long after the test explosion the BWS considered the state of existing intelligence about Soviet biological weapons and concluded:

> While it was fully realised that the Joint Intelligence Committee could not be more specific in their appreciation of the threat . . . the general feeling was that the USSR had considerable leeway to make good before that country would be in a position to employ this weapon offensively; nevertheless in the light of recent events, the Joint Intelligence Committee's broad conclusion could not be lightly disregarded, namely that the USSR could achieve the United States level of production of 1945 by about 1952.[26]

Here, as elsewhere, such assessments were frequently based on the assumption that Soviet capabilities were similar to those of the US and UK. In direct response, the BWS recommended a "firm and consistent policy for research."[27] The subcommittee also felt that there was no need to change priorities from those adopted by the DRPC in 1947. Rapid detection and "development of, and production research on, selected agents and weapons for strategic use" remained of "supreme" importance for completion by 1955 and were classed as "projects which are so important as to call for continuous effort regardless of date of completion."[28]

A year later, in a broad strategic review that characterized the Cold War enemy as an expansionist Soviet Union, the Chiefs of Staff also noted that in the future there should be "some form of supersonic unmanned bomber or other vehicle which will take atomic and other weapons to the heart of Russia."[29] In this scenario, "other weapons" almost certainly included CBW as complements to the nuclear bomb.

The Microbiological Research Department

With BW assuming such a high priority, the scientists at the Biology Department Porton, now renamed the Microbiological Research Department (MRD), enjoyed a period of relative prosperity. David Henderson, the new superintendent, presided over the building of expensive new facilities and an expansion in the scale and scope of the research program. The MRD recruited sufficient staff, despite general personnel shortages, to carry out an ambitious program of basic microbiological research. A fermentation plant to mass-produce nonpathogenic organisms was built, and plans were soon in place to build an expensive new experimental plant to study the bulk production of pathogenic organisms. Research on the design of the biological bomb was passed to the neighboring Chemical Defence Experimental Establishment (CDEE) and, in a few recorded instances, was designated by the code name Project Red Admiral.[30] Beyond the laboratory, Operation Harness, a large-scale field trial at sea using pathogenic organisms, was approved and carried out in 1948 off Antigua. A series of further "hot" trials followed.

Operation Harness

Prior to Operation Harness, several remote sites were surveyed for their potential suitability for trials, and the BWS eventually chose a site on Antigua, Parham Sound, as a suitable headquarters. Although the conduct of the trials remained entirely in the hands of the British scientists and military, the US agreed to provide additional supplies and personnel, and the Canadians supplied additional scientific officers.[31] The experiments, which ran from December 1948 to February 1949, involved some 450 personnel and a menagerie of experimental animals. In each trial, experimenters floated a "trot" of up to 35 inflatable rubber dinghies, each mounted with an aluminum animal crate and glass sampler, out to sea. Each crate could accommodate a sheep in the main body and a monkey in a separate box on top. In addition, side arms could slide onto the main body, each containing three guinea pigs. The entire apparatus was then towed into position by two motorboats. A bomb for each test was mounted on a float and fired by remote control, thus releasing an infectious cloud over the trot. The pathogens used in the Harness trials were *Bacillus anthracis, Brucella suis, Brucella abortus,* and *Francisella tularensis.*

Two parent ships were responsible for entry and exit into the test. HMS *Narvik* acted as the "clean" ship from which animals and equipment were floated out to sea, and the "dirty" ship, HMS *Ben Lomond,* received infected animals and contaminated equipment after exposure to agents. The dirty ship also contained accommodation for the scientific staff, an

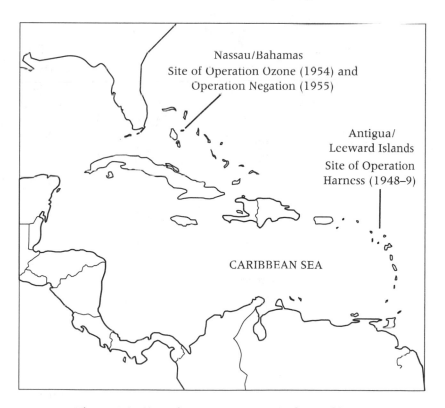

Figure 3.2 Sites of UK open-air tests in the Caribbean.

enlarged sick bay, and laboratory space. Still another ship operated downwind of the trot and performed "clear range" duties, warning off any shipping that strayed within range of the trials. After a nine-day series of tests using nonpathogenic organisms as simulant agents, the experimenters proceeded to trials with pathogens. The first agent tested was *Brucella,* although the initial layout of animals on dinghies provided attractive meals for sharks.[32] Once special shark-repellent bags had been hastily procured from the US Navy, the scientists and their naval assistants continued trials throughout January and February. The full apparatus, involving all 35 dinghies stocked with sheep and guinea pigs, was used in just two anthrax trials.

The main findings in the technical report of the trial remain closed. The open sections stress that "the operation was not primarily designed to test weapons or weapon design." Instead, the trial had aimed to locate a suitable base and to refine research techniques for future tests. The report concluded that the logistics of the operation were too complex, at times exposing "an excessively large number of men to 'risk.'"[33] This said, the authors still favored future sea trials. Of the 22 toxic trials attempted, 15 were judged by the experimenters to be completely successful, having been carried through to completion without being disrupted; 3 were partially successful, and 4 were a "complete failure."[34] From the successful trials, the report noted, the higher toxicity of biological over chemical warfare agents could be cautiously accepted.

A further report by the BRAB to the BWS reveals the optimism with which this first sea operation was viewed:

> The operation has confirmed the value of BW. The laboratory work since the previous field trials has been confirmed. It has provided information regarding the behaviour of new types of bacterial suspensions with certain experimental animals, and brought out the value of monkeys in this type of research. It will be unnecessary, in future, to rely on such clumsy animals as sheep in trials with bacterial clouds. The trials have brought out many administrative and scientific problems.[35]

Operations Cauldron and Hesperus

Although there was some talk of establishing a permanent Tripartite base for trials, nothing materialized.[36] The next series of British trials, codenamed Operation Cauldron, took place from May to September 1952 off the Scottish island of Lewis, in the Outer Hebrides. Scientists tested the agents responsible for plague and brucellosis during the trials, and now employed a far more manageable experimental apparatus than that used in Harness.

The *Ben Lomond* was the only ship involved, acting as both the clean and dirty vessel. A 200-by-60-foot floating pontoon, described as "little more than a floating box with 24 compartments, 9 of which had to be converted to house diesel generators, pumps, 'clean' and 'dirty' animal stowage, change rooms, &c," was the focus of the experimental tests. The experimenters placed their guinea pigs and monkeys in cages on the surface of the pontoon, with a few naval and civilian personnel in protective gear retreating into the compartments below, and the test bomb (or in some cases spray) was detonated from a boom holding the munition some 25 feet away. Timing the detonation proved to be no easy task, and the report of the trial noted that "determination of the exact moment of release is important and difficult . . . a couple of seconds can make the difference between success and failure."[37] The exposed test animals were then taken on board the *Ben Lomond*, where, after several days, scientists carried out postmortems on animals that had died during the holding period and on some that were sacrificed at the end of this period. During the trials 3,492 guinea pigs and 83 monkeys were used in this manner. Humans as well as other animals ended up being exposed in these trials. On the last day of Cauldron, a fishing vessel, the *Carella*, strayed into the path of a trial using plague.[38] The trawler was tailed by two naval vessels for 21 days, waiting for any distress call. When none came, almost all records of the incident were burnt.

The trial report concluded that good results had been obtained for *Brucella suis*, "new ground had been broken with plague trials," and the pontoon technique had been a success. A year later, this assessment was contradicted by the BRAB. Apart from the utility of the pontoon, other results were now seen as less promising. Fildes, with the general agreement of his colleagues, declared that the test of plague had been a "failure" and that brucellosis "had not increased its reputation as a dangerous agent."[39]

The following year saw a return to Lewis for Operation Hesperus. This series of trials encountered terrible weather and was eventually judged to have been "productive of experiments and less productive of positive conclusions," with a "fair number" of negative results among the 72 toxic runs.[40] *Brucella suis* and *Francisella tularensis* were used as the test agents, although there had been some discussion among Porton's scientific advisers to include a virus, Venezuelan equine encephalitis (VEE). In their discussion of the preliminary results of Hesperus, BRAB members noted that tularemia had proved to be a major disappointment, with very few organisms surviving dispersal in trials over longer distances. The scientists also reported that their trials using spraying techniques produced far better results than those with prototype bombs. In the BRAB annual report for 1953 a briefer assessment attributed the poor success of the long-range trials to bad weather and nearby cliffs. The report remarked, however, that there was evidence to "suggest that organisms are less damaged by half a mile travel than by explosive dispersal."[41] This passing remark appears to be an early endorsement, albeit tentative and conceptually embryonic, of the idea that spraying a cloud of organisms toward a target might show more promise than using a bomb directly against that target.

Operation Ozone

Following Operation Cauldron the prime minister, Winston Churchill, had approved a decision to return to the Bahamas for trials. Following a three-week delay, due to damage to the pontoon, Operation Ozone commenced early in 1954 some 60 miles south of Nassau. The thinking behind these trials had changed, according to the scientific report on Operation Ozone, which noted that this shift had been guided by changes in policy that will be discussed later in this chapter:

> In accordance with higher policy, and with the co-ordination of our work with that of our North American allies, the emphasis of these seaborne operations has continually shifted away from *ad hoc* trials of weapons to more fundamental experiments on the behaviour of pathogens in natural conditions. There is an intentional distinction between terms in this statement: we are no longer concerned with "trials" which may amount to no more than a demonstration that a contrivance works, but with "experiments" similar in kind and quantity to those done in the laboratory . . . The most interesting question which the field team has been asked is: what happens to an airborne infective particle?[42]

And, indeed, the focus of Ozone was on how agents survived after dispersal, with the objectives of studying how airborne travel affected *Brucella suis* and *Francisella tularensis;* undertaking preliminary studies with a virus, VEE; and examining the influence of various dispersal methods on the pathogens. These dispersal methods included high explosive, propellant explosive, and spraying. The familiar pontoon-and-bomb method was employed in a small number of trials, in which scientists measured effects on the agents over short distances. The major change with Operation Ozone involved investigating the behavior of agents over distances up to 1,200 yards. In these trials, experimenters placed agents in a flask "fitted with a device which introduced compressed air into it and emitted the liquid in the form of a spray on firing." One significant result emerged from these spray tests: the vulnerability of the organisms in daylight. The trial report, referring to the three test organisms, pointed out that "we can confidently say that their offensive use in such conditions would lose a great deal of its potential effect through aerosol decay and lack of persistent contamination. This would appear to apply only to seaborne attack using onshore winds, which would generally have to be made in daylight: other forms of attack would more probably be made at night or in enclosed spaces."[43]

Operation Negation

Approval for the next series of trials, Operation Negation, scheduled for the Bahamas between October 1954 and March 1955, came from the Chiefs of Staff in August 1954. Permission came with an ominous proviso: the Chiefs of Staff had "stipulated . . . that the continuance of trials beyond 1955 must be subject to further consideration at the appropriate time."[44]

During Negation, Henderson intended to use the same organisms as in Ozone and possibly also eastern equine encephalitis, "as it was more lethal."[45] He proposed to supplement the virus work by using vaccinia virus, related to cowpox and smallpox virus and proposed as a simulant of smallpox, in order "to break new ground." In the event, Negation employed *Brucella suis, Francisella tularensis,* and vaccinia. By June 1955, BRAB members had summarized the main conclusions drawn from Negation. As with Operation Ozone, they emphasized the "extreme dependence of survival of airborne micro-organisms on external conditions: the life in unfavourable circumstances was very short, in favourable circumstances very long."[46] Prior to this discussion Fildes had noted pessimistically that "it was becoming clear that the hazards of biological warfare, as distinct from sabotage, were not as great as has been thought. A statement should be made to this effect."[47] Although this opinion was minuted, no official statement of it was made by the board in its recommendations from the meeting.

Special Operations Trials

On land during 1955 another series of trials took place away from the MRD to ascertain the vulnerability of Whitehall to a germ warfare attack. There is very little detail about these Special Operations trials in the public domain. With assistance from the Post Office and MI5 (Military Intelligence Department 5, responsible for domestic security), at least two trials took place. The first simply investigated the air currents in the building and found leaks throughout the system. The following trial was "more ambitious" and involved MRD staff distributing spores of the simulant *Bacillus globigii*.[48] Two weeks after releasing the bacteria, experimenters took samples at various points within the building and concluded that the Air Ministry Citadel and the Whitehall telephone exchange had been contaminated. The BRAB was also informed that a later test had succeeded in contaminating the Cabinet Office.[49] After this initial investigation, which seemed to have confirmed fears about sabotage, no immediate follow-up trials were planned. These do not appear to have been the only trials connected with a potential sabotage attack. Henderson elsewhere referred in passing to "contamination work" that had been carried out on the London Underground, suggesting that this form of attack could be "applied to the whole country."[50]

Policy Changes: "A Change of Wording and a Rearrangement of Order"

Over the course of the biological warfare sea trials, the broader economic and policy context changed. In 1952, as an already economically exhausted Britain began to feel the cost of involvement in the Korean conflict, the Chiefs of Staff formulated an extensive strategy review at the behest of a new Conservative government.[51] War, they judged, would most likely be a short and intense affair, and all preparations outlined in the document were geared toward this possibility.[52] All WMD remained as likely deterrents, but now a greater emphasis was placed on nuclear deterrence, though without neglecting the additional role of conventional forces.

BW still featured as an integral part of this new strategy, but was now no longer on a par with nuclear weaponry. In March 1952, a month before their serious deliberations for a new Global Strategy Paper commenced, the Chiefs of Staff issued a detailed appraisal of biological warfare. In it they noted that British research had attempted primarily to gauge the danger of biological warfare. The US researchers, on the other hand, were being pushed toward offensive research, weapon development in particular. The report predicted that a strategic germ weapon with "uncertain capabilities" would be available to the US Air Force within a year. A number of suggestions for British policy were also made. First, resources should be devoted to the "immediate defensive problems" facing civil defense and the military services in the UK. Next, some appreciation should be made of the "military and strategic consequences of the probable United States weapons." A third proposition was to "diminish our effort on short term offensive problems but we should strengthen them on research aimed at very long range offensive possibilities." Finally, research should be directed at "learning all we can of the ultimate dangers of BW, and dealing with short and long term defensive measures."[53]

When the Global Strategy Paper was issued in June, the Chiefs of Staff adopted a strong retaliatory position on biological and chemical warfare: "The Allies should not take up a position which would deprive them of the ability to use Bacteriological Warfare or Chemical Warfare in retaliation, if such were to their advantage."[54]

Yet despite this endorsement, the emphasis in British defense policy was now firmly on nuclear deterrence. And, with the first British atomic bomb tested in October 1952, the importance of biological warfare in defense policy began to decline. In the wake of the Global Strategy Paper, the Chiefs of Staff drafted a thorough report for the Cabinet Defence Committee on "Biological and Chemical Warfare Research and Development Policy." In a draft version, which clearly drew on the March assessment of biological warfare, the Defence Committee was asked to agree on a threefold research policy: "Research and trials to determine the true risk of biological warfare should be continued; Research to establish the best defensive measures should be continued; offensive research should be concentrated mainly on the study of long range possibilities."[55]

Despite, or possibly because of, these developments during 1952, the details of policy remained far from clear at all levels of discussion. The scientific advisors, in particular, felt ostracized by secrecy. Fildes reported to the BRAB that he had been involved in disagreements with Ministry of Supply officials, which he attributed to his "lack of knowledge of changing policies."[56] In response, the chief scientist at the Ministry of Supply, Owen Wansbrough-Jones, pointed out that the Chiefs of Staff were not scientists, and so could "give broad direction only and had agreed on a policy that research and trials to determine the true risk of BW should be continued, the best defensive measures should be established, and that we should concentrate mainly on the study of long range offensive possibilities."[57]

Here, the chief scientist was simply quoting the Chiefs of Staff recommendations from earlier in the year, and which were condensed into a formal directive on biological warfare a few months later. BRAB members approved the directive in February 1953, although the minister of supply, Duncan Sandys, had not yet commented on it. Matters were complicated because Sandys was also preparing his own general directive on BW policy.[58] By June, however, the Chiefs of Staff directive, which had not been altered further, reached a stage where no major modification by any of the advisory committees was expected before it was approved at ministerial level.

The minister of supply issued his separate directive on BW research in November 1953, and although much of its content appeared to consolidate the various deliberations from the previous year, there were some significant changes of emphasis.[59] Certainly, the directive had not gained a particularly easy passage past the Chiefs of Staff. In July a draft version suffered delays as doubts were expressed in various departments about its "timing and suitability."[60] Another draft was issued in August and a few months later was submitted to the Cabinet Defence Committee for approval. From the Admiralty perspective, the new draft had successfully shifted emphasis toward a defensive policy. Moreover, because the Navy was the service "least concerned with the offensive and defensive aspects of this form of warfare," the new draft had advantageously reigned in the scope and economic scale of activities about which it had only marginal concern:[61]

> A change of wording . . . and a rearrangement of the order in which the objects of research are set out . . . combine to put the emphasis on the defensive aspects of this work rather than on offense and retaliation. Further a paragraph in the earlier directive which might have been taken as authority for scientists engaged in biological warfare research to range rather widely over their subject, has disappeared entirely. As the draft now stands, it does not appear to be much more than an orderly statement of what is in fact going on, and it does not contain instructions which could be held to cover the Minister of Supply in demanding an unreasonable share of resources for this purpose.[62]

The change of emphasis did not escape the notice of scientists on the BRAB, who complained that "priority appeared to be given to the defensive and research aspects rather than to the building up of BW offensive potential by weapon development and production investigations."[63] The revised version of the directive nonetheless echoed the Global Strategy Paper by announcing that, in the event of war, Britain should be able to "protect her civil population and Service personnel, as well as crops and livestock, against attack by biological methods, and to retaliate by those methods against the enemy should the Government of the day decide to adopt this course."[64]

Research was to concentrate on six areas. Protection and treatment were the first two items on the list, followed by "practical potentialities of biological methods of warfare" relative to the cost and effectiveness of other means of warfare, obtaining suitable agents, bulk production techniques, and the determination of appropriate weapons. Close collaboration with the US was encouraged for the purposes of "economy and efficiency." Complementarity was also mentioned at this point, specifically with respect to weapon R&D: "Since the Americans are concentrating mainly on the development of weapons capable of early introduction into the service, our programme of weapon development should put somewhat greater emphasis on the study of long term projects."[65]

The two directives help to explain the shift from "trials to experiments" noted in the introduction to the Operation Ozone trial report. While this shift was interpreted in the Ozone report as a positive move away from ad hoc testing, it could equally be construed as implementing the policy shift to longer-term projects and, significantly, as a mark of the decline of military interest in BW. Accordingly, the BW scientists did not benefit in the longer term. In March 1954 a DRPC review of defense R&D reported that work on WMD had been fruitful only with respect to nuclear warfare.[66] The review continued in the same negative vein: "It is still too soon in so new a field as biological warfare to forecast with any certainty what developments may yet arise . . . Development of offensive biological weapons has been largely disappointing . . . Apart altogether from the political issues which must be faced before biological warfare could be initiated, the use of such weapons must be limited even in the strategic role to attritional forms of warfare."[67]

A few months later the DRPC produced a final version of this review, which opened with a statement of "new" defense policy. This was interpreted as meaning that "to prevent global war we need a deterrent that includes nuclear weapons and their means of delivery; we must be prepared for warm wars, which we define as peripheral wars of the Korean or Indo-China type."[68] Biological warfare received only brief attention in this final version, including a recommendation that "with regard to offensive BW, it is stated that basic research that is necessary for a proper assessment of the problem of defence is also necessary for work on the offence. On the other hand, it was considered that we could not afford to undertake any work that had not primarily a defensive bias. Our effort should be limited to the defence aspect of BW."[69]

This recommendation had already been acted on by the Air Ministry, which in July 1954 cancelled its requirement for an antipersonnel biological bomb. Although the memorandum cancelling the weapon noted that the Air Staff would "ultimately require toxic biological weapons," it would not need them just yet:

It is now apparent that because of the magnitude of the problem a great deal of research still remains to be done in both the agent and weapon fields before a satisfactory weapon can be recommended to the Service. Moreover, it appears that the storage, transportation, testing and preparation for use of the weapon and its agent will present new problems to the Service and may necessitate the provision of special skills and equipment. The Air Staff considers that until problems associated with the toxic biological weapons are better appreciated it would be unwise to state definite requirements for weapons and other equipment. This A.S.T. [Air Staff Target] is therefore issued to replace ASR [Air Staff Requirement] No. OR/1065 which is hereby cancelled.[70]

The new aims of the Air Force were then spelled out in broad and general terms: "The Air Staff requires research to determine effective agents and suitable weapons for waging biological warfare. They also require investigation of the associated problems connected with the storage, handling, transportation, testing and preparation for use of such agents and weapons with a view to determining what is required in the way of skills, equipment and procedures in the Service."[71]

Defense: "The Committee Wanted to Vote against BW but Didn't Know Quite How to Do It"

Changing policy soon filtered through to the MRD's 1957 successor, the Microbiological Research Establishment (MRE), where the move to a defensive outlook had already resulted in several reversals of fortune for Experimental Plant No. 2, the fermentation plant intended to produce pathogenic agents in bulk. The series of sea trials with pathogens also ground to a halt. Following Operation Negation in 1955, further sea trials for 1957 and 1958 had been agreed in principle by the Ministry of Supply's Scientific Advisory Council, but the committee noted that the trials might have to be deferred because of difficulties in providing a crew for HMS *Ben Lomond*.[72]

In parallel with these developments, the Air Force's goal of a massive nuclear retaliatory capability had moved closer. The decision by Churchill's government to produce a British hydrogen bomb was taken in 1954,[73] and the first delivery of nuclear bombs was made to the Royal Air Force at the end of the same year. During 1955 the first squadron of Valiant bombers became operational. Meanwhile, the Chiefs of Staff now regarded BW as suitable only for the very global war that nuclear weapons were meant to deter. It is therefore unsurprising to find the chief scientist,

Wansbrough-Jones, writing of the 1956 decision to discontinue sea trials: "At the DRPC yesterday the Committee wanted, I think, to vote against BW in general but didn't quite know how to do it so the Chairman was more inclined to be helpful . . . On present notions of priority (global war) it would really come out as priority six with no effort being allocated. I accepted this without argument, because there really was no point in arguing."[74]

The upshot of this maneuvering was that the trials were to be included in the formal defense research program, but no resources would be devoted to them. Within the Ministry of Defence, Admiralty representatives had already stated categorically that HMS *Ben Lomond* would not be made available for future biological warfare field trials.[75] Negation was the final British trial at sea with pathogens. The only concession for germ warfare was that the DRPC recommended a review of both BW and chemical weapons (CW) policy "against the background of the use of the megaton bomb and the possible limitation of nuclear tests."[76]

The status of biological warfare had thus diminished radically during the 1950s, from being placed on an equal research priority with nuclear warfare, and assuming a complementary strategic deterrence role, to a low status in defense policy and supported by a largely defense-orientated research program. This change was, in part, a consequence of the UK's increasing reliance on a nuclear deterrent and, in part, the result of cutbacks in defense expenditure. One might have expected the gradual shifts in program and policy to culminate in a firm Cabinet decision to adopt a defensive biological warfare policy. Instead, the decision was subsumed under a separate policy decision concerning chemical warfare. In July 1956 the Cabinet took a highly secretive decision to abandon an offensive CW capability.[77] Specifically, it ended the large-scale production of nerve gas and the development of nerve-gas munitions and resolved to destroy the residue of the World War II stockpile of other chemical agents and weapons.

There was no mention of biological weapons in the record of this Cabinet decision on chemical weapons. Nonetheless, the DRPC staff, who serviced the committee, interpreted the Cabinet decision over chemical warfare in broader terms, contending: "The arguments which led to the cancellation of weapons for the offensive use of chemical warfare agents largely apply to BW weapons."[78] And this position was underlined in December 1956, when the minister of supply was provided with a reappraisal of the 1953 directive. He was informed that defense research was "going pretty well" but that the offensive aspects of the directive were in doubt, in particular, production "of agents and charged weapons. Tacitly, mainly through the pressure of economy and partly through some belief that the BW policy on weapons should approach the CW one, less and less work is being done under these heads."[79]

The same memorandum stated that "if Government policy permitted," it would be more accurate to delete those parts of the directive relating to determining suitable biological agents, to bulk production and storage of agents and filling weapons, and to suitable forms of weapons. Instead, the directive was maintained by the Ministry, but defensive parts had been assigned a higher priority than offensive aspects. In all practical respects, the British biological warfare program was now defensive only.

Renewed Threat: "Fully Utilizing the Insidious Nature of Biological Agents"

The move to a defensive program did not, as might have been expected, amount to a retreat from an ambitious research program at the Microbiological Research Establishment. Whether as a deliberate strategy or through more indirect means, it was imperative for those involved in the area that the status of biological warfare be revived within the ambit of a defensively orientated policy regime. Scientists, in the new policy environment, began to reconceptualize the potential threat, no longer as an on-target attack with a bomb, but in terms of sabotage and coverage of very large areas with pathogens.

British scientists put the case for revising how the threat was to be construed to the Americans and Canadians at their annual collaborative conference in 1956. This presentation is worth quoting at length, as it provides an insight into the thinking behind the large area threat. The extract also demonstrates how the divergence in policy between the UK and the US, which was still wedded to an offensive program, translated into differences in conceptualizing the goals of their respective research efforts:

The United Kingdom has led Tripartite thinking in emphasising the special properties of BW which not only distinguish it from other weapons
but also make it of great potential effect. These properties are the delay
between exposure and clinical symptoms, and the difficulty of achieving
any form of rapid detection. It was stated plainly by the UK representatives at the 10th Tripartite Conference that the present emphasis (especially in American work) on overt on-target attack with conventional
aircraft BW cluster bombs was, in their opinion, quite wrong and was
missing the whole point of the covert potential of BW. There are two aspects to this covert use. Firstly, what is loosely called sabotage—the undetected release of an airborne BW agent in or near an important installation . . . Secondly, the possibility of covering a very large area and long
distance. This is an unproved idea, depending on the biological, mechanical and meteorological considerations which are far from fully worked
out.[80]

So, although the US still held out hope for a workable BW bomb, the
attention of the UK scientists had now turned to sabotage and large area
coverage. While large area coverage was unproven, according to the authors it still merited further investigation as potentially a credible threat.
Referring to US tests, they argued that in

American trials as long ago as 1952 simulant material has been spread
over areas of up to 34,000 square miles. It is notable that the American
emphasis on conventional BW munitions is in despite of their pioneer
work in both "sabotage" and large area coverage. The point of importance to the present argument is that either form of attack depends on
the ability to disseminate the agents in the form that will survive exposure to natural conditions, in the airborne state, for substantial periods.
And so indeed, does the conventional attack, depending in large measure on penetration into buildings to establish concentrations that are
relatively low but will persist for a long time. It is therefore of fundamental importance that as much as possible may be learned about the dissemination and survival of agents under natural conditions.[81]

A year later, Henderson, the director of the MRE, was still attempting
to refocus everyone's attention on sabotage. Early in 1957 he reported to
the BRAB that he had attended the 11th Tripartite conference and argued

"that the manufacture and dropping of bombs was not the best way to employ BW, and that sabotage in its various forms was probably the most effective method."[82] Henderson, however, remained skeptical about whether this opinion would even be minuted in the conference proceedings.

A few months later, in the June meeting of the BRAB, mention was made of new possibilities for waging biological warfare. During the meeting, Henderson reported to the board that collaboration with the US was being threatened by shifting policy in the UK, noting that "the Americans had begun to regard us as poor allies owing to their belief that we were retrenching in BW policy." The US was continuing work in an offensive direction, and according to the chief scientist, Owen Wansbrough-Jones, US "production was keyed to policy and was directed at the manufacture of bomb fillings." The British scientists were less enamored with bombs, and, taking the chief scientist's lead, the BRAB recommended "that work on the development of bomb clusters should cease."[83] Although there is no open record of whether this recommendation was endorsed at higher levels, the BRAB did not discuss bombs again. It is not even clear how much work was being performed on bombs at this time, since the Air Staff Target had been watered down in July 1954 from a specific requirement for a bomb to a general request for long-term research on the topic.

Instead of bombs, the topic of large area coverage soon received serious attention from scientists and advisors on other committees. In June 1957 a new Offensive Evaluation Committee held its first meeting within the Ministry of Supply. At the inaugural meeting "consideration was given to proposals that direct attack by conventional weapons limits the effectiveness of BW and that clandestine, off-target methods fully utilizing the insidious nature of biological agents would possibly enable a single aircraft to attack effectively tens of thousands of square miles."[84]

This new threat, now dubbed the large area concept (LAC) by scientists, envisaged an aircraft or ship spreading a line of pathogenic biological agent some miles away from an area (off-target) and thus spreading a deadly cloud across an entire region.

In a later report, John Morton of the MRE, who had essentially been in charge of the earlier sea trials, claimed that interest in the LAC had originated in the US some ten years previously but "did not catch on" because its success depended on meteorological conditions that would require a "lid" of air to prevent upward diffusion of a large cloud. According to Morton, interest had been "re-awakened" during 1955 because of three developments. First, calculations at the CDEE "suggested that the meteorological requirements need not be so stringent." Second, work in the laboratory had demonstrated that some agents could survive for long periods while airborne. Finally, Morton pointed out that there had been "growing dissatisfaction with the inefficiency of on-target attack by cluster bombs."[85]

Researchers from both the chemical and biological sections at Porton put forward the case that the UK was especially vulnerable to the new threat. In a detailed report to the Offensive Evaluation Committee they discussed possibilities and problems concerned with maintaining an aerosol cloud of organisms over a long distance, keeping the organisms alive and also their means of dissemination. The report concluded that "in general, the feasibility of effective attack of very large areas with BW agents is far from proven, but evidence is available which would make it dangerous to assume that it is not possible."[86]

This recommendation carried with it an implicit call for further research. Whether or not this was deliberately implied in order to revive the trials program, a proposal to investigate and evaluate the large area threat would have aligned with the new defensively orientated regime in biological warfare policy. There are, at the same time, more distant echoes of wider concerns associated with the concept. In particular, hydrogen bomb tests had provoked a great deal of unease and public debate about fallout over large areas and long periods.[87] Although there is no explicit link between nuclear fallout and the biological "fallout" of the LAC, the parallels remain tantalizing.

Within a short time, scientists at Porton had embarked on a renewed series of open-air trials underpinned by the rationale of the LAC as a threat to UK defenses. The first series of such trials used aircraft to spray fluorescent tracer particles of zinc cadmium sulfide to simulate a biological agent cloud.[88] In the tests, the planes released 300 pounds of zinc cadmium sulfide along a 300-mile line over the Irish Sea, with samples taken at meteorological stations in England and Wales.[89] According to a report on the trials by the Offensive Evaluation Committee, "if the samplers gave a true picture," then 28 million people would have received a dose of 100 particles.[90] Volunteer trials in the US had indicated that such a dose would have been effective for causing tularemia and Q fever.

Insignificance and Negligible Risk

The trials progressed but within an increasingly unfavorable policy climate. A DRPC Staff (DRPS) ad hoc working party was established in 1957 to carry out the review that had been requested by the committee a year earlier. Their report quickly became an official DRPC report, which was later quoted by the Chiefs of Staff. The report had been written soon after the UK had successfully exploded its first test hydrogen bomb, so it is not too surprising to read the verdict that, in comparison, BW possessed a "negligible additional deterrent effect." BW was further deemed to have the disadvantages of delayed effects, susceptibility to weather conditions, and lack of precise control, and was potentially subject to premature discovery. The DRPS concluded that "because of these disadvantages, BW is unsuitable for the initiation of war, counter attack or tactical use, although it could be very useful to increase dislocation in the follow-up stages if effort were available." Finally, the committee conceded that some military interest in defensive BW research was still justified in order to keep abreast of new technical developments.[91]

The Chiefs of Staff Planning Committee considered the DRPC paper on chemical and biological warfare together with a report on intelligence and dismissed both the threat from, and potential for, biological warfare: it attributed "the insignificance of BW as a deterrent to global war . . . not only to the existence of nuclear weapons, but in particular to the practical difficulties of delivery and the uncertainty and delay in its effects as well as to the legal aspects of its use." Given these considerations, the deterrent effect of BW could "not be regarded as credible." Strategic uses of CBW against the UK were likewise dismissed. Soviet CBW offensives against nuclear retaliatory bases would, the report judged, serve "no strategic advantage." Against civilian targets, effects would depend on the "conditions of post-nuclear devastation." The high concentration of UK targets in a relatively small area meant that it was "not likely that BW and CW follow-up to nuclear attack would be considered profitable," presumably because of the marginal additional effect compared with the massive destruction that would already have been wrought by hydrogen bombs. Tactical biological warfare was also dismissed as unpromising. In summary, the Chiefs of Staff's report concluded with several telling points against BW, the key items being:

> The strategic value of BW in present known forms is insignificant.
>
> There is negligible risk of BW or CW being used against the United Kingdom which ought to be accepted when deciding upon the scope of defensive measures.
>
> The West must continue to possess an offensive capability in BW and CW, but this does not require the United Kingdom itself to possess such a capability.
>
> The degree of risk does not justify defensive measures against BW or CW being taken by units of armed services, based in the United Kingdom, or by ships at sea.
>
> The West as a whole must continue some co-ordinated research and development effort into the offensive and defensive aspects of both BW and CW.[92]

Once the Chiefs of Staff had all but dismissed BW from the defense agenda, the possibility of abandoning all UK research in both chemical and biological warfare was raised. The DRPS issued a report on this matter early in 1959, which considered both the financial savings and the political benefits flowing from close collaborative links between research scientists in the UK and US.[93] The DRPS recommended that the DRPC not support the closure of Porton. Instead the staff suggested that the committee inform the Chiefs of Staff Committee that, if all work ceased at Porton, a maximum of £1.6 million per annum would be saved, probably less. In addition, they noted, there were too many scientific and political advantages arising from current collaboration to terminate all CBW research.

Despite being consigned to "insignificance," work at the MRE continued, and five more long-distance trials of fluorescent particles were carried out from October 1958 to August 1959.[94] Toward the end of this series, however, the BRAB was informed that resources for future trials should be obtained on an informal basis; the DRPC "would not allot a high priority to the work, and the official priority might in fact be less effective than the present loosely defined arrangements."[95] The BRAB, in response, continued to protest about the slow progress of aircraft trials in the UK. The board also reaffirmed "its belief that large area attack with BW agents constituted a major threat to the country's defences, and its profound dissatisfaction with the low priority accorded to this threat."[96]

The trials did continue, and by July 1960 a total of 12 tracer trials had taken place, the last 2 from ships rather than aircraft.[97] These sea trials involved releases of fluorescent particles, "simulating a breathable BW cloud as regards particle size"—one in the English Channel and the second in the Irish Sea.[98] The results were taken to demonstrate that an attack from off the coast was a potential threat. At around the same time, the idea that Britain would be vulnerable to a large-scale attack spurred new efforts at the MRE to produce the means to detect such an attack.

A Reawakening of Interest

The constant struggles by the BRAB and others to raise the profile of biological warfare started to meet less resistance in the early 1960s. The new chief scientific advisor at the Ministry of Defence, the zoologist Solly Zuckerman, ordered a review of biological (and chemical) weapons policy in 1961.[99] The Chiefs of Staff had already asked for an operational assessment of CBW and had not entirely dismissed the possibility of manufacturing them at some point.[100] This move had already been interpreted positively by one BRAB member as an "awakening of interest" in the area. Reporting to the board on his recent visit to the US, another BRAB member, the virologist Wilson Smith, said that he "had been impressed by the increased interest in bacteriological warfare that had resulted from the nuclear stalemate."[101]

Zuckerman's panel was chaired by Sir Alexander Todd and given the task of considering "the potentialities in warfare of biological and chemical agents, and to make recommendations about the scope of the programme devoted to their study in the UK."[102] The papers for these assessments were completed by the end of the year but were deemed "too complex" for the Chiefs of Staff.[103] In order to remedy the situation, the War Office chief scientist, now Walter Cawood, established a subcommittee of the DRPC to prepare a shorter and simpler paper.

Meanwhile the Chiefs of Staff in their operational assessment of BW had concluded that "while we cannot foresee the need for this country in isolation to acquire an offensive strategic potential, BW if further developed might be used" as an incapacitating weapon in limited war, as a clandestine weapon, as "a weapon in the exploitation phase in Global war, and on rear target areas in all phases of global war." The Todd Panel had concluded that the USSR was not a suitable target for BW attack and had ruled out the potential of BW in tactical situations. Cawood's ad hoc group, drawing on these conclusions, agreed that "an expansion of effort on investigations of the offensive aspects of CW and BW is warranted," although they added that no strategic offensive capability in either chemical or biological warfare was required. They also complained that the Todd Panel's statement that Russia was not a suitable target for a germ warfare attack was too categorical and might change, given future technical advances.[104]

The change in fortune for biological and chemical warfare rests in part upon the threat from the LAC and, on the chemical side, the potential of incapacitating agents and the discovery of a new and especially lethal series of chemical agents, the V-agents. Broader changes in the political climate, suggested by Wilson Smith's earlier comment on the nuclear stalemate, also contributed to a more favorable context for biological warfare policy. Continuing threats from Khrushchev over Berlin from late 1958 into 1961 and the building of the Berlin Wall did "not shift the British view that only nuclear war could prevent the Russians from strangling West Berlin if they chose to do so and that, despite the rhetoric, West Berlin was not worth a holocaust."[105]

And if Berlin was not worth the holocaust, neither was Cuba the following year. In this light, while still emphasizing an independent nuclear deterrent, UK policy began to shift and, following the US, the notion of a graduated deterrence emerged.[106] Nuclear deterrence was to be a last resort, and thus a new role for CBW was suggested on the way to this ultimate response.

Policy did not immediately adapt to these changes in strategic outlook. At a meeting of the Chiefs of Staff in 1962, Cawood pointed out that the defensive policy for CBW still dated from 1958, when "it was considered that if the West were subject to an attack by the USSR the response would be immediate retaliation with nuclear weapons; it was not therefore considered profitable to develop an offensive strategic BW or CW capability."[107] The changed situation was, however, noted in a Joint Planning Staff report on CBW that drew together the operational assessments, Todd's report, and Cawood's synthesis.

Chemical warfare, in particular, fitted the new thinking about a graduated response to the Soviet threat. According to the Joint Planning Staff, in a global war "CW could be an effective means of delaying the enemy in a period before nuclear weapons are used. Although tactical nuclear weapons might be more effective in producing delay, there is risk of escalation. We agree that the use of CW might well keep the battle under control and provide a means of delaying the enemy and gaining time for negotiations."[108]

Moreover, CW were envisaged as having the potential to prevent escalation from limited to global war in areas such as Southeast Asia, because CW were deemed more "politically acceptable" than tactical nuclear weapons. In the final amendments to the report, the recommendation on CW read: "Further study is required of the tactical employment of lethal and incapacitating CW agents in a global war in Europe before nuclear weapons are used and their possible subsequent use in conjunction with nuclear weapons."[109]

BW, on the other hand, were taken to be more of a threat than an opportunity. The Joint Planning Staff of the Chiefs of Staff Committee conceded that BW did have potential in a global war, but CW would be "more effective in the period preceding the strategic nuclear exchange, and thereafter attack with either CW or BW would be irrelevant." In a limited war, BW might be used as incapacitants or for secret operations, but once again chemical agents were deemed more effective for these purposes. Against these possible opportunities, the planning staff considered that the UK was especially vulnerable to both chemical and biological attack because "the United Kingdom may be covered in one attack, the Sino-Soviet Bloc could not." Moreover, there was thought to be no advantage in initiating a biological attack:

> Political implications apart, we would not wish to be the first to use BW against the Soviet Bloc because of the relative vulnerability of this country should they retaliate strategically against the United Kingdom. However, we agree with the DRPC that there should be increased research

into the offensive use of BW both in order to ensure that defensive measures are properly developed, and to be aware of any break-through increasing its potential effectiveness.[110]

Because of the nation's vulnerability, the planning staff also noted that the DRPC had recommended that "as a matter of the highest priority" the UK should carry out a military and scientific assessment of the threat to the UK.[111] This was to include "large scale BW dissemination trials." Additionally, a five-year increase in the MRE budget was requested to develop early warning systems and immunization techniques.

The case for renewed political and military interest in biological warfare continued to mount, with pressures coming from both East and West. In 1962 NATO's SHAPE (Supreme Headquarters Allied Powers Europe) had requested that "all nations in Allied Command Europe should equip their shield forces defensively and with a chemical and biological retaliatory capability."[112] The British authorities, in the light of this request, had no wish to reveal that they had neither a retaliatory capability nor up-to-date defensive equipment. Looking east, a lengthy Joint Intelligence Committee report provided a less-than-certain but nonetheless worrying appraisal of Soviet biological warfare capabilities:

> We must assume Soviet parity with the West in basic knowledge and technical know-how. Soviet familiarity with the broad lines of current Western outlook on biological warfare as revealed in publications is clearly established. With wide powers of direction at all levels, with tight security, and with vast remote areas in which field trials could be carried out, the Soviet authorities are in a favourable position to conduct biological warfare research and development unknown to the rest of the world. Some areas are strongly suspected of being concerned with biological warfare trials, but details of the nature of any such trials are lacking.[113]

As a result of these deliberations and various reports on CBW the Chiefs of Staff recommended to Minister of Defence Peter Thorneycroft renewed efforts to "increase somewhat the level of our activity in this area" if politically and financially acceptable. This increased activity was to amount to about £1.294 million additional expenditure over five years. In addition a political decision was called for on the proposal to conduct large-scale dissemination trials using microbes.[114] The proposals were approved by Harold Macmillan's Cabinet in May 1963 and were summarized as follows in a brief for new ministers: first, and most generally, "The UK should develop a limited chemical retaliatory capability"; second, "Research should be carried out on biological agents with an authorised expenditure of £0.47 million over 5 years"; and finally, "production of equipment for defence against chemical and biological agents" should proceed "at a rate of £17.5 million for the three services over the next 5 years."[115] In the event, the amount allocated to biological warfare research was split between £0.37 million over five years for research on biological agents and £0.1 million over three years for outdoor trials.[116]

Field Trials with Simulants

The open-air trials now had two main aims: assessment of the threat from the LAC and the early detection of biological warfare agents.[117] They were accompanied by a "ventilation trial" using the simulant *B. globigii* on the London Underground in July 1963.[118] Scientists at Porton were informed that further approval for use of living organisms in trials by the secretary of state for war was not required, although the chief scientist would keep him informed of progress.[119] The chief scientist justified the trials to the BRAB on the grounds that the release of "harmless micro-organisms presents no special hazards," especially in comparison with activities such as brewing, sewage disposal, and agricultural operations.[120]

A progress report on field trials undertaken in 1963 and 1964 noted that 10 trials had been carried out, mainly at night, using a ship as a source vehicle along the south coast of England. Initially these trials had used dead, stained organisms.[121] Most of the remaining trials used a mixture of live *Escherichia coli* (162) and *B. globigii* spores, the latter being used as a tracer.[122] From October 1964 into May 1965 13 trials took place at night in Lyme Bay and Weymouth Bay on the south coast.[123] Scientists and naval personnel released suspensions of *E. coli* (162) and *B. globigii* along a line between 5 and 20 nautical miles long and between 5 and 20 nautical miles from the shore. The cloud, which was tracked by releasing balloons carrying radar reflectors, was generated by four spray heads spraying bacterial suspension for between 55 and 113 minutes. Sampling of the cloud was performed on land at distances of up to 37 nautical miles downwind as the simulated biological attack spread over southern England.[124] Scientists drew various conclusions from this series of trials—primarily, that *E. coli* survived better when airborne in large particles than in small ones.[125]

The later trials also employed a new technique that involved "microthreads" made from spider's web to hold the microorganisms during studies of their viability in a simulated airborne state.[126] Alongside the spraying trials, sets of *E. coli* held on microthreads were exposed to the atmosphere in exposed and sheltered sites. The scientists concluded that *E. coli* on microthreads survived better than airborne cells. Even with this disparity, they argued that the microthread technique was still in better agreement with field results than laboratory methods for estimating survival.[127]

Prudent Preparation or Expensive Insurance?

Throughout these trials, BW policy remained unchanged, despite a change to a Labour government in 1964. A note from the secretary of state for war to Prime Minister Harold Wilson in November 1965 confirmed that policy, inherited from the previous government, was to develop defensive measures.[128] By February 1967, and with a review of defense expenditure R&D under way, the prime minister had not replied to the memorandum. At this stage, one of the successor committees to the DPRC, the Defence Research Committee (DRC), decided to allow the review to run its course before returning to the prime minister. With the MRE employing 120 staff and costing around £1 million per year, the establishment was a prime target for possible defense cuts. A Ministry of Defence briefing note reported that the DRC was convinced that BW could be made and stored easily and cheaply. On the other hand, the committee "was divided on its views about the likelihood of an attack and was therefore unable to reach a decision about the whether the work of the MRE should be regarded as a thoroughly prudent preparation for a real danger or as a rather expensive insurance against a remote contingency."[129]

Members of the committee had argued that an enemy might regard BW as more humane than nuclear warfare, adding that BW might also be used either to divert attention to internal matters from events elsewhere in the world or to take over the country as a "going concern" with its industry intact. Other committee members had dismissed these ideas as unlikely while the UK remained within NATO. The DRC had, nonetheless, expressed the view that tactical use of BW in a limited war was increasingly feasible, with new combinations of agents able to produce effects within four hours. Hence, the committee reported: "The tactical value of BW may thus be greater than was previously thought, and it may be that, irrespective of doubts about the likelihood of strategic attack against this country, MRE's work on defensive measures should nevertheless continue on account of this tactical possibility."[130]

At around this time the Joint Intelligence Committee also reported that the "likelihood of a full scale BW attack on the UK is small . . . although the Russians possess a delivery capability, their preparations appear to be defensive rather than offensive."[131] This view was supported in an earlier review of the BW program, which noted that a BW attack was feasible but unlikely because "it is difficult to see convincing military or political motives for such an attack."[132]

A later note to the minister of defense carried these assessments forward, explaining that the UK continued to adhere to the Geneva Protocol and would not initiate chemical or biological warfare. With respect to retaliation, the author noted that there were no plans to retaliate in kind against a biological attack, and although the previous Conservative government had sanctioned research on a limited chemical retaliatory capability, the prime minister had since maintained a "masterly silence" on the subject. The author did, however, endorse research for "passive defence." With this in mind, the memorandum recommended that the Ministry of Defence should "try and shift the main load of running it [MRE] onto someone else's back—the Medical Research Council, or the Minister of Health, or the Department of Education and Science . . . I suggest our slogan should be 'out of the MRE by 1970.'"[133]

This view was endorsed by Roy Mason, minister of defense for equipment, who argued that although the MRE could be shifted to civil departments, the decision needed to be made in the light of the spread "of knowledge about chemical and biological warfare among middle-class Powers who will quite soon be able to arm themselves with highly effective weapons of this sort at a fraction of the cost in money or facilities of a nuclear armoury."[134]

In the wake of this discussion, a variety of funding sources for the
MRE was considered, such as the University of Southampton Medical
School, the Home Office, Ministry of Health, Ministry of Technology, the
National Research Development Council, the Science Research Council,
the Ministry for Overseas Development, and the Medical Research Council (MRC).[135] Of these, the proposal to the Medical Research Council appears to have been most vigorously pursued within the Ministry of Defence, although there is evidence that this move was regarded by the
Chiefs of Staff as a "bluff" that they thought would fail and thus allow them to maintain control of the establishment.[136] The chief scientific
advisor, Solly Zuckerman, made a number of unsuccessful approaches
to the secretary of the MRC, Harold Himsworth. Zuckerman conveyed
Himsworth's reasons directly to the secretary of state for defence, Dennis
Healey:

> Himsworth's general line is that it is not the policy of the MRC to dabble
> in affairs that are not designed to improve health, and that they are also
> in general averse to secret work . . . they would not only be going against
> one of their principles, but might also be damaging their "image" in
> countries overseas, which assume they have nothing to do with matters
> relating to defence and, in particular, with subjects like microbiological
> and chemical warfare.[137]

Other efforts to "civilianize" the MRE in the UK continued unsuccessfully alongside cutbacks in defense expenditure. Furthermore, at this
time Britain took a lead role in proposing a new Biological Warfare Convention to the Eighteen Nation Committee on Disarmament. The details
of the UK's involvement are discussed elsewhere.[138] The Ministry of Defence was less enthusiastic about the convention than the Foreign and
Commonwealth Office. Probably as a result, the ministry predicted that
the impending treaty would have little impact on BW research. In 1968
the secretary of state for defence reported to the Cabinet Defence and
Overseas Policy Committee that "we have stated clearly on a number of
occasions that, in the field of BW, we are concerned only with devising effective means of defence. Our position is not affected by the recent UK
initiative in the Eighteen Nation Disarmament Committee . . . The terms
of any such agreement consistent with the UK initiative, would not pre-

clude work of a defensive nature, nor would the need to continue research on BW defence be reduced."[139]

Indeed, although there had been and would continue to be a steady increase in the amount of civilian research performed at Porton, the establishment continued research in "early warning and detection devices, physical protection, prophylaxis, therapy, decontamination and training aids."[140] Two years later a report on the future of Porton concluded that the UK "should have the capability of providing for the defence of our armed forces in the face of the existing threat of a biological attack" and that the budget for this work "was unlikely to be affected for some years by the current international discussions on banning the use of biological agents." Classified work also continued on microorganism production, aerobiology, early warning, and outdoor trials to assess the vulnerability of British troops to a BW attack.[141]

Further Trials

Trials entered a new phase toward the end of the 1960s and into the 1970s, when assessment of risk moved away from the entire population and became focused on naval crews.[142] It is not clear why this shift took place, although a 1966 review of BW research noted that "the central scientific staff . . . take the view that the present swing in the research effort away from BW attack on the United Kingdom toward its use in small scale operations should receive more emphasis."[143] Additionally, there is little doubt that senior scientists at Porton would have been aware of the Project 112 tests being undertaken in the US (see Chapter 2). However, only CW tests at Porton have been directly linked to Project 112.[144]

Shipboard tests were initially carried out using bacteria held on microthreads in order to compare their decay rates in different parts of a ship.[145] These trials graduated to a larger scale as a frigate, HMS *Andromeda*, was sailed through a cloud of *E. coli* and *B. globigii*.[146] In December 1970 the ship was exposed on three successive nights to clouds released from around three miles away, remaining in the cloud for between two and a half and eight minutes. Dosages measured in the ship were used to estimate the dose that would have been inhaled by crew members, and the trial report noted that the results indicated that, had the simulant been the pathogenic agent *Francisella tularensis,* enough cells would survive to present a hazard.[147]

The scientists next investigated the fate of bacterial cells on the clothing of crew after a ship had passed through a cloud of bacteria.[148] This trial, on HMS *Achilles,* took place over four days in January 1973. On the fourth day the vessel spent about an hour in the bacterial cloud. The trial investigators concluded that a "ship in the normal ventilation condition is vulnerable to biological operations, and showed that the outer clothing of men exposed on the upper deck and in the machinery spaces was invariably contaminated with microorganisms."[149]

From indirect exposure of crews to clouds of bacteria, the scientists moved on to assess direct contamination of individuals with microorganisms. Tests were carried out mainly at Porton, although a few took place on ships.[150] An initial series of tests in 1974 demonstrated that the spores of *B. globigii* could penetrate inner and outer layers of clothing and settle onto hair. A second series of tests, under the codename Gondolier, took place in 1976 in Portsmouth, where "the trials subjects were all young males with about three months of basic naval training."[151] In six batches of five people, the men were sprayed for about four and six minutes with *B. globigii*.[152] Sampling from air and clothing confirmed that there was contamination of the outer clothing and from secondary aerosols, which arose during undressing.[153] Similar trials continued for another year, after which, according to the open literature, the series was terminated.

Other aspects of the British program were also drawing to a close at this time. In 1976 plans were laid to close the MRE within two years.[154] The BRAB was dissolved in 1977. Work at the MRE did continue, but now under successive reviews of its future, and in 1979 the Establishment was reconstituted.[155] A small Defence Microbiology division was created within the Chemical Defence Establishment (CDE) to focus on defense research, while the majority of staff from the MRE remained within a new Centre for Applied Microbiology and Research. During the 1980s and 1990s military leaders regarded the threat from biological warfare as increasingly prominent, and consequently defensive research received a boost. In 1991 the CDE became the Chemical and Biological Defence Establishment (CBDE), falling in 1995 under the auspices of an umbrella organization for several defense research establishments, the Defence Evaluation and Research Agency (DERA). Most recently, in July

2001, the government split the DERA into two organizations, one publicly and one privately owned. The public organization was renamed the Defence Science and Technology Laboratory (Dstl), and it contained the CBDE, now known as Dstl Chemical and Biological Sciences.

Conclusion

The British BW program stands as one of the most significant of the 20th century in its scale, scope, and degree of integration with the state. The program, commenced in order to prepare against a similar attack from Germany, continued throughout the Cold War with the constant threat of the Soviet Union providing a *raison d'être*. It would, however, be over-simplifying to claim that the only, or even the main, influence on the development of both policy and program was a straightforward reaction to the perceived threat or, more accurately, to uncertainty over the threat. The gradual change to, and subsequent maintenance of, a defensive policy was undoubtedly informed by the various, but nebulous, assessments of the Joint Intelligence Committee. It was also clearly a result of changes in the relative priority of biological, chemical, and nuclear weapons in the minds of the military and political authorities, as they witnessed national and international tests demonstrating the power of nuclear weaponry and concurrently struggled with an increasingly stretched defense budget.

Changing policy was also accompanied by various and changing interpretations of how to implement policy and also of the manner of the threat. After World War II, the term *biological weapon* referred primarily to bombs, and here scientists and policymakers could draw close parallels with nuclear weapons. As the race to develop a biological bomb was called off, scientific—and only later political and military—attention shifted to the LAC. Large-scale outdoor trials to determine the extent of this threat were readily justified as conforming first to the new defensive policy, and then as research on "offensive aspects" of biological warfare. By the late 1960s, when a Soviet attack on the UK was deemed feasible but unlikely, the term *offensive aspects* was dropped from Whitehall discussions in favor of emphasizing that British research was aimed at developing defensive measures. Throughout all of these changes, the threat from sabotage remained ubiquitous but relatively disregarded.

CHAPTER 4

The Canadian Biological
Weapons Program and
the Tripartite Alliance

DONALD AVERY

On 14 March 1970 the Canadian delegate to the United Nations Conference of the Committee on Disarmament (CCD) declared that "Canada has never had, and does not possess biological weapons (or toxins) and has no intention to develop, acquire or possess such weapons in the future." This statement was, and is, technically correct: Canada has never acquired an independent biological weapons (BW) capability. But it ignores the important support role Canadian scientists assumed in facilitating American and British offensive BW programs between 1939 and 1969. This involvement ended abruptly in November 1969, when President Nixon curtailed the US offensive BW program, paving the way for the 1972 Biological Weapons Convention (BWC), which was supposed to do away with these deadly instruments of war. Significantly, the ban occurred at a time when allied scientists had demonstrated that biological warfare could actually become a powerful strategic force—given sufficient scientific, technological, and financial resources. And for this reason the BW arms race did not end after 1972; it only went underground.

Since the early 1970s the expertise of Canada's military scientists has been utilized in defensive biological research and in international efforts to develop effective CBW verification measures. Recently, as the threat of BW has increased, with "rogue states" and terrorist organizations being of particular concern, there has been a dramatic increase in the demand for CBW protective equipment and detection systems, both for the Canadian forces and for their NATO allies. Most of the military research has been carried out at the Defence Research Establishment Suffield (DRES), while civil biodefense programs have been primarily the responsibility of Health Canada and federal law enforcement agencies, with increasing cooperation from their provincial counterparts. Yet despite Canada's important historical and contemporary role in responding to the threat of biological warfare and bioterrorism, this subject remains virtually unexplored in the scholarly literature. One reason is that a veil of secrecy has shrouded this important aspect of Canadian national defense and international relations; another is that historians have concentrated their attention on the threat of nuclear weapons.[1]

This chapter examines the central question of whether Canada had an offensive BW program between 1945 and 1972; and, if so, when it was curtailed. It will also assess how Canada's military, scientific, and political leaders viewed BW, and the relationship between BW and other WMD.

Figure 4.1 Major sites involved in the Canadian BW program.

Issues of secrecy, accuracy of intelligence sources, and dependency on scientific and technological support from its two major allies are other parts of the Canadian BW experience. More specifically, an analysis of Canada's role within the Tripartite system, with its annual meetings and specialized working groups, reinforces the argument that most of the major BW projects were based on cooperative research rather than being solely national initiatives. Records of the Tripartite meetings also provide a useful "window" of analysis for understanding the US and UK offensive and defensive biological programs at the operational level, where scientific rather than political priorities predominated. The chapter also surveys NATO's planned responses to a possible Soviet BW attack, particularly during the 1960s, and how these concerns influenced Canadian defense policies.

Any analysis of Canadian involvement with BW must consider four factors. First is the fact that Canada, unlike its two allies, did not have an independent nuclear deterrent; instead, it was dependent upon the protection of the US nuclear umbrella. This meant that Ottawa's contribution to Tripartite WMD capabilities was restricted to the fields of biological and chemical warfare, where it made significant contributions. Second, although CBW systems were viewed with revulsion, as inhuman forms of warfare, at least chemical weapons (CW) had a proven battlefield function. In contrast, BW were often labeled as "uncontrollable and unpredictable" WMD, with only minimal strategic potential.[2]

Another important factor was the unwillingness of successive Canadian governments, between 1945 and 1969, to acknowledge that Canada's armed forces were involved with the offensive biological warfare research of its two allies, either as a deterrent or for retaliatory purposes.[3] In turn, this refusal raises the question of whether Canada had an official policy on biological warfare, aside from the mantra of denial. The answer is no. Rarely was the issue discussed in the Cabinet, the only exception being during the late 1960s, when Prime Minister Pierre Trudeau took a personal interest in the creation of the BWC. But for the most part, discussion in Parliament or coverage by the media was infrequent and superficial: a situation that continues to the present day.

Finally, in understanding Canada's BW priorities it is important to appreciate that most of the crucial decisions were made by bureaucrats rather than by elected officials. This was particularly the case during the offensive phase (1945–1969), when the hierarchy of the Defence Research Board (DRB) had virtually a free hand in the operation of Canada's four BW research facilities (Suffield, Ottawa, Kingston, Grosse Isle), as well as their interaction with US and British defense facilities. While this system was efficient, it was not conducive to either public disclosure or debate.[4]

Post–World War II Planning and the Cold War

During World War II Canada was actively involved with various British and American projects to develop new biological/toxin warfare munitions and delivery systems. Although there are various reasons for Canada's participation, looming largest was the assumption that these terrible weapons would be used by Nazi Germany and imperial Japan, and therefore a retaliatory capability was required. As a junior partner in this CBW alliance, Canada's military had neither the capability nor the responsibility of actually using these weapons in a retaliatory attack. This heavy burden rested with its allies.[5]

Yet by the end of World War II, Canadian bacteriologists, biochemists, and veterinarians had made a number of significant contributions to the allied BW program. On the defensive side, there was a unique version of the botulinum toxoid (A & B), which could have been used to immunize Canadian troops, if intelligence reports about possible German use of BW had materialized during the June 1944 D-Day landings. In addition, there was the joint US-Canadian project at Grosse Isle, Québec, which produced an effective vaccine against rinderpest, a deadly animal virus. From 1942 into 1944 the Grosse Isle site was also used to produce virulent anthrax spores that were subsequently analyzed by Canadian researchers at the Suffield, Alberta, military testing facility and by their American counterparts at Camp Detrick, Maryland.

On the other hand, Suffield scientists were among the first to recognize that anthrax *(Bacillus anthracis)* munitions had serious limitations, and to encourage the weaponization of other pathogenic agents and toxins. High on their list was botulinum toxin, for use in both 4-pound bombs and in 500-pound cluster/dart bombs. They also supported the use of *Brucella suis* (brucellosis)—an agent with high infectivity but low mortality—as well as carrying out research on a variety of other antipersonnel and anticrop agents. Although the biological and toxin munitions tested at Suffield were not used during World War II, Canada's military and political leaders were sufficiently impressed with their potential that they decided to retain most of the CBW wartime facilities, and the services of key university and military BW scientists, under the auspices of the newly created Defence Research Board.[6]

Whereas during the war the civilian National Research Council had coordinated Canada's CBW activities, the DRB was a unit of the Department of Defence, with its own director general, Dr. Omond Solandt. A respected medical scientist and skillful administrator, Solandt recognized the importance of reestablishing the wartime Anglo-American-Canadian BW system that would give Canadian scientists access to top-secret American and British BW research while, in return, making Suffield available for large-scale US and UK field tests with chemical agents and BW agents or simulants.[7] In pursuing this goal, he utilized the services of the special DRB Biological Sub-Committee, composed of scientists who had been involved with wartime research either at Suffield, Ottawa, or in the university laboratories of McGill, Queen's, or Toronto. This small expert group had another important asset: they were personally acquainted with the scientists who were directing the US and UK programs during the postwar years.[8]

The founding meeting of the Tripartite Conference took place in August 1946 in the US. It began with David Henderson, director of Porton Down's biological warfare research center (the Microbiological Research Department, or MRD), providing an overview of Britain's priorities and problems, notably the limited number of qualified scientists, minimal research facilities, and difficulties of reconciling military and scientific goals. A similar message came from the DRB team, who gave assurances that Canada's BW wartime operation would continue, though at a reduced level, "for continuing fundamental and basic research at Kingston and field experiments at Suffield." Priority would be given to three projects: insect vectors, the problems of ground contamination, and field trials with BW munitions. In summary, Solandt called for "close cooperation in the research and development of BW between Canada, Great Britain and the United States . . . [with] close scientific cooperation more important than cooperation at the policy level." General Alden H. Waitt, head of the US Chemical Warfare Service (CWS), echoed this viewpoint, expressing particular interest in having US munitions tested at Suffield, since the Granite Peak, Utah, and Horn Island, Mississippi testing facilities were scheduled for closing. He also indicated that the CWS regarded the postwar development of BW as essential to US national security "on a basis comparable to the Atomic Bomb."[9]

An important dimension of Canada's BW planning was direct linkages with the scientists at the UK Microbiological Research Department. These linkages were in part an extension of wartime cooperation, when key scientists (David Henderson, Paul Fildes, Lord T. C. Stamp) had been closely associated with the Suffield operation, and in part a result of British interest in developing an offensive BW capability.[10] Within the hierarchy of Canada's Department of National Defence (DND), there was certainly an awareness of the top-secret Operation Red Admiral, which called for the mobilization of Britain's scientific expertise "to bring into service by 1957 biological weapons comparable in strategic effect with the atomic bomb, and defense measures against them."[11] Moreover, during their frequent North American forays, Porton scientists made a point of visiting Suffield and Ottawa, as was the case in 1947, when David Henderson attended the first Tripartite meeting at Suffield. On this occasion he noted that Canadian scientists retained "a very keen interest in BW Research and are anxious that the facilities at Suffield should be fully utilized for co-operative effort with USA or United Kingdom, or both." On the research front, Henderson found DRES scientists of high quality, and was particularly anxious to recruit University of British Columbia scientist Dr. Alex Wood, professor of animal husbandry, as director of field operations for Operation Harness, the first of a series of British BW sea trials.[12]

This ambitious undertaking was a continuation of wartime attempts to determine the threat of BW in terms of agents, delivery systems, and possible defensive measures—building upon previous wartime trials carried out at Suffield, Detrick, Dugway, and Porton Down itself. The maritime exercises, however, had the great advantage of being able to use "hot" pathogens—*Bacillus anthracis, Yersinia pestis,* and *Francisella tularensis*—in order to test "the generally accepted belief that the most effective way to distribute pathogenic bacteria was to produce airborne particles containing the virulent organism . . . [and] by the direct inhalation of such clouds a percentage of casualties in man or animals would follow."[13] Many of the results obtained from Operation Harness were reassessed during follow-up trials at Suffield during the summer of 1950. Further confirmation about agent performance was recommended at the fifth Tripartite meeting (1952), when it was agreed that munitions tests using *Brucella suis* and *Francisella tularensis* would take place at DRES in the summer of 1953. Scientists from Detrick and Porton agreed to provide agents and equipment.[14]

Between 1945 and 1950 there were a number of field tests at Suffield with various BW munitions, using both the 4-pound bomb and the 500-pound cluster bomb.[15] The extent of Canada's activity was revealed in a January 1949 report of the DRB Bacteriological Warfare Research Panel (BWRP), which described and prioritized the various BW projects. On the assumption "that each agent would require its own munition," DRB scientists decided to assign first place to botulinum toxin, followed by "non-sporulating species such as *Brucella, Tularaemia* or Melioidosis (Whitmore's Bacillus)."[16] There was also a consensus that the DRB should establish a biological pilot plant for the production of "a number of agents with the same equipment."[17] In addition, there was considerable discussion about the dissemination of agents either through aerosols or by use of insect vectors, preferably those "found in the larger populated areas of likely enemy countries"—obviously either the Soviet Union or China.[18] In this regard, Guilford Reed's sophisticated work on the possibilities of using plague-infected fleas was regarded as quite promising, particularly since the project also involved US researchers.[19] In keeping with Canada's arctic image, many of the DRES trials attempted to determine the effect of cold-weather conditions on bacterial and viral aerosols, as well as exploring the possibilities of using CBW in the arctic environment.[20]

With the escalation of the Cold War, BW cooperation between Canada and the US increased substantially. In December 1951, for instance, Omond Solandt was informed that the US Air Force was "giving serious consideration to special weapons" and was preparing an inventory of possible "areas where user tests of live agents can be undertaken." The US Air Force was particularly interested in the Royal Canadian Air Force base at Cold Lake, Alberta, "since the climate there should be similar to many parts of European Russia." Suffield was its second choice, with the joint US-Canadian arctic research center at Churchill, Manitoba, another possibility. However, the DRB was insistent that the US military could not "carry out BW trials in Canada, elsewhere than at Suffield, without very good reason." In a subsequent report, conveyed to the US Air Force, the advantages of Suffield's 700-square-mile operation were carefully outlined, notably the special "mile-square area fenced with rodent proof fence for the conduct of BW trials . . . a bursting chamber of about 200 cu.m. [cubic meter] capacity which will handle shells up to 25 pd. [pound] size. It has a vacuum system and lines to allow sampling of dispersed agents from six positions in the chamber . . . [with] an animal exposure chamber attached."[21]

These discussions coincided with major upgrades in the US BW system, notably the establishment of a huge manufacturing plant at Pine Bluff, Arkansas, with an initial capability to produce four pathogens (brucellosis, tularemia, anthrax, and botulinum toxin); along with improvements in delivery systems, with a range of new bombs, sprays, rockets, and insect vectors. In addition, there were apparent changes in the US position toward the offensive use of BW, a trend that Porton's David Henderson noted during his March 1952 visit to the US: "American colleagues of long standing had become very offensive minded . . . emphasis was now entirely on anti-personnel and anti-crop weapons . . . [and] the Services had gained complete control of BW matters in the US."[22]

From the Korean War to the Cuban Missile Crisis

During the Korean War, Canada's involvement with BW came under intense public scrutiny at home and abroad, particularly after the 1952 Communist-bloc allegations that the US had used germ warfare in North Korea and northeastern China. In part, this crisis reflected growing international concern about BW, fueled by a variety of doomsday predictions. One of the most graphic declarations came from Dr. Brock Chisholm, the Canadian-born director general of the World Health Organization, who gave the following warning in a September 1949 address to the World Union of Peace Organizations: "Biological warfare is not a new kind of war, it is just the latest step . . . Some seven ounces of a certain biological agent, if it could be effectively distributed, would be sufficient to kill all the people of the world . . . Large armies, navies and air forces, including the Atomic bomb, which have been regarded as the symbol and fact of Power, are now obsolete."[23]

The Korean "germ warfare" controversy erupted in January 1952, when Soviet officials accused the US military of using BW against Chinese and Korean soldiers and civilians (see Chapter 13). The charge was followed by a series of North Korean reports about specific American BW crimes, which were endorsed by various western procommunist organizations, including the Canadian Peace Congress. This public relations campaign sought to portray the US as a barbarous, warmongering nation. Much of the propaganda was directed at the nonaligned Asian members of the United Nations, who were bombarded with "irrefutable" evidence of American guilt.[24]

To assist its beleaguered ally, the Canadian government used its influence at the UN to try to convince Commonwealth countries such as India, Pakistan, and Ceylon (now Sri Lanka) that the Communist charges were nothing more than malicious propaganda, with no scientific validity.[25] This "educational" program profited greatly from a special June 1952 report, "Statement Concerning Charges of the Practice of Bacteriological Warfare by United States Forces in Korea and North-East China," prepared by three eminent Canadian entomologists. After examining available evidence, these experts dismissed the Chinese and Korean charges as nothing more than "biological absurdities," pointing out that if the US military had really wanted to use BW for strategic purposes, they would not "have adopted such inept, infantile and altogether stupid methods in a field in which they are supposed to be masters."[26] Throughout the next six months Canadian diplomats made good use of this report at the deliberations of the UN General Assembly and at the meetings of the Conference on Disarmament in Geneva. In 1953 the germ warfare controversy abruptly ended following the death of Josef Stalin, and the Korean armistice.[27]

Significantly, the Korean War experience increased rather than decreased Canadian support for US military containment policies. This trend was evident in the 1958 creation of the North American Air Defense (NORAD) Treaty and the Defense Production Sharing Agreement of the following year. Even Prime Minister John Diefenbaker, who would later oppose Canada's having tactical nuclear weapons, was at this stage in his political career prepared to support the Eisenhower/Dulles doctrine of nuclear containment.[28] On the other hand, there is no evidence that the Canadian leader was aware of the 1959 US presidential decree authorizing possible first use of CBW.[29]

Within Canada's defense establishment the Korean germ warfare controversy provided additional incentive for increased BW research, on the assumption that it had been designed to conceal accelerated Soviet work in this field. These concerns were evident in the 1952 deliberations of the BWRP involving the various ongoing programs at the four BW research centers: Suffield (DRES), Ottawa/Shirley's Bay (DREO), Kingston (DREK), and Grosse Isle (GIR).[30] Two projects were regarded as most important: expansion of the work at Kingston in producing "more virulent strains of *B. mallei*, and *C. botulinum* toxin, along with continued work on insect vectors"; and production and storage of rinderpest vaccine "of sufficient quantities for use by US and UK as well as Canada."[31]

The Grosse Isle research station had been reactivated in 1948 because of DRB concerns, shared by the US Chemical Warfare Service, that the Soviets might attack the North American cattle industry with rinderpest or other agrobiological weapons.[32] The task of creating an effective "R" vaccine was, however, complicated by the fact that the original "seed stock" activated virus could not be used for large-scale immunization. As a result, two years were lost searching for "fresh virulent" sources, and it was not until December 1950 that DRB scientists could announce "that they could fully protect cattle against highly virulent (R) material . . . [and] prevent an exceedingly grave attack."[33]

Ironically, it was at this juncture that Grosse Isle scientists became involved with the CWS's "basic research in the field of anti-animal agents." More specifically, the US wanted an opportunity to carry out BW field tests at the station—a request that was quickly granted with one reservation: Dr. Charles Mitchell, of the Department of Agriculture, would represent the DRB, "with full authority over the trials." In short order, scientists from Camp Detrick were busy acquiring new "R" viral strains based on their weapon potential, ease of production techniques, storage capacity, stabilization as an aerosol, "and methods for dissemination." During the next seven years Grosse Isle carried out an extensive offensive antianimal program—maintaining an inventory of more than 16 different viruses. These included 22 strains of rinderpest, along with samples of African swine fever, Western, Eastern, and Venezuelan equine encephalitis, Newcastle disease, fowl plague, hog cholera, rabies, and Rift Valley fever. In the case of rinderpest, scientists were successful in developing strains (the Turkish) transmissible not only through inhalation but also through saliva on the feed trough.[34]

In October 1954 the BWRP reviewed the Joint US-Canadian antianimal program, noting that three viruses were presently available as "efficient agent[s] for the destruction of . . . food-bearing animals. 1. Cattle (Rinderpest); 2. Swine (African Wart Hog disease) [African swine fever]; 3. Chickens (Fowl Plague)." Despite these successes, four years later the Grosse Isle program was cancelled, apparently because of the US decision to discontinue its involvement in antianimal BW research. For their part, DRB authorities were only too happy to get out of this field, particularly after the unfounded 1952 allegations that germ warfare experiments at Suffield had been responsible for the devastating epidemic of foot and mouth disease that had ravaged western Canadian cattle herds.[35]

DRES scientists had been extensively involved with a range of other Tripartite projects calculated to improve the quality of BW munitions and delivery systems for lethal agents, and also in developing strategies for greater utilization of agents such as *Brucella suis* and *Coxiella burnetii* (Q fever), "used to incapacitate rather than to kill." As a result, there was a continual flow of defense scientists between DRES, Camp Detrick, and Porton Down as part of an elaborate division of labor that usually centered upon US priorities and policies.[36] The scale of this cooperation was outlined in a Canadian report submitted to the 1958 Tripartite meetings which highlighted four major trends that promised to transform BW into a powerful strategic weapon: "a) The feasibility of large area coverage with inert particles has been demonstrated both in the UK and US, even under random weather conditions; (b) Considerable success has been achieved in the US in spray devices for producing the fine aerosols necessary to penetrate the lungs; (c) Work in Canada on the mechanism of death of airborne organisms led to a biochemical explanation which in turn led to significant success in decreasing the death rate of airborne bacteria; (d) Canadian work established the feasibility of developing new diseases by adaptation of viruses from one species to another." Within the Tripartite alliance there was a consensus that most attention should be devoted to "the large area coverage concept of BW with a view to its evaluation under military conditions." In order to consolidate large area concept (LAC) R&D activity in one place, it was decided that previous US work on in this field should be transferred to Suffield "because of the availability of field facilities side by side with laboratory facilities." This move would also expedite future tests for the dispersal of bacterial and viral slurries from aircraft, "to produce fine aerosols."[37]

This change in testing venue did not concern Dugway scientists, largely because of their ongoing collaboration with DRES "to test present candidate CW and BW munition-agent systems, to provide field data on munition and component functioning . . . and the estimation of target effects." Indeed, throughout the 1950s semiannual informal conferences between the two facilities had been institutionalized, providing an opportunity for scientists to discuss ways of achieving greater standardization in their respective BW testing techniques.[38]

This cozy relationship was somewhat altered by the 1961–62 changes in the US Defense Department's approach to BW, engineered by its new boss, Robert McNamara. In an attempt to rationalize existing weapon systems, McNamara established the Project 112 Working Group, which subsequently recommended that the Army's CWS carry out BW agent research for all the military services. To facilitate this added responsibility, it was decided to place Fort Detrick's BW R&D activities and Pine Bluff's production facilities within the Army's Munitions Control Division, while BW testing was assigned to the Army's Testing and Evaluation Command, located at the Deseret Test Center (DTC), Utah, which was also responsible for Dugway. In addition, the DTC was responsible "for planning, organizing, executing and reporting all extra-continental field work for the US forces," as well as carrying out negotiations with Canadian and British research centers.[39]

During the next nine years the Deseret Test Center was involved with the planning of a large number of CBW munitions tests outside the US, many involving live pathogens. These occurred at tropical sites such as the Marshall Islands and at the Gerstle River cold-weather site in Alaska, as well as a series of maritime trials to determine how the US Navy and its NATO allies could withstand CBW attacks. At least three of these involved Canadian scientific and military personnel: Elk Hunt, Phase II CW (June–July 1964); Copperhead, January–February 1965, (BW) off Newfoundland; West Side II, (BW) Suffield, January–March 1965; and Rapid Tan, (CW) July–August 1968.[40]

Nuclear War, CBW, and NATO, 1962–1968

Canada's close association with US military policies was brought into clear focus during the Cuban Missile Crisis of October 1962.[41] In both countries there was, for example, renewed interest in civil defense. Ottawa's worst-case scenario was based on the assumption that if 16 Canadian cities were attacked with nuclear weapons, the combined effects of blast, thermal energy, immediate radiation, fallout residual radiation, and subsequent epidemics would result in approximately 4.5 million deaths, or 25 percent of the Canadian population. The related threat of BW was also part of the Canadian civil defense agenda, as outlined by one 1963 report:

> The form of overt biological warfare most likely to be used against us in the event of war, would be the creation of aerosol clouds of critically sized particles containing aggregates of pathogenic agents. Urban areas and important military targets might be subjected to this form of attack. Biological warfare agents could be conceivably used by subversive methods in a variety of fashions and a saboteur might introduce pathogenic agents into the air of localized but strategically important communities.[42]

Although the nuclear Limited Test Ban Treaty of 1963 marked some progress in controlling one WMD, the BW threat was not necessarily diminished. On the contrary, the nuclear stalemate appeared to increase the possibility that the Warsaw Pact might launch a series of limited wars, using CBW. If such a situation materialized, the crucial question was whether the US or its allies should retaliate in kind or immediately resort to nuclear weapons. In this debate over the advantages of a flexible response, one group of military planners called for the deployment of biological and chemical incapacitants both for military reasons and because "public feeling against the use of BW and CW would be largely eliminated."[43]

Between 1963 and 1969 Canada redefined its position on *all* WMD, under the watchful eye of the US. Ottawa's belated decision to accept tactical nuclear weapons, as part of its NATO and NORAD obligations, has been the focus of many scholarly studies. Less well known was the May 1963 secret policy statement on CBW issued by the Canadian Chiefs of the General Staff, which established three operational principles: "a)

Canada will in no instance initiate nuclear, biological or chemical warfare; b) Canadian Armed Forces may be committed to participate in a war in which the use of N, B or C is initiated by an enemy; c) The Canadian Armed Forces will develop the knowledge and the capacity to ensure that protective measures are adequate, and that a capability for retaliation in kind could be quickly instituted if so directed."[44]

Canadian BW policies were also affected by its active role in NATO, particularly during the 1960s. The Cuban Missile Crisis had forced NATO to consider seriously the chemical, biological, radiological, and nuclear civil defense capabilities of its members through a myriad of standing committees and working groups. In the case of nuclear and chemical weapons, efforts had been under way since the early 1950s to devise defensive and survival tactics. In April 1962 biological warfare was added to the list as part of a comprehensive survey carried out by the NATO Military Committee on Civil Defense Preparedness. In its final report the committee lamented "the lack of progress in Allied Command Europe in acquiring a CB defensive and retaliatory capability . . . due in part, to the sensitive political nature of this type of warfare."[45] This was deemed a serious military deficiency because of growing evidence that the USSR had dramatically increased its chemical and biological arsenal of nerve gases, incapacitating agents, and the means "for spreading biological warfare agents over large areas, thus constituting a new major weapons threat."[46]

The specific threat of biological warfare was discussed in an April 1965 report of NATO's Standing Group on Science and Technology: "The Soviets appreciate the potentialities of biological warfare and have given consideration to its use . . . [but] There is no evidence to indicate a tactical offensive biological or radiological warfare capability in the Soviet armed forces." The following year, however, this relatively optimistic assessment was revised on the basis of more accurate intelligence. Now the Soviet biological warfare threat was regarded as real and imminent: "Advances in biological warfare offensive programs will depend upon Soviet intentions . . . there is no doubt the Soviets have the necessary background and experience to develop a complete array of BW munitions systems for a variety of strategic and tactical uses . . . The agents for use could already be standardized individually or in combination with other biological or chemical agents."[47]

Given this ominous prognosis, NATO's Military Committee decided to adopt a more aggressive stance toward the perceived Soviet CBW threat. On 22 September 1967 it issued the controversial "Overall Strategic Concept for the Defence of the North Atlantic Treaty Organization Area": "It is not evident to what extent BW or CW capabilities might affect deterrence. However, there is a danger that Soviet leaders might come to believe that their capabilities in these fields would give them a significant military advantage. NATO should rely principally upon its conventional and nuclear forces for deterrence, but should also possess the capability to employ effectively lethal CW agents in retaliation, on a limited scale; passive defensive measures against CW; and passive defensive measures against BW.[48]

Significantly, this new NATO policy emerged at a time when Canada's Department of Defence was reconsidering its own CBW priorities. Normally this was a frustrating process for the DRB, since the Chiefs of the General Staff usually downgraded CBW requirements in favor of nuclear-weapons-related programs—despite the fact that Canada was not part of the nuclear-bomb club. All this changed during the July 1968 CBW review, when DRB officials were finally able to convince their superiors that BW represented a serious threat to Canadian national security. In building its case, the DRB first pointed out that advances in the biological sciences not only increased the threat of the traditional bacterial, viral and toxin agents, but also facilitated the emergence of frightening new pathogens. Reference was also made to ongoing DRB projects of special military importance, notably "warning detection systems . . . development of test agents (use in urban area, use in troop trials), evaluation of defensive systems . . . and evaluation of microorganisms, suitable for use as incapacitating agents."[49] Other arguments focused on the magnitude of the Soviet threat, NATO's new policy of CBW deterrence, and Canada's obligations to its now Quadripartite partners (Australia joined in 1964).[50]

In keeping with this latter commitment the Defence Department announced that Operation Vacuum, a full-scale field trial designed to test the effectiveness of CBW defensive equipment and operational tactics, would take place at Suffield during the fall of 1968. The actual exercise lasted three weeks and involved more than 2,000 military personnel, most seconded from the Canadian forces, although there was a small British and US military and scientific component. While most of the emphasis was on troop performance within a CW environment, there were also secret plans to stage a mock BW attack, using the simulant *Bacillus globigii,* which was cancelled at the last moment because of concerns about negative public opinion. Indeed, throughout the late 1960s the Canadian peace movement and leftist groups consistently portrayed Suffield as an evil and dangerous place, a symbol of Canada's connection with US "militarism and imperialism."[51]

In addition, there was strong pressure from the Department of External Affairs (DEA), which was outraged that Operation Vacuum was being held at the same time that Canada was involved with important CBW arms control negotiations in Geneva. But defense officials remained adamant that Canada should fulfill its alliance commitments. This position was outlined in December 1968, when Brigadier General H. Tellier, on behalf of the Chiefs of the General Staff, issued an official statement about its future CBW policies. Although he admitted that nothing had been done to implement the secret directive of May 1963, Tellier argued that the time had now arrived for decisive action:

> The new policy proposes that agreements should be reached with our Allies whereby suitable weapons can be made available . . . Discussions with the US Army are in the very early stages. They are aimed at determining the costs involved in providing a retaliatory capacity. I would emphasize that the Canadian Forces have no intention of holding B or CW munitions in Canada or in Europe. Our requirements would be held in British or American stockpiles, to be supplied in the event B or CW is employed against NATO forces.[52]

A related development was the October 1968 Third Tripartite Intelligence Conference on CBW in Ottawa. Its agenda focused on three major problems: "the potential impact of new scientific developments on the threat to the Western world from biological and chemical warfare . . . the capabilities of those nations whose intentions we have reason to suspect, and [the need] to define those areas . . . where more collaboration between the participating intelligence organizations is needed." Since Canada had "virtually no facilities for the collecting of scientific intelligence," there was great concern in Ottawa that nothing should threaten access to US and UK sources, including unfavorable publicity about the proceedings.[53]

Throughout the remainder of 1968 the Departments of National Defence and Foreign Affairs continued their internal struggle over the proposed munition exchange system, although both sides agreed that little consideration should be given "to offensive capability in BW, in part because Canada did not now have and probably would not in future have the necessary (BW) equipment, such as long-range aircraft."[54] In addition, DEA officials convinced their Cabinet colleagues that the CBW munition exchange system was a difficult and risky endeavor: "if it were discovered that Canada were engaged in exploratory discussions of this sort, the public outcry would be loud, and the international embarrassment considerable." In these deliberations, reference was also made to the fact that the British government was prepared to renounce all offensive BW research, remove Porton Down from the jurisdiction of the Ministry of Defence, and intensify its efforts to obtain international CBW disarmament.[55]

Arms Control and Canada's BW Program, 1968–2004

Yet at the beginning of 1968 there seemed little chance for such a ban. Locked in a ferocious war of attrition in Vietnam, the US continued the massive use of "nonlethal" tear gas and herbicides despite growing international criticism that this violated the Geneva Protocol, and which rejected US claims that these nonlethal weapons could actually be seen as serving a "humanitarian" purpose. In addition, the fact that the US was one of the few countries that remained outside the Geneva Protocol reinforced the Soviet propaganda campaign to portray the US as a "rogue nation," determined to flout international standards. This negative image of "American militarism" gained credence even among the US's closest allies.[56]

In Canada, the 1968 election of Pierre Trudeau as prime minister, with his more critical view of nuclear weapons, Canadian-US relations, and the rigid approach to the Cold War, resulted in a comprehensive and controversial reassessment of the country's foreign and defense policies. Although Trudeau's role in determining Canada's position in CBW matters during this crucial period remains somewhat obscure, it appears that unlike his predecessors the prime minister took great interest in the unique threat of BW. This involvement was evident in his instructions that the Cabinet Committee on External Policy and Defense should carefully examine all of Canada's CBW commitments, and in his personal belief that the time was propitious for pursuing aggressive CBW arms control discussions at both the United Nations and the Geneva-based Eighteen Nation Committee on Disarmament (ENDC). After all, he reasoned, the US and the Soviet Union had already embarked upon the important Non-Proliferation Treaty, which in turn facilitated the SALT process.[57]

The linkages between these various arms control agreements were greatly enhanced by the active involvement of UN Secretary General U Thant, who in July 1969 submitted a special scientific report on CBW to the UN General Assembly. In turn, the earlier British proposal for a separate Convention for the Prohibition of Biological Methods of Warfare, as the first stage in a more comprehensive CBW arms control regime, became the consensus document for the Western members of the UN Conference of the Committee on Disarmament (formerly the ENDC). While Canadian diplomats supported these initiatives, they realized that Washington's endorsement was crucial to the success of any CBW arms control measure. As a result, they carefully cultivated their counterparts in the US State Department and the Arms Control and Disarmament Agency (ACDA) even while trying, with limited success, to ensure that CBW discussions at the ENDC and General Assembly did not turn into anti-US polemics. Bu their greatest concern was the direction US policies would assume after the November 1968 US presidential elections. Would the Pentagon, for instance, adopt a hostile position toward any significant CBW disarmament proposal, and therefore thwart any important initiatives? Would Congress heed the advice of scientists such as Matt Meselson that BW were "useless and foolish" and should be removed from the US arsenal? And was it possible that as president Richard Nixon, the Cold War warrior, would actually commit the US to a unilateral renunciation of BW?[58]

During the summer of 1969, with ENDC discussions reaching a critical stage, DEA officials speculated whether they should become involved with the internal US debate, since Canada's "close and widely known cooperation with the United States in CBW research implicates us politically in USA CBW policy. We therefore have a legitimate and direct national interest in seeing that USA policy is as internally consistent and solidly based as is possible." It is not clear, however, to what extent DEA officials were aware of the interagency investigation that had been established by the US National Security Council in May 1969. Did they know that Secretary of Defense Laird, the State Department, the ACDA, and the President's Scientific Advisory Committee all disagreed with the Joint Chiefs of Staff that biological weapons had a future in US strategic planning? And did they anticipate that after eight months of intense negotiations, the Nixon administration would adopt three bold initiatives: to ratify the 1925 Geneva Protocol, with certain qualifications; to pledge "no first use" of "lethal" chemical weapons; and to abolish the US offensive BW program? Unfortunately, answers to these questions remain difficult, given restrictions on key US and Canadian documents for this period.[59]

Consequently, it is not clear why Nixon's momentous unilateral renunciation of 25 November 1969 caught the Canadian government so completely by surprise. But it did.[60] Indeed, Minister of Defence Marcel Cadieux, when grilled in the House of Commons, had to rely on an article published in the *Montreal Gazette* for details about Nixon's statement; and even then he got it wrong. But an official Canadian response was soon forthcoming, and it came in two parts.

First was the Cabinet policy statement of 11 December, which adopted an even more sweeping BW disarmament position than the US—since it also included toxin weapons (which the US later added in March 1970). The second stage was Canada's official declaration about CBW, which was issued on 24 March 1970, not by Prime Minister Trudeau but by the veteran diplomat George Ignatieff.[61]

In his eloquent speech on 24 March at the CCD meetings in Geneva, Ignatieff stressed that "Canada never has had and does not now possess any biological weapon (or toxins) and does not intend to develop, produce, acquire, stockpile or use such weapons in the future." He quickly added, however, that until a CW convention became a reality Canada would reserve the right to use CW in retaliation and would carry out defensive CBW research activities. Thus both the DRES and DREO operations would be continued, albeit on a reduced basis. More specifically, Suffield scientists would continue their work in developing CW and BW detection systems and protective equipment for Canadian troops facing CBW, as well as carrying out research "towards resolving problems associated with the verification of a comprehensive ban on chemical and biological warfare that may be concluded."[62]

In September 1971 Canada enthusiastically endorsed the draft BWC, which had emerged after intensive US and Soviet negotiations, with unanimous consent in the House of Commons and Senate. Although there was a general concern about the BWC's lack of adequate provision for verification, and regret that it was not part of a broader CBW arms control system, Canadian diplomats and their defense advisors recognized the advantages of having achieved the first "true disarmament of a weapon category to have occurred since 1945." They also recognized that obtaining a separate ban on BW had been possible for a number of important reasons: they had unproven military capabilities; they were morally repugnant; and they had a reputation for being "dangerous to produce, difficult to handle and unreliable in their operations." On the other hand, DEA officials remained hopeful that a CW convention was imminent and that, when it came into being, serious deficiencies in the BWC could also be rectified.[63]

After 1975, Ottawa's greatest concern was that international tensions between the two superpowers would undermine the entire international arms control system. This became a matter of special concern during the early 1980s, when the new Reagan administration cancelled the nuclear SALT II treaty process, accelerated the development of binary CW, and questioned the effectiveness of the BWC. This last crisis gained momentum during the First Review Conference of 1980, when the US delegation claimed that the anthrax outbreak near the Russian city of Sverdlovsk demonstrated clearly that the USSR was "pursuing the development and probable production of biological weapons." This image of Soviet aggressive behavior was reinforced by the "Yellow Rain" controversy, associated with 1981 US allegations that "toxins and other chemical warfare agents were developed in the Soviet Union, [and] provided to the Lao and Vietnamese"[64] (see Chapter 13 for more detailed discussion of both allegations). Canada's response to these two incidents varied. In the case of Sverdlovsk, the DEA did not become directly involved with the US campaign against Soviet noncompliance, in part because of concerns about the quality of US intelligence, and in part because of fears that the incident might scuttle the BWC, thereby leading to another superpower biological arms race. On the other hand, Ottawa did send two scientific fact-finding missions to Kampuchea and Laos, in response to an invitation from the UN secretary general, and to prevent the Soviet bloc "from using toxins and other CB weapons in a variety of surrogate wars."[65]

With the end of the Cold War, Canada's efforts went beyond trying to save the BWC. Now the emphasis was on enhancing its effectiveness through an elaborate compliance and verification system. But progress was slow—at least until the 1994 Special Conference on BW Verification, attended by 79 interested States Parties, which created the specialized Ad Hoc Group of Experts with a mandate "to develop proposals for a legally binding verification protocol in time for the Fourth Review Conference in 1996." By July 2001 there was considerable optimism within the Department of Foreign Affairs that the proposed Verification Protocol of the BWC, based on nine years of continuous consultations between scientists and diplomats, would finally be accepted by the 145 member states gathered at Geneva for the Fifth Review Conference. These hopes were dashed in August, when US arms control delegates informed the Geneva meeting that the proposed verification system was both inoperable and dangerous. The terrorist attacks of 11 September 2001 have made further discussions of the verification process even more difficult.[66]

Preventing Bioterrorism after 9/11

In September 2001 the Canadian government was quick to respond to the threat of bioterrorism. First on the agenda was the Anti-Terrorist Bill (C-36), which dramatically increased the federal government's capabilities to deal with terrorism in its many forms. This was followed in October by the even more sweeping Public Safety Bill (C-42). Administratively, a number of security programs had already been implemented by the Department of National Defense and Health Canada's Centre for Emergency Preparedness and Response, originally established in July 2000 "as the country's single coordinating point for public health security in Canada," with special responsibility for enforcing regulations on the importation of human pathogens and biosafety in laboratories, administering the National Emergency Stockpile System, training and coordinating provincial and local public health officials, and tracing disease outbreaks globally. In addition, provision was made for the creation of the National Advisory Committee on Chemical, Biological, Radio-Nuclear Safety and Security.[67]

Most of the Canadian government's initiatives have had dual purposes: to deal with the possibility of a bioterrorist attack on Canada, and to reassure American officials that the Canadian government is prepared to adopt stringent and effective methods in preventing terrorists from launching a biowarfare attack on the US. As part of this coordinated effort, in November 2001 there were meetings in Ottawa between Tommy Thompson, the US secretary of health and human services, and Allan Rock, the minister of Health Canada, to discuss the possibilities of collaboration in the large-scale development of smallpox vaccines. These consultations were broadened later in the month when Rock hosted a meeting in Ottawa of health ministers from G-7 countries, consultations that resulted in the formation of the Global Health Security Initiative (GHSI) "to strengthen the public health response to the threat of international biological, chemical and radio-nuclear terrorism." Since its founding in November 2001, the GHSI has held four ministerial meetings, prepared a risk assessment system for possible chemical, biological, radiological, and nuclear attacks, improved communication between high-containment laboratories in the various countries, and sponsored one major multinational exercise (Global Mercury) to evaluate "international assistance and collaboration in the case of a smallpox incident." In May 2003 Canadian authorities were also involved in another transboundary counterterrorism exercise, TOPOFF 2, which involved key US and Canadian political and law enforcement officials.[68]

Conclusion

Was biological warfare on the verge of becoming a controllable and predictable form of WMD in 1969? At the time, the verdict of most Western politicians and arms control specialists was a resounding no. Quite a different verdict emerges, however, if one examines contemporary reports emanating from the scientists at Suffield, Detrick, and Porton, which stressed both the level of technological innovation and operational possibilities. And it was this message, that the "stuff is too damn good to go away," which motivated Soviet leaders to secretly expand their BW program after 1972.[69]

Within the framework of the international BW arms race, the Canadian experience has many important dimensions. First, its involvement with the important Tripartite exchange system was primarily an extension of its World War II experience, when prominent medical scientists such as Nobel laureate Sir Frederick Banting and associates made significant contributions to the Allied offensive and defensive BW capabilities, as well as serving as a linchpin between UK and US scientists. Significantly, Canada did not abandon its involvement with CBW after 1945, unlike the case of its nuclear bomb research. If it had, Suffield would have disappeared as a major military facility, and Canada would not have been part of the Tripartite CBW system.

Not surprisingly, it was the US program that most influenced Canadian R&D priorities. Indeed, DRB scientists were convinced that they could not "carry on an effective program in biological defense if we were cut off from the constant flow of US information, particularly information on weapon effects." This dependency was also revealed by the DND's attempts in 1968–69 to acquire a CBW retaliatory capability—through access to American stockpiles. Even more telling was the enormous impact of President Nixon's November 1969 declaration on all aspects of Canada's BW operation.[70]

But was the Canadian BW program merely an appendage of its more powerful allies' programs? In terms of the British connection, there is no evidence, as some writers have claimed, that Canada's postwar CBW priorities were determined by Whitehall. Similarly, the DRB's relations with the key US military and scientific BW agencies, notably Camp Detrick and the Dugway/Deseret test centers, operated on the basis of mutual respect and reciprocal exchange, a relationship that continues today. In terms of offense-related research, the Canadians made a number of important contributions, such as the insect vector program at Kingston and the antianimal research at Grosse Isle. But it was the many Suffield projects, often coordinated with the Dugway and Deseret testing centers, that were most highly regarded by Canada's Tripartite allies. These included munitions tests with lethal and nonlethal agents, the LAC trials, studies on the decay of aerosolized bacteria and viruses, and early warning detection systems. Equally valuable were defensive measures related to battlefield CBW sensors, decontamination techniques, protective equipment; along with new vaccines and antibiotics, particularly since these undertakings continued after the BWC. In fact cooperative defensive work has been a central feature of DRES research priorities during the past 35 years under the auspices of NATO, the Tripartite Technical Cooperation Program (TTCP), and the more exclusive 1980 US/UK/Canada

Memorandum of Understanding, which calls for Tripartite cooperation in defining BW problems, achieving interoperability, and developing new technologies.[71]

Increasingly, scholars have analyzed the unique culture of weapons establishments. While most of the focus has been on nuclear scientists, a recent study of Porton Down provides valuable insights into the dynamics of a BW laboratory. Unfortunately, there is no comparable history of Defence Research and Development Canada–Suffield, although it has an equally long institutional record (over 60 years) and an impressive list of accomplishments. The documentary record does, however, demonstrate that Suffield had its own set of values and rituals. Of central importance was its role in defending Canada against the threat of CBW during both World War II and the Cold War. In both cases, defense was related to effective deterrents, or, more specifically, a retaliatory BW capability. From this perspective, it is not surprising that Suffield scientists believed that in supporting the US offensive BW program they were also safeguarding Canada's national security.[72]

CHAPTER 5

The French Biological
Weapons Program

OLIVIER LEPICK

Numerous monographs have been written on arms policy in France since World War II, particularly on its nuclear weapons program, yet the issue of biological weapons (BW) in France is completely absent from the nation's historiography. Indeed, it is impossible to find a reference work that devotes more than a few lines to the subject.[1] It seems that the secrecy surrounding the study of these matters, in France and elsewhere, is a result of the sensitivity of the subject at a time when the risk of BW proliferation continues to threaten international security. This secrecy largely explains why historians have found it impossible to obtain access to sources and archives that, despite the time elapsed, contain scientific information which is still valid and which could contribute, if it were not supplied with extreme caution, to the proliferation of these armaments. Owing to the largely technical nature of the source materials and archives on which this chapter is based, what follows is essentially a description of the principal phases, technical parameters, and structural organization of the French BW program (and more specifically its offensive program) since the end of World War II. As it was not possible to gain access to political sources (the Secrétariat Général de la Défense Nationale and the Conseil de Défense), this study lacks the political dimension that would have placed the biological military program in France within the wider perspective of French defense policy since 1945. For this reason, the following pages are limited to a factual history of the BW program in France.

It is possible to identify two distinct periods in the history of the French BW program between 1947, the date of its inception, and 1972, when

France officially terminated all activities connected with offensive biological armaments. The first period, from 1947 to 1956, was characterized by a large-scale military program of biological R&D, including the establishment of France's biological arms systems. The second period, from 1956 to 1972, saw budgetary redistribution in favor of nuclear armaments, relegating BW to a subsidiary position. The program declined in political and budgetary significance throughout the second period. This progressive abandonment was interrupted only in 1963, when French government interest in incapacitating BW briefly reinvigorated the program.

1947–1956: Development of the Program

It appears that the first steps toward, or consideration of, the resumption of military activity in the field of BW in France took place early in 1947. This is the date of a memo, emanating from the Bureau Scientifique de l'Armée and addressed to the chief of staff of the Défense Nationale, General Juin, which stated that "biological warfare is as formidable as nuclear warfare" and went on to recommend that "systematic studies" be undertaken into these matters.[2] The memo proposed an initial meeting and contained a tentative outline of a future BW program. This five-point program proposed that, following preliminary analysis of the prevailing situation at home and abroad, France should immediately embark upon:

- Studies of dispersal devices (bombs, aerosols, atomizers)
- Studies of diseases caused by exotoxins and endotoxins and of diseases capable of being transmitted by insects
- Studies on the detection of germs in the atmosphere
- Studies on protection
- Studies on biological aggression (contamination of waterways and foodstuffs, livestock, and crops, and investigations into the possibility of using insects as a vehicle of biological warfare)[3]

From the agenda for this launch meeting, which took place on 11 March 1947 at Val de Grâce military hospital, under the aegis of the Section Armement et Etude of the Army General Staff and within the framework of the activities of the Commission Médicale de Défense contre la Guerre Moderne, it seems that the decision to reembark upon a BW program had already been made, since "the aim of the meeting is to take stock of studies concerning germ warfare and to lay the foundations of an organization that will undertake such studies."[4] The meeting was chaired by General Devers. Also present were Lieutenant Colonel Ailleret; a representative of the Service de Santé des Armées; the surgeon general, Costedoat; and Colonel Krebs, representing the Comité Scientifique de la Défense Nationale. In the preliminaries to the discussion General Devers said that it was "a duty to undertake studies in this field."[5] The conclusions of the meeting mentioned the need to "propose to the minister of war the creation of a commission that will be his technical advisory body" on matters of biological warfare. This commission would have as its missions the establishment of a research program and the implementation of a plan, which would be entrusted to the French intelligence agencies, to gather information on the military biological activity of other principal nations. Among the military and civil research organizations capable of undertaking the work necessary for the creation of a BW program, participants at the meeting identified the prophylaxis laboratory of the Centre d'Etudes du Bouchet (CEB), the laboratories of the Service de Santé des Armées, the Army veterinary research laboratories, and the laboratory of the Chemical Weapons Section of the Service Technique de l'Armée (STA).

During the meeting, Surgeon General Costedoat drew up a list of French initiatives in biological warfare and detailed the knowledge acquired by France during its previous biological program between 1921 and 1940.[6] He described material and documents seized from Germany by French forces which proved that the Germans were interested in this subject, although it would not have been possible for him to specify the nature and scale of these seizures. He also detailed the scope of the American program, citing the Merck Report of the US War Department as his principal source, as well as the British and Soviet programs, in particular the Aral Sea installation, Vozrozdeniye Island. Costedoat added that the only organization working on these matters in France since 1945 was the Commission Médicale de Défense contre la Guerre Moderne, the creation of which had been authorized by the military cabinet on 24 December 1946. Forming part of the Direction Centrale du Service de Santé des Armées, this commission, chaired by Costedoat, was subdivided into three sections: chemical, microbiological, and nuclear.[7]

Referring to the principal recommendations of this meeting, Colonel

Krebs, head of the Permanent Secretariat of the Comité Scientifique de la Défense Nationale, wrote on 21 March 1947 to Lieutenant Colonel Ailleret, who was overseeing the matter on behalf of the General Staff,[8] proposing that the commission be composed only of military personnel, given the significant hostility that civilian groups were likely to show toward biological warfare. He suggested that members of the commission should include representatives from the Etat-Major Général de la Défense Nationale, the Army General Staff, the Office of the Minister of Defense, the Service de Santé des Armées, the STA, the Laboratoire de Prophylaxie du Bouchet, and the Services des Poudres. Regarding the program of study, Colonel Krebs recommended, taking into account the

Figure 5.1 Major sites involved in the French BW program.

lack of information on the designated means and objectives, a two-stage process: first, to draft a precise description of the situation, of work previously undertaken and results obtained; and second, to establish a BW study program.

In anticipation of the establishment of the commission, two meetings of the Commission Médicale de Défense contre la Guerre Moderne devoted specifically to germ warfare were held on 17 and 25 April 1947.[9] A 21-point program of study was adopted at the 25 April meeting. Many of the topics discussed dealt with the danger of aerial transmission of microbes and their toxins. On the basis of a model of the situation as it existed before the war, the document advocated the creation of a special laboratory within the Services des Poudres to study means of attack. This research, it was suggested, would take place at the Bouchet facility. The program would continue work undertaken before 1940 on microbial aerosols (analysis and determination of particular size, spontaneous and induced flocculation). These studies would be led by the CEB. The document recommended that microbial toxin aerosol dispersion trials begin in February 1948.

Simultaneously, the military intelligence agencies were responsible for gathering information on biological warfare from their military attachés on the ground.[10] The fruits of this mission between 1946 and 1949 are to be found in the archives of the Service Historique de l'Armée de Terre in Vincennes.[11] The file contains few classified documents, being more of a compilation of open source material (witness statements from the Nuremberg trials describing German activities, nonclassified American sources, and Allied military sources detailing the program in Japan) than a true intelligence mission. In the same way, the first studies at the STA laboratories were undertaken "in accordance with a very reduced and specific program as indicated a priori by the areas of study that appear to be the most profitable."[12]

Costedoat, the surgeon general, was responsible for drawing up a list of the work undertaken before 1940 by the CEB and, more specifically, the prophylaxis laboratories of Veterinary Colonel Velu. The list features in the report written by Costedoat in 1953 and sheds light on the progress of the French military biological program between the two world wars:

- Trials on the infection of guinea pigs by spore-carrying projectiles: the trials showed that fine metallic particles (a waste product from the manufacture of metal spikes) carrying spores of *Bacillus anthracis* and dispersed by a bomb with a 15-gram explosive charge could cause anthrax septicemia in guinea pigs exposed to the spores.

- Resistance to explosion of anthrax spores: this research showed that explosion did not completely destroy the spores and that they could be kept in stable suspension in a semisolid gel that, at the moment of explosion, would coat the projectiles within the missile as well as the shrapnel of the missile; experiments also allowed researchers to calculate the minimum infectious doses according to the missile employed.
- Spore dispersion using real bombs: two types of bombs (grenades and conventional bombs) were used to demonstrate the capacity of spores to contaminate all bodies in contact with the shrapnel, to spray all objects in the vicinity with a fine mist, and to produce microbial aerosols.
- Effect of chemical agents on infection: the effect of chloroform on the triggering of anthrax infection.
- Virulence of nonsporulated bacteria via the respiratory tract: experiments showed that it was possible to transmit *Brucella abortus* to guinea pigs using an aerosol.
- Dispersal within a room by explosion of nonsporulated bacteria.
- Dispersal of bacteria by inserting glass ampules into the explosive device.
- Production of microbial aerosols using explosive missiles.
- First real trials of the dispersal of anthrax spores by standard aerial bomb: coated projectiles caused death in 91 percent of the injured guinea pigs.

It was upon these technical foundations, results of the pre-1940 BW program, that the French program was to be relaunched at the start of 1948.

In keeping with tenets enshrined in France's ratification of the Geneva Protocol in 1926, it seems that the French authorities, despite resuming their examination of offensive BW, intended to have recourse to this type of armament only "in case of retaliation."[13] This position was continually reiterated by military leaders responsible for piloting the biological military program and is mentioned frequently in the preamble to the minutes, which are classified, of meetings concerning the French biological warfare program. This position was adhered to unswervingly throughout the whole of the 1947–1972 period.

In April 1948 the Comité Scientifique des Poudres et Explosifs approved the proposals put forward by Surgeon General Costedoat at the meeting of 11 March 1947, and allocated a budget of five million Francs to allow "the launch of a germ warfare program."[14] At the same time a decision was taken to prioritize the dispersal of toxins (specifically botulinum toxin, which French intelligence reports claimed had been the subject of promising work in the US, resulting in the manufacture of a pure poison). During the initial phase, the biological agents maintained in the scope of the French program were botulinum toxin (produced by *Clostridium botulinum*), ricin toxin, and *Bacillus anthracis*. Beginning in early 1948, the first trials of animal infection in enclosures, using microbial aerosols, were successfully completed at Aubervilliers by the biological division of the STA's Chemical Weapons Section.[15]

Between 1948 and 1952, in the absence of an effective administrative structure to steer the BW program, the laboratory of the biological division of the STA's Chemical Weapons Section led the ambitious project to test the BW data acquired by France. This undertaking, covering both offensive and defensive aspects, appeared in its technical schedule to be a continuation of interwar biological activity.[16] Research carried out between 1948 and 1953 focused on the following areas:

- Trials on the infection of animals by microbial aerosols: experiments showed that fine aerosolized delivery of bacteria was an excellent propagator of infection, and that the pulmonary tract was the best point of entry into the host; that a pathogenic microbe *(Salmonella typhimurium)*, when administered deep into the bronchioles and alveoli, rapidly became "a pneumotropic pathogenic germ."[17]
- Trials on animal infection through the air and dispersion on the ground in the form of a bacterial aerosol using dried bacteria *(Salmonella typhimurium)*.[18]
- Trials on animal infection through the air and dispersion on the ground in the form of a liquid aerosol using fresh cultures *(Salmonella typhimurium)*.[19]
- Trials on animal infection through the air and dispersion on the ground in the form of a bacterial aerosol created from bacteria *(Salmonella typhimurium)* kept in dried form and rehydrated at the moment of use.[20]
- Trials on animal infection through the air and dispersion by aircraft in the form of a bacterial spray created from fresh cultures.[21]
- Various tests on aerial spread with dried bacteria made from fresh cultures.[22]

- Direct contamination of an area using anthrax spores dispersed by means of frangible devices without the need for explosive charge. Direct contamination of the ground using anthrax spores dispersed by bouncing bombs with a weak explosive charge.[23]
- Direct contamination of an area with anthrax spores and with a persistent vesicant agent (mustard) dispersed together by bouncing mines.[24]
- Susceptibility and modes of infection of cattle to rinderpest virus and pneumonia virus at the Laboratoire Militaire de Recherches Vétérinaires (LMRV) at Maisons-Alfort.

The results obtained during this period remain limited from a technical point of view. Many of the areas explored did not yield conclusive data. Without underestimating the findings of trials conducted between 1948 and 1953, it is nevertheless appropriate to put the scope of the tests into context and to conclude, as did Surgeon General Costedoat, that the results were "mixed," given the "difficulty of artificially creating sweeping epidemics."[25] These results, rightly characterized as "mixed," were largely explained by the relative inadequacy of the human and financial resources dedicated to the BW program. This dearth of resources is attested in a letter from General Ailleret, commandant of special weapons,[26] on 22 January 1952, emphasizing the need "to create a relatively significant mass of executive officers and assistant officers trained in chemical, bacteriological, and atomic techniques . . . who would serve in the services responsible for research and experimentation. At the present time, this mass of educated operatives does not exist: we estimate that the number of officers truly competent in the fields of chemistry and bacteriology is no greater than 20 and 15, respectively (of which 90 percent are military veterinarians)."[27]

Nevertheless, the knowledge acquired should not be underestimated, especially in the case of lyophilization of aerobic and anaerobic bacteria and the aerosolized dispersion of bacterial suspensions. These tests also enabled scientists to document the serious infection of war wounds through the use of contaminated artillery, infection by pulmonary inhalation using microbial aerosols with very weak doses of chemical agents, and the triggering of enzootics through the dissemination of pathogenic germs.

On 26 August 1952 the minister of defense, René Pleven, along with the secretary of state for war, approved the creation of the Commission des Etudes et Expérimentations Chimiques et Bactériologiques de l'Armée de Terre (CEECB).[28] The CEECB was to coordinate the efforts of the armed forces dealing with CBW. Its role was to provide a rapid exchange of information among the different bodies regarding their respective activities in CBW and to coordinate experiments, manufacture, protection, and employment connected with such armaments. Among the members listed were the commandant of special weapons, who was to chair the CEECB, the director of the CEB, the head of the bureau Armement et Etudes of the Army General Staff (ARMET), a representative from the Services des Poudres, a representative of the Army Veterinary Service (Services Vétérinaires des Armées), and a representative of the Comité d'Action Scientifique de la Défense Nationale (CASDN).[29] The creation of this commission gave France's BW program a dedicated administrative body that until 1956 continued the research begun in 1948.

In order to establish the terms of reference for this new body, Surgeon General Costedoat drafted a report whose conclusions and technical recommendations created a broad framework for the continuation of work on France's BW program. After providing an inventory of pre-1953 activity, the report concluded that "these findings would endorse the continuation and development of experimental research and the establishment of practical trials on the ground." The report went on to address the broad outline of the future program, stating that "it must be conceived in terms of the results already obtained and in the light of working hypotheses on the use of biological armaments," thus sketching France's embryonic biological doctrine. In the case of tactical weapons usage, the objective must be "to increase the seriousness of wounds from exploding devices, thus overloading hospitals, prompting the general issue of medical prophylaxis (vaccines, specific serums)"; in strategic instances, the targets were "sensitive civilian and military zones, training camps, maritime bases, large towns, supply centers, livestock farms, and industrial centers, with the aim of causing panic and creating multiple infection hotspots that, in certain favorable though unforeseeable circumstances, could trigger epidemics or epizootics which would weaken the target's military potential, reduce the output from arsenals and from industry, disrupt supply lines, and deal a severe blow to livestock rearing." The report proposed delivery using 105- and 155-millimeter shells, 120-millimeter mortars, 50- and 250-kilogram aviation bombs, bouncing mines, and aerial dispersion.

It also recommended that methods of delivery by aircraft be perfected (research on large-capacity reservoirs, dispersion by pressure generator, emission ramps, rain diffusers, mist, aerosol). Criteria for the choice of agent included high virulence, resistance (notably to explosion), long storage life, and potential for large-scale production. Costedoat proposed the use of the following in munitions: *Bacillus anthracis, Clostridium tetani* (tetanus), *Clostridium oedematiens* (gangrene), *Clostridium histolyticum* (gangrene), *Clostridium sporogenes* (gangrene), *Clostridium perfringens* (gangrene), and *Clostridium botulinum* (botulism). The biological agents to be used in explosive vectors or dispersed in strategic situations were *Burkholderia mallei* (glanders), *Burkholderia pseudomallei* (melioidosis), and *Brucella melitensis* (Malta fever). The report recommended rinderpest and peripneumonic virus as the biological agents suitable for delivery by explosive devices or dispersal equipment used to infect livestock-rearing areas.[30]

Undoubtedly bearing in mind the budgetary constraints that would limit the scope of the program, the report concluded by recommending the adoption of a "minimum" program whose principal activities would be

- Selection and preparation of agents
- Study of the preservation of vitality and virulence of lyophilized agents
- Study of the best storage conditions (suspension in the form of liquid, gel, dust, ampules)
- Study of the antiseptic properties of certain materials (including metals and plastics, used in explosive devices)
- Study of resistance to explosion of germs and determination of minimum active doses on wounds
- Determination of minimum active doses in aerosolized delivery
- Study of microbial aerosols in enclosed spaces and under varying conditions of humidity, temperature, light, pressure, and other factors
- Study of the stability of microbial aerosols in the open air (dispersal by aircraft and the monitoring of stability under varying meteorological conditions)
- Protection and disinfection

It was upon the recommendations contained in Costedoat's report that France based its BW program from 1953 into 1956. The genuine influence of the report is clear from a 1955 memo titled "Program of Biological Studies,"[31] which takes up Costedoat's key recommendations. The memo allocated the tasks among the research bodies involved in the BW program. The CEB undertook the selection and preparation of biological agents (studies of complex toxins), storage of biological agents, studies on the preservation of the vitality and virulence of lyophilized microorganisms, resistance of bacteria to explosion, potential to infect areas by primary and secondary aerosols, the improvement of explosive devices (study of dispersion by 150-millimeter self-propelled projectiles and cluster bombs once the Air Force had chosen its preferred delivery means), and protection (studies on the effectiveness of gas masks vis-à-vis bacterial aerosols and viruses). The research laboratories of the Army Veterinary Service were entrusted with the study of pathogenic animal viruses and the study of the behavior of viruses when subject to freeze-drying.

1956–1962: Decline

The years 1955 and 1956 marked a turning point in the French BW program, when budgetary and strategic priority shifted to nuclear arms.[32] This change in military policy considerably reduced the budgetary allocation to BW and pared biological warfare activities to a program of scientific monitoring. Despite the lack of written evidence (the subject was mentioned only indirectly in CEECB meetings),[33] it seems likely that it was Minister of Defense Maurice Bourgès-Maunoury who decided to reduce drastically the monetary allocation to the CBW programs in 1956. The decision coincided exactly with the start of France's nuclear weapons development program. The significant budgetary outlay required by the nuclear program caused an immediate drain on the resources allocated to the biological and chemical programs. The years 1956–1972 were marked by uncertainty and continual challenges as France abandoned research into offensive BW. On several occasions, decisions were made to recommence activities and even to create an operational biological arsenal, but these were not implemented owing to insufficient funding.

At a meeting of the CEECB on 20 November 1957, General Fleury (director of the Services des Poudres) opened by noting that "significant budgetary cutbacks are hampering the BW program" and leading to a "suspension of bacteriological research." CEECB members condemned this situation and in their conclusions recalled not only the results of their operations since 1947 but also the considerable amount of work still to be done. They emphasized the danger of totally abandoning operations in this field in the light of advances being made in the US and the USSR. In its final recommendations, the CEECB urged that research not be totally abandoned, while acknowledging that it would pose considerable "financial issues." The CASDN representative, who was appealed to during the meeting for extra funding, could only decline such requests on the grounds of "limited means."[34] Most of the minutes of meetings of the CEECB and the entities that superseded it between 1957 and 1972 touch on the chronic suffering of the BW program as a result of lack of financial support. From this point onward the program survived in a form described by those who ran it as minimal.

On 23 January 1958 the supervisory body of the CEECB, the Commandement des Armes Spéciales, which had been answerable to the Army General Staff, became the Commandement Interarmées des Armes Spéciales (CIAS). Directed by General Thiry, the new entity was answerable to the Joint Chiefs of Staff (Etat-Major des Armées), which had the power to confer on the CIAS responsibility for all examination of and assignments regarding special weapons (nuclear, chemical, and biological). This decision turned the CEECB into a body covering all the armed forces, and thus it was renamed the Commission Interarmées d'Etudes et d'Expérimentations Chimique et Bactériologique (CIEECB). In practice, however, CIAS control over the functioning of the CIEECB was limited, as the change of supervisory control was motivated more by considerations regarding the nuclear program than CBW considerations.

At its first meeting, on 6 July 1959, the CIEECB, having again condemned reductions in funding and stressed the need to continue work in progress, drew conclusions from decisions ratified in the previous few months and proposed the adoption of a three-point "basic" program:

- Inventory of potential biological agents, studies of the bacteria and viruses that could be used by the enemy, study of the preservation of bacteria and viruses by lyophilization, studies of viruses that are pathogenic for animals, study of the effect of irradiation on the susceptibility of animals to infections of microbial or viral origin

- Detection devices for microbial aerosols (including evaluation of US technology that made use of fluorescent antibodies)
- Systematic evaluation of physical and chemical means of biological decontamination

Should there be any room for maneuver in the budget, the subgroup would include two additional elements in its program:

- Examination of the adaptation of chemical weaponry to carry biological agents
- Examination of the industrial manufacture of freeze-dried biological agents[35]

Until this "basic" program was approved and funded nine months later,[36] the CIEECB was able only to give the go-ahead to several less significant projects, namely the study of animal pathogens, freeze-drying of viruses, inventory of potential biological agents, and study of the effects of irradiation on animal susceptibility to infection.

In July 1959 the CIEECB also proposed the creation of a working group, the Sous Groupe de Travail et d'Etudes Biologiques (SGTEB), to be chaired by Veterinary General Guillot. This group would follow and coordinate the biological warfare program. It comprised, among others, representatives from the CIAS, the CASDN, the ARMET, the Service de Santé des Armées (Bacteriology Section), the CEB, and the Army Veterinary Service.[37] From then on, the subgroup would meet independently and report regularly to the CIEECB on its activities. The first meeting of the SGTEB took place on 29 November 1959.[38] The BW program, therefore, continued to function at a level close to low-water mark and with ambitions limited to modest advances on earlier progress. This work was focused in the following areas:[39]

- Studies on biological agents. Positive results were obtained in the following areas: use of freeze-drying to conserve and store viruses, bacteria, and toxins; the pollution of terrain using dust or aerosols to disperse biological agents (botulinum toxin in particular); study of the ecology of viruses by observing the behavior of two viruses of opposing characteristics, one fragile and the other very robust, to determine their behavior when put in different exposure chambers; enhancing or creating the pathogenic properties of certain bacteria vis-à-vis species with little or no susceptibility, using irradiated animals; staphylococcal enterotoxin, transferable in food sources (prefiguring an incapacitating agent).

• Biological detection devices. The development of a detection alert was attributed to the microbiological division of the Centre de Recherches du Service de Santé des Armées (CRSSA) in Lyon, which tested detection of aerosols by infrared spectrography. This effort led to the development of a prototype detector. The LMRV developed a method of detecting botulinum toxins in water using ultrasensitive equipment and continued its work on identification of microbes, using fluorescent antibodies.

In 1962 the operational and administrative organization of France's BW program was characterized by fragmentation. Responsibility for the research and operation of the biological programs was dispersed among the Délégation Ministérielle à l'Armement (DMA);[40] the Service Biologique et Vétérinaire des Armées (SBVA), part of the Joint Chiefs of Staff;[41] the Chemical Weapons Section of the STA, part of the Army General Staff; and the Direction Centrale du Service de Santé des Armées. One coordinating body, the CIAS, oversaw all activities but had no direct influence upon either financial or personnel support; thus it became merely a means to centralize the information structure, allowing those governing the programs to catch up periodically on their progress. Financial restrictions and the fragmented structures of research into BW were made it difficult for the technical bodies to formulate precise definitions of goals: type of materials, importance of the programs, time scales for realization, and industrial infrastructure to get the programs up and running.[42]

1962–1964: Incapacitating Agents

At the beginning of 1960, following a visit to France by the chief of the US Chemical Corps, General Stubbs, key figures in France's BW program became aware of the importance that "Americans attach to biological warfare. They consider that it is almost certain to occur during some future conflict and . . . are prepared to manufacture rapidly the necessary biological agents."[43] In November 1961 Veterinary General Guillot presented conclusions from his own visit to the US to a meeting of the SGTEB.[44] He stressed the importance that US military leaders attached to biological agents. At the Fort Detrick laboratories, Guillot noted, the Americans were studying 160 biological agents, including *Francisella tularensis, Bacillus anthracis,* botulinum toxin, staphylococcal enterotoxins, and various rickettsias and viruses. He also drew attention to US interest in biological agents as incapacitants and to studies being carried out on diseases carried by mosquitoes and ticks. The US efforts in the field of biological warfare were deemed to be very impressive. After hearing Guillot's report, the SGTEB decided to alert military commanders to the dangers of falling behind in BW research. Throughout the 1960s US-French contacts took place regularly, all producing the same conclusions: that France's program was significantly behind America's.[45]

The unofficial decision to relaunch a robust BW program can probably be dated to the end of 1961, as would appear from remarks made by the CIEECB chairman at a meeting on 23 January 1962: "for the first time the need for retaliation has been addressed: the concept of static defense (a type of Maginot Line) based solely on protection, would in fact lead to disaster."[46] Information gathered on BW activities in the US contributed significantly to the revival of interest in such weapon programs among the political and military authorities in France. At a CIEECB meeting on 16 May 1963, General Thiry conceded that "since 1961, in the wake of several visits to the United States, it has become obvious to the authorities that [CBW] are not viewed with the importance that they deserve."[47]

In February 1962 the general commander of the CIAS proposed a scheme to advance study, research, and manufacture. On 9 March 1962 Prime Minister Michel Debré informed Pierre Messmer, the armed forces minister, that the government had decided in principle to relaunch a BW program. It was not until 29 January 1963, however, that the Conseil de Défense made clear its stance on this issue. A directive from the ministerial delegate for armaments dated 12 April 1963, referring to decisions taken at the Conseil de Défense, asked the CIAS "to take stock of the advances made relating to chemical and biological warfare, and bring together the aspects necessary to allow the formulation of a plan for chemical and biological armament."[48] For the first time since the end of World War I, France was considering committing the country to maintaining a BW arsenal and no longer limiting its activities to R&D.

The CEB, as instructed by the ministerial delegate for armament, drew up for the CIEECB a summary of past activities in the field of biological aggression and outlined the advances made:

we have carried out studies on production, conservation, and dispersal of biological agents. With regard to production, we have created in the laboratory a *Bacillus anthracis* and botulinum toxin in raw, purified, and lyophilized state. Concerning conservation, work has been carried out on these two agents subject to variable factors (temperature, aeration, pH, and storage methods, including, for example, botulinum toxin was kept in a lyophilized state for one year). Finally, we have studied the dispersal of bacteria in general, using cooled powder gas projectiles of 120-millimeter mortars, antipersonnel mines, self-propelled missiles, explosion in general. Studies on the aerosolized delivery of toxins and potential contamination through ground and particles are in progress.[49]

At the meeting on 16 May 1963, the CIEECB chairman, in response to ministerial directives, laid out a future CBW program, marking out two periods: one short-term, covering 1963–1964, and one longer-term, covering 1965–1969. In doing so he stressed that it was vital that a war doctrine bringing such arms into play be defined before any biological armament plan was enacted. The chairman then declared his opinion that the country should first equip itself with protective materials, including offensive weapons designed to wage tactical war on a European battlefield. It thus appears that only tactical utilization of BW was envisaged. The objective was for the laboratories to prioritize the development of biological incapacitants. It was also decided that a budget for these additional projects should be drawn up as quickly as possible and submitted to the minister. For 1962, the funds allocated to the BW program came to FF5 million; for 1963, and FF5.7 million.[50]

In June 1963 the CIEECB was given new directives by the Conseil de Défense. The commission was instructed to conduct biological studies into incapacitating armaments rather than into lethal agents. The SGTEB president wrote to all the heads of all the collaborating research organizations informing them that from now on they were "to plan to extend their studies in 1964 under the new terms for the program and to participate in it according to their abilities and resources. Given that research on alert, detection, protection, and decontamination is being carried out according to a previously defined program, it would be appropriate in our opinion to consider new studies on the following":

- Creation and maintenance of a prototype biological offensive agent
- Selection and conservation of microbe stocks
- Familiarization with the manipulation of such offensive agents (ecological studies) and with their conservation and production
- Choice of vector and mode of delivery congruent with their military use
- Experimentation[51]

In accordance with Conseil de Défense directives, the CIEECB undertook "to focus research on incapacitating biological agents rather than lethal agents."[52] Veterinary General Guillot explained that aerial dispersion of staphylococcal enterotoxin had already been perfected in the US and suggested that this method also be explored for the agents of tularemia and dengue fever. From this point onward research was carried out in accordance with the new directives of the BW program. The SBVA, at Maisons-Alfort, carried out its experiments with two biological incapacitants: staphylococcal enterotoxin, dispersed by aerosol; and *Brucella abortus*, studied in irradiated animals. The Centre d'Etudes du Bouchet, although attention had shifted to incapacitating agents, continued its work on *Bacillus anthracis*. The CEB intended during 1964 to resume its theoretical studies, abandoned in 1958, into biological ordnance and aircraft reservoirs with aerosolization nozzles and bomblets, so as to be ready to carry out the first actual tests the following year.[53]

The SGTEB meeting in January 1964 was largely taken up with the question of biological armament. The SGTEB wanted to focus on the dispersion of biological agents in dry product form rather than in liquid form because it was easier to conserve agents in that way. During the discussions, Medical Corps Colonel Colobert pointed out that the US was having to replace its stocks of liquid agents annually, which was a very costly undertaking.[54] The SGTEB also spent time during this session on defining the notion of a biological incapacitant: "this type of agent when used in normal conditions should not cause mortality greater than 1 percent or create permanent lesions, in addition to possessing the properties required for a weapon agent (pathogenic character, ease of production, resistance to the release environment)."[55] Besides staphylococcal enterotoxin and *Brucella abortus*, other potential incapacitants were studied in

1964 by the LMRV, in particular *Erysipelothrix insidiosa, Listeria monocyto-
genes, Bacillus cereus, Shigella flexneri, Salmonella typhimurium, Salmonella
enteritidis;* then, in 1966, it studied certain viruses that could be used as
incapacitants (adenovirus types 1, 3, and 5, coxsackie virus A21, and influenza virus APR8).[56] The SGTEB concluded that studies on enterotoxin
and incapacitants should be carried out by the LMRV. The CRSSA microbiology lab would be responsible for studying complications involving
diphtheria toxoid and tuberculin in the context of complications resulting
from mass vaccination.

The study of vectors and of dispersion of simulants was to be carried
out on a theoretical basis by the CEB in 1964. To this end, the CEB transferred to biological experimentation all the technical resources originally
financed as part of the chemical weapons (CW) program. Thus the dispersion division, which had all the necessary apparatus and gas chambers
of different volumes (15–30 cubic meters), was asked to carry out studies
on aerosol generation. It was also decided that work done jointly by the
dispersion and bacteriological services on methods of dispersion should
begin again. Before 1958 this collaboration had furthered the development of various BW. Regarding the choice of carriers, CEB technicians
felt that as far as toxins and bacteria were concerned, problems could be
swiftly resolved, since work had been done and was still under way at
the CEB. However, for viruses and rickettsias, an arthropod carrier was
required, which in turn necessitated a breeding program. Experiments
would have to take place initially in an enclosed space, then outside
(which meant finding and managing a space in which real agents could
be studied, which would then have to be decontaminated).[57]

The various agencies responsible for BW studies over the period 1962–
1965 focused on the following.

Centre d'Etudes du Bouchet at Vert-Le-Petit:

- Contamination of soil and aerosol dissemination: trials on poisoning by
  aerosol delivery of purified botulinum toxin (100 percent of mice exposed were dead within 12 hours), discovery of minimum active dose
  according to terrain (animals) and form of the agent (raw, purified, and
  lyophilized toxin) on healthy or partially poisoned animals.
- Preservation of toxins by lyophilization: results were encouraging, help-
  ing to preserve the vitality and virulence of the toxins.

- Agent production: extraction and concentration of *Salmonella paratyphus* endotoxin.
- Weapons testing: tests on a high-output aerosolizer consisting of a nozzle into which the microbial suspension was delivered, surrounded by a sleeve through which compressed air passed. The two tubes—liquid and gas—could be adjusted, permitting control of output and particle size; examination of the performance of bacteriological aerosols in the lab with *Serratia marcescens;* study of the influence of different factors on microbial solutions (humidity, age, and so on).
- Detection: testing of air sampling equipment.
- Protection: study of the reliability of gas masks.

Laboratory of the Centre de Recherche du Service de Santé des Armées at Lyon:

- Warning detection: trials of a prototype detection device (which took samples of air and detected bacteria using infrared spectrography); construction of a permanent bacteriological air monitoring station; rapid identification of bacteria (a method based on enzymatic reactions in a bacterial suspension in the presence of a concentrated metabolic substance and pH indicator); taxonomic procedures allowing the interpretation of results obtained using earlier methods.

Service Biologique et Vétérinaire des Armées (Maisons-Alfort and Tarbes):

- Weapons testing: conditions in which biological agents were used (always using *Bacillus anthracis* and botulinum toxin); creation of virulence in bacterial strains as they affected resistant animal species with a view to using these strains as biological agents; enhancement or creation of pathogenic or toxigenic potency of certain bacteria used to infect laboratory animals irradiated with X rays, then used on unirradiated animals; conditions in which biological offensive agents were used; ecology of viruses and research on methods of extending survival of agents; establishing the conditions for diffusion or dispersal of biological agents; production of pathogenic clones of biological agents used to infect irradiated animals; studies of a staphylococcal enterotoxin used as a biological incapacitant and of botulinum toxin; production of viruses in tissue cultures; aerosol infection of animals.

- Protection: detection of abnormal levels of microbial agents; identification of specific agents; studies on protection against attack with viral agents; experimental diagnosis of certain ailments in laboratory animals irradiated by X rays; experiments on laboratory animals with a view to using mass vaccination by aerosol; studies of the effectiveness of CF51/33 mask cartridges.
- Detection: immunofluorescent detection of biological agents; studies on the effectiveness of vaccination by aerosol.

The commitment to creating an operational BW arsenal lasted only a few months. In a letter dated 21 August 1964, General Lavaud, ministerial delegate for armaments, informed the chief of the Army General Staff that

> the Conseil de Défense, at the recommendation of the chiefs of staff of the three armed forces, has made the establishment of operational bacteriological and chemical systems a secondary priority behind other programs judged to be more essential but which have not been pursued until now because of a lack of resources. The Conseil de Défense has decided, therefore, that in the field of biological and chemical weapons, priority should be given to researching scientific advances and new technology prior to considering the doctrine for the use of such armaments and the creation of materials accordingly. The only operational activities planned in the long term in the field of biological warfare are those concerning means of protection. This includes limited-scope offensive operations integral to the work.[58]

These decisions were based essentially on financial considerations[59] and marked the end of a fleeting commitment by France to furnish itself with a BW arsenal. This brief period is nevertheless of special interest in that it constituted a clear breach of biological arms policy as it had existed in France since 1922, the date of the first French military activities in this field. The break with policy probably had its origins in the advances in BW being made by the US as perceived by French officials. The increasing international tension that marked the start of the 1960s, following the second Berlin crisis of 1961 and the Cuban Missile Crisis of 1962, probably played an equal part.

1964–1972: Death Throes

In 1964 France's BW program entered a period of incoherence exacerbated by chronic organizational instability. The switch in policy brought about periods of administrative and hierarchical confusion that left research organizations to their own devices, without clear instructions as to where their efforts should be directed. This situation reflected the total lack of concern among political and military decisionmakers regarding the status of BW, which from then occupied a distant third place behind chemical and nuclear armaments. The policy change also signaled a considerable shrinkage in the budget allocated to BW. From the mid-1960s on, France gradually abandoned the offensive element of the program and retained only activities that were strictly defensive. The shift away from offensive measures was made official in 1972 following the signing of the Biological Weapons Convention (BWC), although France did not formally ratify the treaty until 1984 because it made no provision for a system of verification.

In June 1964, the dissolution of the CIAS[60] brought with it the demise of the CIEECB and another administrative reorganization of the BW program. However, the activities of the SGTEB remained unaffected; having been given a new lease on life in January 1965,[61] thereafter it operated under the aegis of the DMA. The latter determined the parameters of research and study in liaison with the Special Weapons Section of the Army General Staff. The DMA thus assumed the role of the ex-CIAS.[62] The missions of the working group remained, as before, "to facilitate the exchange of information between different organizations participating at different levels and within various hierarchies in the execution of these programs in order to coordinate execution, to formulate proposals that may alter the direction of these programs, or to improve the conditions of their execution."[63]

Subsequently the Délégation Générale à l'Armement, which became *de facto* the authority over the military weapons program, began considering future organization in the light of the new objectives assigned to the program. Its considerations took the form of a memo drafted during 1965, addressed to Pierre Messmer, minister of the armed forces, and titled "Organization of Biological and Chemical Operations." The memo noted that "the biological program has never in recent years been granted sufficient priority for it to be regarded with interest by the Armed Forces Staff. In

such a situation the technical organizations have experienced great difficulty arriving at a precise definition of the goals to be attained."[64] The
DMA therefore proposed the creation, within its own structure, of an organization that would govern the entire BW program, in order to bring
together all biological activity, establish liaison with the Armed Forces
Staff, and allow the programs to be defined. At the same time, the Conseil
de Défense agreed on 25 June 1965 that "a person shall be designated by
the minister of the armed forces to assume responsibility for activities relating to biological and bacteriological weapons and protection against
their effects. This person will coordinate the tasks presently distributed
among the Armed Forces Staff, the Service de Santé des Armées, the
Services des Poudres, the Direction des Recherches et Moyens d'Essais,
and other collaborators."[65] The DMA's proposals were finally ratified by
General Ailleret on 20 July 1967,[66] centralizing under the Direction Technique des Armements Terrestres responsibility for research and operations in biological and chemical spheres that had previously been de-
volved to the Services des Poudres and the SBVA. A newly formed
Nuclear Biological Chemical Joint Armed Forces Committee (Comité
Interarmées NBC, or CINBC) would in turn comprise two working
groups: the first, dealing with "biological and chemical armament and
retaliation," would be responsible for drafting a doctrine for utilization,
proposing a concept of use that would determine the equipment required, and, finally, determining the parameters for the programs of research, study, and execution; the second, the Défense NBC, would have
responsibility for all defensive aspects. The CINBC met for the first time in
January 1968.

Although the SGTEB continued its work during this period, there was
widespread uncertainty about the future of the program, which was left
to fend for itself in the absence of reliable direction. At a meeting of the
biology subgroup on 20 December 1966, the chairman remarked that
"the only directives received on this subject for the last three years are
those that came from the former Commission Interarmées des Etudes et
Expérimentations Chimiques et Biologiques" and that the only directives
addressed to the working group were those interpreting the decisions of
the Conseil de Défense. He expressed a desire for further clarification,
particularly with regard to actions altering the instructions received previously.[67] Although no clear evidence of formal notification by the minister of defense or Conseil de Défense survives in the archives of the

Service Historique de l'Armée de Terre, a gradual abandonment of the offensive BW program is discernible from the end of 1966. Beginning in 1967, the minutes of meetings of the SGTEB refer only to CW issues, and the ministerial directives setting out the priorities for that year's research show clearly that offensive CW took precedence, with only secondary consideration given to biological detection. In 1967 the traditional references to "aggression" in reports on progress in BW from the Centre d'Etudes du Bouchet disappear abruptly from the minutes of the SGTEB.[68] It thus appears that France renounced *de facto* the offensive element of its weapons program at the end of 1966. It is likely, moreover, that this situation was not the result of a formal decision taken following an official policy review (which happened only in 1972) but rather a consequence of lack of political interest and attendant budgetary support, accompanied by a move to prioritize nuclear weapons; for in the months that followed, only a very few residual activities relating to offensive BW were maintained within certain laboratories.

Of the working groups established in 1967, only the Défense NBC group continued to meet at irregular intervals from 1969 into 1972. In contrast, in the absence of clear instructions and despite repeated requests, the SGTEB maintained its existence on paper but never actually met. In fact, at the first meeting of the CINBC, in January 1968, "instructions to the biological and chemical armament and retaliation working group failed to be established, to the extent that the group had no mandate and could not operate."[69] In an attempt to resolve this impasse, the army chief of staff in February 1969 had requested that a concept be defined and that a meeting of the CINBC working group on biological and chemical armaments and retaliation be convened to that end.[70] Demonstrating the lack of interest in these issues, these requests elicited no response, confirming *de facto* the abandonment of the offensive element of France's BW program.

On 6 February 1969, more than 15 months after the last meeting of the SGTEB, the Défense NBC group met for the first time. The majority of its members were from the Service de Santé des Armées. As an introduction to the work of the group, the chairman confirmed that "predominance in the field of BW is henceforward given to protection."[71] Although certain residual activities pertaining to offensive weapons were carried on within the CEB,[72] research focused more and more on defensive aspects, and particularly upon detection (use of fluorescent antibodies in warning systems with nonspecific antibodies) and decontamination (trials of tri-ethylene glycol, ethylene oxide, B-propiolactone, and hexylresorcinol).[73] On this date the CEB ceased all manufacture of botulinum toxin and put a stop to studies on dispersion with the exception of research relating to detection. For its own requirements, the CRSSA focused on the production of various simulants (neisseriae, micrococci, Pseudomonad enterobacteria, vibrios, actinomycetes, and especially corynebacteria).[74]

Conclusion

With the exception of an early period of enthusiasm from 1948 to 1956, during which France committed limited human and financial resources to the resumption of its prewar R&D program, France's BW program between 1945 and 1972 is characterized by a lack of strategic planning and of continuity in action. This lack of interest was expressed in turn in budgetary, organizational, military, and political terms. In fact, despite a short period from 1962 to 1964, during which the decision to create a biological arsenal was taken without ever being put into practice, France's program was essentially one of R&D without industrialization of the armament systems, the theoretical study of which was carried out only between 1948 and 1956. As such, the scale of the program in France bore no comparison with the programs of the US and the USSR at that time, and was very much behind that of Britain. The fact remains, however, that the advances made by the program, though described as modest, were substantial and allowed key hypothetical data to be validated. Conversely, in operational applications advances were very limited, particularly after 1956, the year when real trials were partially abandoned. France's decision at the end of the 1960s to obtain a nuclear arsenal, and the substantial expenditure associated with the nuclear program in the years that followed, sounded the death knell in 1967 for France's aspirations in biological matters, at least as far as offensive weapons were concerned. Signature of the BWC in 1972 served to enshrine legally a decision already implicit in fact for six years. Since that date, France has worked unceasingly to ensure the strengthening of the Convention and to promote its implementation.

CHAPTER 6

The Soviet Biological
Weapons Program

JOHN HART

It is generally believed that the Soviet Union had the largest, most extensive biological weapons (BW) program of any country. The highly secret program, which was expanded on the basis of a decision taken in 1973 by the Central Committee of the Soviet Communist Party, continued until at least March 1992, when Russia's President Boris Yeltsin acknowledged a delay in his country's implementation of the 1972 Biological Weapons Convention (BWC).[1] The following month he issued a decree on the implementation of Russia's treaty commitments with regard to chemical and biological weapons (CBW).[2] However, questions continue to be raised regarding the fate of the former Soviet BW program institutions, structures, and personnel.

An authoritative and comprehensive account of the post–World War II Soviet BW program based on archival documents and oral histories by participants has never been published. However, significant works have been produced, including memoirs, academic studies, and partially or entirely declassified intelligence assessments.[3]

The program also reportedly involved the development and fielding of both tactical and strategic BW systems.[4] Estimates of the number of people employed by the program at its height are generally put at between 25,000 and 60,000. It is unclear whether and how the estimates include support staff and the criteria by which military personnel are counted.[5] Concerns and uncertainty also persist about the lack of authoritative, detailed information on the organization of the program, the nature and type of work carried out, and how the structure and work changed after the collapse of the Soviet Union.

Assessing the Program

There is no authoritative and comprehensive account of the Soviet BW program based on oral histories and a systematic study of primary, including archival, documents. Such an account would describe the evolution of people, organizations, activities, and policies and would indicate how official policies and programs were actually implemented. Such an account would also describe the motivations of the Soviet government in its decision to pursue an offensive BW program after it had signed the BWC in 1972.

Two key participants in the Soviet BW program, Kanatjan Baizakovich Alibekov (who later changed his name to Ken Alibek) and Igor Valeryanovich Domaradsky, have published accounts of their work.[6] A third important participant, Vladimir Artemovich Pascchnik, defected to the UK in 1989 and provided much reliable information during his debriefing.[7] However, little of this information is publicly available. A fourth knowledgeable participant who has apparently provided BW-related information is V. S. Koshcheev, a former head of the Third Main Directorate of the Soviet Union's Ministry of Health who now lives in the US.

Some Russian and Soviet journal articles may, in some cases, be viewed as "official" or otherwise authoritative. The discussion of Soviet or Russian activities is almost always confined to those related to defensive aspects, usually vaccine development.[8] Information on vaccine work, including the names of the individuals and institutions involved, is probably reliable. At least three other factors are notable in published Soviet and Russian works on CBW-related matters. One is the emphasis on the external CBW threat to the country. Second, there is little or no discussion about Soviet offensive BW work (in contrast to chemical weapons–related activities, about which there is greater openness).[9] Finally, discussion of BW is sometimes confined to activities in other countries.

Articles have been published in Soviet scientific journals on basic and applied research that has potential offensive BW applications.[10] In such cases, no firm conclusions may be drawn in the absence of information that reveals intent. The consequent uncertainty is reflected in the declassified sections of intelligence estimates by other countries.

Analyses and discussions of civil defense and military doctrine have also been studied, including information that indicates the views of military planners on the role of BW in military doctrine.[11] The US has devoted more resources than any other country since the end of World War II to determine the nature and status of the Soviet BW program. Partially declassified intelligence assessments are available for much of this period.

Origins and Development of the Program

Origins

Soviet interest in BW dates to at least 1928, when Yakov Moiseevich Fishman, the head of the Military Chemical Directorate of the Worker-Peasant Red Army, prepared a report on BW.[12] By the time World War II began, the Soviet Union appears to have developed and tested a variety of BW systems, including aerosol generators and frangible air bombs.[13] Outside assessments generally appear to conclude, however, that any Soviet BW program was limited in scope at the end of World War II and that any production of BW occurred on a small scale.

After the war, the Soviet BW establishment underwent further development and expansion. For example, secret laboratories were reportedly established and attached to most universities and technical institutes.[14] In 1957 the possible use of new types of weapons began to be "actively" discussed by the Soviet Communist Party Central Committee.[15] The so-called Problem No. 5 dates to at least the 1950s and, according to some sources, to the end of World War II. The term was initially used to refer to developing defenses against BW. Its use was apparently ended in 1992, when the Russian CBW defense establishment was reorganized.

In the 1960s Soviet scientists expressed increased concern about the growing backwardness of Soviet science, including fundamental research in experimental and theoretical biology, by writing to the Council of Ministers and the Soviet Communist Party Central Committee. They also cited the country's inability to produce or otherwise obtain modern laboratory equipment, so that they were not always able to replicate experiments conducted abroad.[16]

In order to give biology sufficient political support to overcome these problems, some academicians and Soviet officials argued for increasing support for the biological sciences in terms of their military significance, including the need not to fall behind the West in the field of BW.[17] According to Domaradsky, Yuri A. Ovchinnikov (a molecular biologist and Academician of the Soviet Academy of Sciences) and V. M. Zhdanov (a virologist and member of the Soviet Academy of Medical Sciences) argued that the Soviet military's BW capabilities had been hindered by Trofim Lysenko's scientifically unfounded views regarding genetic inheritance. This argument was decisive in persuading the Central Committee to reorganize and increase financial and political support for the biological sciences. As part of this effort, the Soviet government decided that civilian expertise had to be effectively incorporated into the military's BW-related work.[18]

In 1963 or 1964 the Anti-Plague Department was also reorganized and integrated into Problem No. 5 projects.[19] Other institutes involved in BW-related work (offensive or defensive) included the Institute of Physical Chemistry (Chernogolovka), the Institute of Bio-Organic Chemistry (Pushchino), the Institute of Biochemistry and Physiology of Microorganisms, the Scientific Research Institute of Biological Experiments in Chemical Compounds, the Institute of Highly Pure Biological Preparations (Leningrad), the Institute of Immunology (Soviet Ministry of Health), the N. F. Gamaleya Institute of Microbiology and Epidemiology, the D. I. Ivanovsky Institute of Microbiology, and various other Academy of Medical Sciences and Ministry of Health facilities.

In 1963 the Soviet Communist Party Central Committee issued a decision to strengthen the biological sciences and their practical application. In February 1966 the Soviet Council of Ministers took a decision to strengthen the country's biological sciences. The Main Directorate of the Microbiology Industry (Glavmikrobioprom) was established under the Council of Ministers to implement the decision.[20]

By the 1970s an expanded and more capable civilian-military structure was put into place as a result of the 1973 Central Committee decision to expand the Soviet BW program. This system was coordinated by the Inter-Agency Scientific-Technical Council on Problems of Molecular Biology and Molecular Genetics.[21] The council, which was established in 1973,[22] consisted of representatives of the Soviet Communist Party's Department of Science; the leadership of military-scientific production facilities (NPOs); the leadership of Glavmikrobioprom; and leading microbiologists, virologists, geneticists, and molecular biologists from the Soviet

Academy of Sciences and the Soviet Academy of Medical Sciences. The
council was answerable to the State Committee on Science and Technology and the Presidium of the Soviet Academy of Sciences. It established
and carried through a number of scientific programs that were viewed by
the Soviets in terms of their utility in strengthening the country's military
capabilities.[23] The council drafted the agenda for scientific research work

Figure 6.1 Major sites involved in the USSR BW program.

on BW and coordinated work plans among the various government ministries and took decisions by consensus. The decisions had to be informally approved by the Military Industrial Commission. The council was reorganized in 1975 partly because of decision-making problems, which in turn were related to the fact that its members lacked the appropriate background or knowledge required to oversee a BW program.[24] This entity appears to have been the main mechanism by which scientific research directions were considered and developed and the necessary resources identified and directed toward BW-related activities.

In the 1970s the Military-Industrial Commission of the USSR Council of Ministers Commission on the Problem of Providing for the Development through Fundamental [Scientific] Research of New Types of Biological Weapons[25] also played a major role in linking the scientific community with BW work. Soviet Academician Ovchinnikov played a key role in these efforts. In the early 1970s, Ovchinnikov served as vice president of the Soviet Academy of Sciences and as a consultant to the Military-Industrial Commission.[26]

The Soviet Union signed the BWC in 1972 following the Convention's opening for signature and ratified the treaty in 1975. That year the Soviet Communist Party Central Committee issued a resolution on the creation of advanced military technology, which in turn required measures to be taken to strengthen basic and applied scientific research.[27] The following year a decision was taken by the Politburo to expand the country's BW program on the basis of a proposal made by the 15th Directorate of the Ministry of Defense (MOD).[28]

Structure

Broadly speaking, the Soviet BW program of the 1970s and 1980s consisted of a military component, a political component, and a civilian component. The military component was largely controlled by the 15th Directorate, established in 1973 and first headed by Colonel General Efim Ivanovich Smirnov.[29] From 1985 to 1989 the directorate was headed by V. A. Lebedinsky, and from 1989 until its abolition in 1992 by Lieutenant General Valentin Evstigneev. The 15th Directorate was abolished on 3 January 1992 by an MOD decree (no. 3), and the Directorate of Biological Defense was established within a Directorate of Radiological, Chemical, and Biological Defense Forces Command.[30] A limited number of personnel with BW-related expertise also began work in the Presidential Committee on Problems of the Chemical and Biological Weapon Conventions, established by Yeltsin in 1992.[31] The committee was established at least partly in response to problems associated with the Russian Federation's compliance with the BWC. In May 1999 the committee and its responsibilities were taken over by the newly established Munitions Agency. In 2004 the Munitions Agency was incorporated into the Federal Agency on Industry (which is subordinate to the Ministry of Industry and Energy).[32]

Key political bodies involved in the Soviet BW program included the Central Committee of the Soviet Communist Party, the Politburo, and the Council of Ministers. Perhaps the key body on BW matters within the Central Committee was the Defense Department (not to be confused with the Ministry of Defense).[33] A variety of ministries participated in the program, including the KGB, the Ministry of Agriculture, the Ministry of Chemical Industry, the Ministry of External Trade, the Ministry of Health, and the Ministry of Internal Affairs. Military officers played an influential role, holding seats in the Politburo and Central Committee. Scientific expertise, including that related to BW, was provided by the Science Department of the Soviet Communist Party Central Committee, while the Military-Industrial Commission was answerable to the Council of Ministers.

Numerous civilian scientific and technical bodies, including the Soviet Academy of Sciences, the Soviet Academy of Medical Sciences, the Ministry of Health, the Anti-Plague Department institutes, and university departments were also directly or indirectly involved.

The major BW field-test facility was Vozrozdeniye Island, located in the Aral Sea. This island was identified as the location for the Bacteriological Institute and Proving Ground for Bacteriological Weapons by the 1951 Hirsch Report.[34] A 1965 CIA analysis of whether the island had a BW test facility found that the evidence, while suggestive, was inconclusive.[35] Today the island is the focus of some US cooperative threat reduction program funding.

There were four key research facilities: the All-Union Scientific Research Institute of Applied Microbiology (at Obolensk, near Moscow); NPO "Vector" (in Kol'stovo, Novosibirsk Region; later renamed the State

Research Center of Virology and Biotechnology [Vector]); the Institute of Experimental Hygiene (Kirov, now called Vyatka); and the Institute of Microbiology of the Ministry of Defense of the Russian Federation (Zagorsk, now Sergeev-Posad).

BW production and storage facilities included those of the Main Directorate for Biological Preparations (Biopreparat) at Berdsk, Omutninsk, Sverdlovsk (now Ekaterinburg), and Stepnogorsk.[36] A decision appears to have been taken after the 1979 anthrax outbreak in Sverdlovsk to replace a storage facility at the MOD's Scientific Research Institute of Bacteriology (Military Compound 19, located at Sverdlovsk) with a new facility at Stepnogorsk, Kazakhstan.[37]

The Sverdlovsk facility appears to date to 1949, when a scientific research facility was established in the city on the grounds of an infantry training school. In 1951 work on developing materials and methods for defending against botulinum toxin was carried out at this location. In 1960 the facility was renamed the Military-Technical Scientific Research Institute of the Ministry of Defense of the USSR. The institute developed production methods for a variety of botulinum antitoxins that were later transferred to the Ministry of Health. It also worked on the prevention and treatment of anthrax, including the development and preparation of anthrax vaccines. In 1974 the institute was renamed the Scientific Research Institute of Bacterial Vaccine Preparations of the Ministry of Defense of the USSR. In 1986 the facility was transferred to Military Epidemiology's section of the MOD's Scientific Research Institute of Microbiology. In 1995 the Sverdlovsk facility was renamed the Center for Military-Technical Problems of Biological Defense (and continued to remain a part of the Scientific Research Institute of Microbiology).[38]

Starting in the 1950s, the Sverdlovsk center developed mathematical techniques for modeling the behavior of BW agents in the field,[39] the persistency of aerosols, the effectiveness of re-aerosolization, and ways to maximize human survival in a BW-contaminated environment. The facility also produced a handbook describing the behavior of BW agents, which is reportedly widely used within the Federal Border Service, the Federal Security Service, the MOD, the Ministry of Emergency Situations, and the Ministry of Internal Affairs.[40] It has also developed anthrax vaccines, botulinum antitoxins, and allergens for the detection of melioidosis.[41] It is unclear, however, whether this facility was part of Military Compound 19 and, if not, how the research facility and the military compound might have been connected.

Biopreparat

In 1972 the Politburo authorized the creation of the Ministry of Medico-Biological Industry (MinMedBioProm), which became the Main Directorate for Biological Preparations, also known as Biopreparat. Biopreparat was also the general name given to the civilian component (but directed by the military) of the Soviet BW program after 1972. Biopreparat initially consisted of at least six scientific production organizations: Biomash, Biosyntez, Enzym, FarmPribor, Progress, and Vector.[42] Most of the Biopreparat personnel were initially military personnel. As a rule, the Soviet military occupied the leadership positions. Vsevolod Ivanovich Ogarkov was the first head of Biopreparat. In 1979 General Yuri Tikhonovich Kalinin replaced him.[43] Following the end of the Soviet Union, Kalinin became chairman of the newly established joint-stock company Biopreparat, a position he held until at least the late 1990s.[44]

Biopreparat, which was sometimes referred to by its postal box address (A-1063) and sometimes referred to as Ogarkov's System, or The System, appears to have had at least 20 main locations. US and UK intelligence were reportedly aware of the existence of Biopreparat before Pasechnik's defection.[45] A number of Biopreparat facilities had been flagged in previous Western intelligence estimates; however, its scale and scope were not properly appreciated until after Pasechnik's information became available.

One of the key Biopreparat facilities was the All-Union Institute of Highly Pure Biological Preparations, founded in Leningrad in 1974. It eventually consisted of three sites and employed approximately 3,500 people. The institute initially focused on developing lethal and debilitating strains of tularemia, respectively. However, it eventually focused on weaponizing *Yersinia pestis*. By 1987 the facility reportedly had a manufacturing capacity of approximately 200 kilograms per week. The dried plague strain that was weaponized was referred to as Weapon of Special Designation One.[46]

Another Biopreparat facility, NPO "Vector," was established in 1974 as the All-Union Research Institute of Molecular Biology and is officially under the control of the Russian Federation Ministry of Health. Vector's areas of BW expertise during the Soviet period included Ebola virus, Lassa virus, Marburg virus, and variola virus.[47]

Defensive and Offensive Activities

The Soviet Union carried out R&D on the full range of traditional agents. According to Ken Alibek, the major R&D included work on *Bacillus anthracis, Brucella spp.*, Ebola virus and Marburg virus, Junin virus, Lassa virus, Machupo virus, (equine) encephalitides, *Burkholderia mallei, Burkholderia pseudomallei, Yersinia pestis,* variola virus, and *Francisella tularensis.*[48] The Soviet program also addressed the problems of detection, prevention, and treatment, as well as all major elements associated with the identification, evaluation, and testing of agents for possible eventual large-scale production, storage, or weaponization.

Once an agent became designated as a "military strain" it might be produced for long-term storage (such as freeze-dried *Bacillus anthracis* spores) or be produced on a regular basis to replenish aging stocks as they became less virulent over time (for example, *Yersinia pestis*). It is uncertain how agents were selected for screening and the process by which they were then selected for more extensive laboratory work and testing in order to be eventually filled into weapons.

The USSR placed great emphasis on the principle of maintaining large standby production capacity in case of national emergency. A part of the effort to maintain such a capacity was to ensure ease of convertibility of civilian facilities to military production.

Research Activities

Soviet research activities in the biological area fell into two major categories: basic research with either offensive or defensive applications, and applied work for offensive or defensive purposes.[49]

Determining whether published research indicates an offensive or defensive program can be problematic. For example, some have pointed to research on enhancing the virulence of pathogens and developing antibiotic resistant strains as evidence for an offensive BW program. Additional context is generally required, as some such work may be done for defensive purposes.

Defensive Activities

Much of the open Soviet BW-related work conducted before and during World War II was on vaccine development. Attention was also devoted to securing domestic production capacities for antibiotics, including penicillin and streptomycin. In 1946 a group that included A. F. Kopylov, N. N. Ginsburg, and M. M. Faibich was awarded the USSR State Prize for developing a penicillin production method. During World War II, the Scientific Research Institute of Epidemiology and Hygiene of the Red Army, presently located at Vyatka (formerly Kirov), worked on the development and production of vaccines against anthrax, plague, and tularemia. In 1945 institute researchers M. M. Faibich, I. A. Chalisov, and R. V. Karneev were awarded the USSR State Prize for developing a dried live plague vaccine, partly based on an EB strain obtained from the Pasteur Institute in 1936. The institute workers N. N. Ginsburg and A. L. Tamarin were awarded the USSR State Prize in 1945 for developing and producing an anthrax vaccine. The first Soviet live anthrax vaccine was created on the basis of work done by N. N. Ginsburg, who isolated an avirulent, highly immunogenic strain called STI-1 in 1940.[50]

The Soviets used aerosol immunization for both humans and animals.[51] Soviet scientists developed aerosolized vaccinations for a variety of agents, including *Bacillus anthracis*.[52] They also developed vaccines to be delivered orally and through skin creams. From 1962 to 1973 the Kirov institute workers, including P. A. Katyrev, V. I. Ogarkov, Yu. S. Pisarevsky, Valentin Ivanovich Evstigneev, V. V. Simonov, and N. Yu. Polonskaya, developed an inhalation method for vaccinating against pneumonic plague "using a small-sized particle aerosol of a rehydrated culture of an EV *Yersinia pestis* vaccine strain."[53] Whereas the use of aerosolized vaccines in the West has been limited (although there has been recent interest), the Soviets reportedly vaccinated animals using helicopter-borne aerosols in Kazakhstan and other areas of the former Soviet Central Asian republics. At least some of the field trials for these vaccination campaigns were carried out at Stepnogorsk, a facility where BW-related work (such as vaccine production) was carried out.[54]

A semiofficial history of the Directorate for Radiological, Chemical, and Biological Defense Forces states that the "start of work in the USA in the area of offensive biological weapons strengthened the apprehension of the Government of the USSR with regard to their possible employment." According to the history, the Scientific Research Institute of Sanitation of the MOD was therefore established in Zagorsk, incorporating a previously existing institute belonging to the Ministry of Health. In the 1960s and 1970s this facility developed mass vaccination techniques (apparently an aerosol vaccine) against smallpox. In the 1970s the facility developed a live oral smallpox vaccine. The oral vaccine work was carried out under the direction of A. A. Vorobyev and V. A. Lebedinsky. Vaccines in tablet form were also developed against Venezuelan equine encephalitis (VEE). In 1986 this facility was transferred to the virology section of the MOD's Scientific Research Institute of Microbiology. The Zagorsk facility was renamed the Virology Center in 1995.[55]

Offensive Activities

Soviet scientists conducted BW-related research on a wide variety of antipersonnel, antiplant, and antilivestock agents. They reportedly weaponized and produced on a large scale a number of agents, including *Bacillus anthracis*, Marburg virus, *Yersinia pestis*, and variola virus. According to Alibek, the Soviet Union had four "major" *Bacillus anthracis* production facilities, located at Kurgan, Penza, Sverdlovsk, and Stepnogorsk.[56]

A significant amount of attention was devoted to manipulating the genetic properties of bacteria and viruses, including the transfer or modification of peptides to destroy the immune system,[57] attempts to genetically modify pathogens to induce the production of endorphins, and the transfer and modification of genes for lethal factors into other bacteria or viruses as part of attempts to create genetically engineered pathogens. Research was also carried out on the mechanisms by which autoimmunity could be induced.[58] In the 1980s Obolensk scientists reportedly genetically modified *Legionella* by inserting genes that triggered autoimmune responses against myelin.[59] When tested on laboratory animals, the altered organism caused brain damage and paralysis and proved nearly 100 percent lethal.[60]

Biopreparat worked to develop pathogenic strains that were resistant to multiple types of antibiotics. It also carried out work on modifying the antigenic structures of bacteria and viruses to evade the body's immune system. The System attempted to obtain strains that were resistant to multiple types of antibiotic treatment and did not lose their virulence in the process. Domaradsky promoted the "binary concept," in which two or more strains were employed simultaneously. One strain of *F. tularensis,* for example, would be developed primarily for its antibiotic resistance, while another strain would be developed primarily for the retention of its virulence. The concept was reportedly used as a basis for developing a plague strain resistant to approximately 10 antibiotics.[61] It is not clear whether this concept was adopted as MOD policy or was instead applied in a more ad hoc manner by low- or mid-level personnel, such as scientific research staff.

At least two projects to genetically engineer variola virus have been reported, one by combining it with VEE virus, and one by combining it with Ebola virus. There is disagreement among intelligence analysts and others about whether such work was actually carried out, as well as about its technical feasibility.[62]

In 1988 Nikolai Ustinov, a Vector employee, accidentally injected himself with Marburg virus and died approximately three weeks later. His blood was used to grow a strain of the virus called Variant U that was subsequently weaponized.[63] It also appears that field tests of variola virus on Vozrozdeniye Island resulted in at least three civilian smallpox deaths in Aralsk in 1971.[64]

Evaluations and Understanding of the Program

Intelligence

The sources of information available to outside intelligence organizations on the Soviet BW program included U-2 airplane overflights of Soviet territory,[65] interviews with defectors and other individuals with firsthand contact with the Soviets, World War II German assessments, reviews of the scientific published literature, and statements on military doctrine. The US understanding of Soviet BW following World War II and through the 1960s appears to have been largely speculative and uncertain (see Chapter 2). US assessments were based partly on the requirements developed for the US BW program and on perceived US vulnerabilities. The US was also concerned about possible BW sabotage operations on its territory. In addition, the US and, to a lesser extent, other countries have systematically collected information by, among other things, talking with individuals formerly associated with the Soviet BW program.

There appears to have been no positive proof until the 1970s that the Soviets had an offensive BW program. The 1979 anthrax deaths in Sverdlovsk strongly suggested an offensive Soviet program. Pasechnik's defection was perhaps the most significant event, convincing skeptics in the US and the UK. The information he provided showed that the Soviet offensive BW program was continuing on a large scale, in contravention of the BWC. He also described work meant to make BW agents resistant to environmental stresses and medical treatment. For example, he reportedly described a powdered form of antibiotic-resistant *Yersinia pestis* strain produced for filling warheads. He maintained that the USSR had a 20-ton stockpile of *Yersinia pestis* and was periodically replenishing it. He also confirmed that Vozrozdeniye Island had been used for large-scale field testing of BW agents.[66]

A 1965 CIA study concluded that there was "no firm evidence of an offensive Soviet BW program." At the time, however, a presumption existed within US intelligence that the Soviets had such a program. This presumption was based partly on the fact that the USSR was undertaking defensive measures and partly on a belief that it was logical for the USSR to have an offensive program in view of a range of factors, including Japan's possession of an offensive BW program during World War II and the widely known US commitment to BW. US intelligence was forced to rely on indirect methods to try to determine whether the Soviet Union had an offensive BW program. The CIA study noted: "Analysts have used speculation, analogy, and parallels with other nations' BW research, development, and practice in recent times and in the historical past. They have analyzed Soviet, Satellite, and Chinese propaganda charges of US germ warfare for clues as to the Communists' sophistication and familiarity with BW hardware and agents." Indirect methods included a literature review of "military-related activity in the field of biology and medicine, all technical publications which appeared to be censored by security considerations, and all biomedical studies which did not jibe with Soviet public health requirements as we know them."[67]

A major consequence of the then recent availability of overhead imagery (the first U-2 overflight was carried out in 1957) appears to have been a renewed focus on Vozrozdeniye Island, which had been identified as a BW facility in the 1951 Hirsch Report on the basis of information developed by the German military before and during World War II. Partly for this reason, the island was the "foremost suspect as a biological warfare center."[68]

Several factors drew attention to the question of Soviet compliance with the BWC starting in the 1970s, in particular a suspicious anthrax outbreak in Sverdlovsk in April and May 1979 that resulted in the deaths of at least 64 people, allegations made primarily by the US that the Soviet Union was using mycotoxins ("yellow rain") in Afghanistan and Southeast Asia (see Chapter 13), the 1989 defection of Pasechnik to the UK, and the information provided by Alibek to the US starting in 1992.

Almost immediately after the signing of the BWC, the CIA and Defense Intelligence Agency reportedly concluded that satellite imagery indicated that the Soviets were not dismantling their offensive BW program.[69] According to former Secretary of Defense Melvin Laird, the US did not raise the issue with the Soviets, in part because nuclear arms control issues were seen as more important. Apparently there was also disagreement among US government officials as to the degree of certainty necessary before a perceived noncompliance issue should be raised and whether broader national interests were not better served through a more gradual approach to seeking clarification.[70]

Beginning in 1975 Arkady Shevchenko, a senior Soviet diplomat and member of the Soviet delegation to the UN in New York City, began providing information to the US, including information that the Soviet Union was violating the BWC.[71]

The names of many of the major Soviet BW facilities were correctly identified between World War II and the mid-1970s. For example, Vozrozdeniye Island, Gorodomlya (located at Seliger Lake), and Sverdlovsk were identified in the Hirsch Report as having known or likely BW facilities. In August 1975 the CIA reportedly leaked information that "questionable activities" were occurring at Kirov, Zagorsk, and Sverdlovsk. The identities of three then recently established sites—Berdsk, Omutninsk, and Pokrov—were also disclosed.[72]

Partly on the basis of satellite thermal imaging that showed the warheads were refrigerated, in 1988 US intelligence tentatively concluded that the Soviet Union had mounted BW warheads on intercontinental ballistic missiles. Between 1984 and 1988 the US reportedly issued six démarches against the Soviet Union on BW. The first three featured concerns about activities at Zagorsk; the rest concerned the 1979 Sverdlovsk anthrax outbreak.[73]

The Sverdlovsk outbreak, together with the yellow rain allegations, provided a major impetus toward the decision by the Second Review Conference in 1986 to agree on annual data exchanges, including outbreaks of infectious diseases and "similar occurrences caused by toxins that appear to deviate from the normal pattern . . . of occurrence," to serve as a confidence-building measure (CBM).[74]

The Trilateral Process

Pasechnik's 1989 defection provided the main impetus for a series of secret meetings among UK, US, and Soviet officials to clarify the status of Soviet compliance with the BWC. The information he provided was key to the identification and selection of Soviet sites the UK and US wished to discuss with Soviet authorities and to visit.

This trilateral process consisted of preliminary informal discussions and visits, discussions and visits within the framework of the Trilateral Agreement, formalized in 1992, and follow-up discussions that began in mid-1994 and effectively ended in 1996, when, according to a former UK official and BW technical expert involved in the process, a letter from Russian Foreign Minister Evgeni Primakov to US Secretary of State Warren Christopher went unanswered for lack of a "collective resolve" by the parties to try to continue to overcome the unresolved issues.[75] In other words, the participants saw no further utility in continuing the process.

In January 1991 US and UK teams were allowed to visit the Institute of Immunology (Chekhov), the Institute of Applied Microbiology (Obolensk), the Institute of Molecular Biology (Kol'tsovo), and the Institute of Highly Pure Preparations (Leningrad).[76] Among the important discoveries made at the Obolensk facility were an explosive containment chamber, extensive physical security and biosecurity measures, and a large-scale fermentation capacity.[77] The most significant event during the visit at Kol'stovo was the admission by a worker that the facility was doing variola virus work. Among other things, the Soviet Union had not declared to the World Health Organization that it was doing variola virus work at this facility.

In Moscow on 10–11 September 1992 a joint UK-US mission discussed BW matters, including the nature of activities at the St. Petersburg Institute of Highly Pure Biological Preparations. On 14 September 1992 the three parties issued the Joint Statement on Biological Weapons by the Governments of the United Kingdom, the United States and the Russian Federation (10–11 September 1992) (Trilateral Agreement), in which the states reiterated their commitment to the BWC and agreed to host reciprocal visits at selected facilities in order to enhance confidence in treaty compliance.[78]

Under the terms of the agreement, Russia "confirmed the termination of offensive research, the dismantlement of experimental technological lines for the production of biological agents, and the closure of the biological weapons testing facility [apparently Vozrozdeniye Island]." It also agreed to reduce the number of personnel "involved in biological programmes" by half, to reduce "military biological research" by 30 percent, and to dissolve the MOD department responsible for the offensive BW program (15th Directorate). The agreement stated that access to nonmilitary biological sites would be "subject to the need to respect proprietary information on the basis of agreed principles" and that access to any military biological facility would be carried out on a reciprocal basis and be "subject to the need to respect confidential information on the basis of agreed principles." It did, however, also state that access to military and nonmilitary biological facilities would include "unrestricted access."[79]

A set of US government talking points about the agreement stated that Russia had again admitted during discussions held in Moscow on 10–11 September 1992 that it had violated the BWC. The points emphasized that an elaborate and extensive "cover story" was "in many respects still functioning."[80]

In 1993–94 there was a second round of trilateral visits to the All-Russian Scientific Research Institute of Veterinary Virology and Microbiology (in or near Pokrov), the Chemical Plant (Berdsk, near Novosibirsk), the Chemical Plant (Omutninsk), and the All-Union Scientific Research Institute of Microbiology (Obolensk).[81] At Pokrov the visiting team viewed hardened underground bunkers capable of holding several hundred thousand chicken eggs.[82] At Berdsk the visiting team reportedly saw 4 operational 64,000-liter fermenters and an uncompleted building capable of holding 40 64,000-liter fermenters. When the team visited Obolensk in January 1994, it noted that the previously inspected explosive test chamber had been removed.[83]

The trilateral process had a number of consequences. According to Alibek, the UK-US visits resulted in a Soviet decision to develop a "completely new type of mobile biological weapon facility."[84] Some offensive work at some facilities was curtailed or suspended. One result of Pasechnik's defection was that Kalinin ordered all offensive work at the Leningrad facility to be halted and incriminating evidence removed or destroyed.

The US (and UK) did not publicly discuss the Soviet BW program during the trilateral process (except for periodic references to yellow rain and Sverdlovsk) because they believed that quiet diplomacy would be more effective in promoting transparency and appropriate follow-up steps. The process remains suspended.

BWC Data Exchanges

On 8–26 September 1986 the Second Review Conference of the States Parties to the BWC agreed to submit annual, politically binding data exchanges on biological-related information to serve as CBMs. On 9–27 September 1991 the Third Review Conference agreed that information would be provided in additional areas, including "past offensive and/or defensive biological research development programmes." The Soviet Union (and then the Russian Federation) has submitted information every year since late 1987, when the first exchange of information and data occurred.

The quality and completeness of Russia's CBMs have been questioned, especially with respect to past programs. The US publicly criticized the Soviet Union's submission in 1991 during the Third Review Conference of the States Parties to the BWC.[85] And the British ambassador to Russia, Sir Rodric Braithwaite, and James F. Collins (the deputy chief of mission at the US embassy in Moscow) reportedly warned Yeltsin in 1992 to "reveal the full extent of the former Soviet biological weapons program or face public denunciation" at the UN.[86] In June of that year Russia reportedly showed a draft declaration to the US listing 4 (Kirov, Sverdlovsk, Vozrozdeniye Island, and Zagorsk)[87] of 20 facilities the US and UK knew or suspected of having been involved in producing or stockpiling BW.[88] No mention was made of the Sverdlovsk *Bacillus anthracis* leak (which some Russian officials have still periodically maintained was a natural disease outbreak) or Soviet work with hemorrhagic fever viruses.[89] A second list was reportedly provided to the US by Russia,[90] which unnamed US government sources characterized as "marginally better." US and UK officials told the Russians, however, that if the data were submitted to the UN as a CBM declaration, they would publicly "attack it as seriously inaccurate";[91] among other things, neither of the drafts provided a "detailed account of the allegedly extensive work with mycotoxins."[92] A third and final draft was also judged inadequate by the US and UK:[93] among other omissions, like the previous two drafts it failed to acknowledge stockpiling of BW.[94]

In 1992 Russia declared that the Soviet Union (and then Russia) had had an offensive BW program from 1946 to March 1992; that the Soviet Union had begun a program in the late 1940s to develop BW for retaliatory purposes; that work had been carried out with *Bacillus anthracis, Francisella tularensis, Brucella spp., Yersinia pestis,* VEE virus, *Rickettsia sp.,* and *Coxiella burnetii* at facilities located in Kirov, Sverdlovsk, and Zagorsk in the 1950s; that models of BW-filled air bombs and rockets had been tested at Vozrozdeniye Island; and that work had been done to determine the threat posed by *Burkholderia mallei* and *Burkholderia pseudomallei.* In addition, "In the late 1960s, industrial facilities with storage capabilities were, by a government decision, established in Glavmikrobioprom for the production of medicinal and other protective preparations, which could also be used for the preparation of biological agents during a crisis." Although "investigations with dangerous pathogens" were carried out in 1982 and 1983 at Glavmikrobioprom (at Kol'tsovo, Obolensk, Chekhov, and Leningrad), the declaration stated that an insufficient level of "scientific-methodological level of work" and a lack of equipment and reagents "did not permit practical significant results in the military field." Russia also declared that a multistep review of the "military biological program" had been begun before the Second Review Conference (held in 1986). Finally, Russia stated that it had not stockpiled BW.[95] In short, the declaration described work that was essentially defensive, or at worst preparatory for a possible full-scale offensive program. There was no clear or straightforward admission of offensive work, in contrast to the declarations provided by, for example, the UK and US. Since 1992, Russia has declared that it has no changes to make on this part of the declaration.

Rationale for the Program and BWC Violations

Before the BWC was opened for signature in 1972, the Soviet Union most often gave the following three reasons for pursuing a BW program: the US agreement at the end of World War II not to prosecute participants in Japan's BW program, in exchange for BW information; alleged US use of BW against North Korea during the Korean War; and a more general charge that the US was an aggressive "imperialist" country intent on dominating the world.[96]

Soviet threat perceptions were heightened by Japan's World War II BW program.[97] However, the extent to which the threat was perceived as being actual rather than a justification for strengthening Soviet BW-related capabilities is unclear. On 25–30 December 1949, in Khabarovsk (USSR), Soviet authorities tried 12 Japanese military personnel for "preparing and employing" BW.[98] The case was based in part on Japanese BW documents and materials captured by Soviet forces in Manchuria. A commission of experts was assembled to evaluate these materials and to provide testimony at the trial. The commission was headed by Nikolai Nikolaevich Zhukov-Verezhnikov, a key figure in the postwar Soviet BW program.[99] It is unclear whether or how Zhukov-Verezhnikov's postwar experience on the commission influenced his later views.

It has also been suggested that the Soviet Union did not wish to give up its offensive BW program after 1972 because the military found attractive the possibility of eliminating the personnel of factories, research facilities, and the like located deep in an enemy's heartland, far from actual combat, while preserving intact the infrastructure and equipment. It is also possible that the military simply did not wish to lose part or most of its existing BW establishment. Another factor in the decision to violate the BWC may have been the close personal ties enjoyed by key supporters of maintaining the program (such as Zhukov-Verezhnikov and Smirnov) with those at the highest levels in government. Smirnov, who was still head of the 15th Directorate at the time, was reportedly a close friend of

Leonid Brezhnev.[100] Another rationale for the BW program reportedly given to Gorbachev by the Soviet military was to help counter the military threat posed by China.[101]

Shevchenko's memoirs provide some insight into both the decision-making process and possible reasons why the USSR wished to retain an offensive BW program. Shevchenko, referring to personal discussions with Soviet Ministry of Defense officials, said that in the early 1970s the military was strongly opposed to any arms control or disarmament agreement on chemical or biological weapons partly because such agreements "could reveal the extent of the development of these weapons and would show Soviet readiness for their eventual use." Shevchenko has said that General Aleksei A. Gryzlov informed him that Defense Minister Andrei Grechko had instructed the Soviet military not to stop production of BW. Shevchenko also believes that the Politburo must have known about this directive.[102]

Some sections of the Soviet government appear not to have believed that the US had in fact abandoned its offensive BW program. This skepticism may have stemmed in part from the deception programs reportedly run by the US in the 1960s and 1970s to encourage Soviet research into unproductive, costly research directions in CBW.[103]

When considering the rationale for the post-BWC offensive BW program, it is also important to take into account the role played by compartmentalization of activities and information—both information specific to the work and more general information regarding the outside world. According to Pasechnik, workers in The System were, depending on their level of security classification, given one of four "legends." The first-level, "open legend" denied there was a BW program. The second-level, "closed legend" acknowledged BW work but said it was defensive. The third-level legend involved providing limited information about some aspects of offensive work. Finally, individuals cleared for the fourth-level legend were permitted to know the true nature and scope of the program.[104]

A lack of outside information probably facilitated the justification for carrying on an offensive program. Many of those involved in the program were apparently unaware of the BWC's existence. The charge that the US used BW against North Korea has appeared in some official and semiofficial Russian-language publications,[105] and Chinese and North Korean government officials continue to make it (see Chapters 2, 4, and 13).[106] Finally, participants have noted that most scientists focused on the technical aspects of their work, giving little, if any, thought to possible moral or legal considerations.[107]

Biology-Related Developments

Since the early 1990s, a number of European countries and the US have implemented cooperative R&D programs with facilities and personnel previously involved in the Soviet BW program to help ensure that the latter remain employed in work for peaceful purposes. The bulk of such assistance has been provided by the US within the framework of the 1991 Nunn-Lugar Cooperative Threat Reduction Program and, since 2002, the Group of Eight (G8) Global Partnership against the Spread of Weapons and Materials of Mass Destruction. Assistance is also provided through the International Science and Technology Center (ISTC) program, established in November 1992 by the European Union, Japan, Russia, and the US.[108] Some European assistance is provided within the framework of Technical Assistance for the Commonwealth of Independent States (TACIS), an EU program that will be ended by 2007.[109] Such programs have provided greater transparency on former BW-related activities. Many of the outside cooperative efforts are directed toward cataloging and securing pathogenic strains at all biological facilities in the former Soviet Union.

Russian Developments and Programs

In the 1990s the Russian government undertook a number of measures to preserve or strengthen the country's biological sciences (for example, domestic production of high-quality laboratory and diagnostic equipment, vaccines, and medicines) and to improve defensive capabilities against CBW attacks, including actions by nonstate actors (often described in the context of the international "war on terrorism"). Some of these efforts have featured cooperative projects between Russian ministries and scientific research establishments. Associations have also been established involving the participation of both state and private (or semiprivate) entities. For example, the Russian Federation Ministry of Science and Technologies has proposed cooperation projects with Ministry of Defense scientific research bodies in the fields of biotechnology and genetics.[110] The Russian Ministry of Industry, Science, and Technology has several projects designed to promote high-tech research in Russia.[111] Some of the financial support for these projects comes from government-funded venture funds, including the A. V. Bortnik Fund (also known as the Foundation for Assistance to Small Innovative Enterprises, or FASIE) and the Russian Fund for Technological Development.[112]

The Russian government has identified a wide range of areas and goals for improving defenses against BW. These include promoting comprehensive, integrated legal statutes; supporting biological monitoring facilities and infrastructure; maintaining reference strains for diagnostic, treatment, and research purposes; clearly delineating institutional areas of responsibility; maintaining and enhancing international cooperation; protecting human, plant, and animal life; implementing measures to meet a perceived bioterrorist threat; and developing and maintaining modern detection equipment and detection systems infrastructure. Valentin Evstigneev, now a retired lieutenant general of the armed forces medical service and the current first deputy director general of the joint-stock company Biopreparat, has argued that such a system be controlled by the State Committee on Problems of Biosecurity and by the Biotechnology and Security Agency. They would oversee industrial facilities, laboratories, medical facilities, and scientific research organizations at the national, regional, and local levels. The extent to which this plan has actually been implemented is not clear. However, the Center for Special Laboratory Diagnosis and Treatment has been established at the Russian Federation MOD's Virology Center (part of the Scientific Research Institute of Microbiology) as part of such an effort.[113] Reportedly the institute is currently working on the further development of anthrax and plague vaccines, antianthrax immunoglobulin, diagnostics, and sanitation measures for the Russian armed forces and civilian population.[114]

Unresolved Concerns

Concerns persist about a continued lack of responsiveness by Russian officials to requests by other governments for clarification regarding the fate of the former Soviet program,[115] the fact that a number of high-level officials in the current Russian CBW defense establishment are known or suspected to have been a part of the Soviet BW program, and the fact that outside access to several Soviet BW military R&D facilities has never been allowed. There is also continued concern that individuals formerly involved in the Soviet BW program could be recruited by countries believed to be interested in pursuing illicit BW programs.

There are five military facilities to which outside access has been either sharply limited or disallowed: the Center for Military-Technical Problems of Anti-Bacteriological Defense (Ekaterinburg, formerly Sverdlovsk), the Center for Virology (Sergeev-Posad, formerly Zagorsk), the Scientific Research Institute of Microbiology (Vyatka), the Scientific Research Institute of Military Medicine (St. Petersburg), and a facility located in Strizhi, near Kirov (Kirov-200). (According to Russian officials, the Strizhi facility is no longer under the control of the Russian MOD.)

Several countries have also expressed concern about residual capacity. The 2001 US assessment of the former Soviet BW program states:

> serious concerns remain about Russia's offensive biological warfare capabilities and the status of some elements of the offensive biological warfare capability inherited from the FSU [former Soviet Union] . . . Many of the key research and production facilities have taken severe cuts in funding and personnel. However, some key components of the former Soviet program may remain largely intact and may support a possible future mobilization capability for the production of biological agents and delivery systems . . . work outside the scope of the legitimate biological defense may be occurring . . . the United States continues to receive unconfirmed reports of some ongoing offensive biological warfare activities.[116]

Part of the concern about a standby capacity relates to a Soviet government decree that reportedly reorganized Biopreparat as a civilian organization, but also instructed that Biopreparat was "to organize the necessary work to keep all of its facilities prepared for further manufacture and development."[117] More generally, maintaining a standby production capacity and the option of converting civilian to military production were major goals in Soviet military planning. According to a former researcher in the Soviet BW program, at least some of the offensive BW research results were preserved. He has also said that research published by former colleagues is not of the highest quality, a fact that suggests that they are continuing to carry out classified research.[118]

On 4 March 2003 a US Department of State official testified to Congress: "We believe, based on available evidence, that Russia continues to maintain an offensive biological weapons program."[119] In 2004 a US Department of Defense official estimated that approximately 40 institutes that were formerly part of the Soviet BW program still exist.[120]

It is unclear how much specific information the US, UK, and others have about recent or current Russian activities that cause concern. The Russian government is generally reluctant to discuss the former Soviet offensive program and subsequent developments, feeling that further discussion would result in no benefit to the Russians and might prove embarrassing.

Conclusion

Further changes in the political and scientific leadership in Russia are necessary before a more definitive account of the Soviet BW program and its legacy can be produced.[121] A sign that such changes have occurred would be the publication of studies on offensive aspects of the Soviet BW program by Russian scholars. (Almost all published scholarly works on the Soviet BW program have been produced by people living outside Russia.) International perceptions of Russian activities in the biological sciences will inevitably change and evolve with time, perhaps someday resulting in an international climate that facilitates disclosures about past activities.

CHAPTER 7

Biological Weapons in Non-Soviet Warsaw Pact Countries

LAJOS RÓZSA

KATHRYN NIXDORFF

In 1944 Allied bombers destroyed Hungary's advanced bio-warfare institute in Budapest. However, the staff survived and after the war expressed a willingness to serve the new Hungarian Communist regime. Yet it appears that the new regime declined. Why did a cruel Communist dictatorship pass up this opportunity to augment its power? Were the other non-Soviet Warsaw Pact countries also free of biological weapons (BW)? Or did they participate in the Soviet BW program? This chapter aims to answer these questions.

Traditionally, Central Europe contained an array of nations exhibiting a diversity of ethnic origins, languages, and religious denominations. Ethnic tensions, traditional in this part of the world, easily develop into international military conflicts here, and, indeed, both world wars originated from this region.

At the close of World War II Poland, Czechoslovakia, Hungary, Romania, Bulgaria, and the eastern part of Germany were occupied by the Soviet Army. The Soviets manipulated political life to ensure that Communist regimes assumed the power. Communists took power in Yugoslavia and Albania as well. In 1955 Poland, Czechoslovakia, Hungary, Romania, Bulgaria, Albania, and, in 1956, the German Democratic Republic (GDR) signed the Warsaw Pact Treaty (WP), and thus formally became allies of the USSR. Attempts to leave this alliance provoked brutal reprisals by the Soviet Union, as in Budapest in 1956 and in Prague in 1968. Thus, these diverse nation states found themselves in a quite uniform historical situation. Their religious and cultural differences, including their ethnic tensions, went into hibernation. Their governments, and the outside world as well, viewed them as a monolithic "Soviet bloc."

Hungary

The first known BW program in the region was in Hungary.[1] In 1936 the Hungarian Highest Defense Council authorized the Headquarters of the Hungarian Royal Defense Forces to establish a BW R&D project in 1936. This decision was said to be a response to reconnaissance information indicating that neighboring states had already started similar activities. A team including a medical bacteriologist, a veterinary parasitologist, a chemical engineer, and two laboratory technicians was organized and led by Colonel Dezso Bartos, a medical bacteriologist/epidemiologist. All employees were unmarried males who lived within the institute.

The institute was properly named the Health Control Station of the Hungarian Royal Defense Forces, and it was situated in an artillery equipment warehouse in Budapest. Surrounded by a high rampart equipped with a wire fence, it housed eight microbiological and one chemical laboratories, a library, an animal house, and storage. Research started in August 1938. The project involved explicitly offensive goals. Colonel Bartos viewed military personnel, civilian populations, agricultural crops, and livestock as potentially vulnerable targets of biowarfare. His team investigated three types of biological attack: by means of bombs and artillery shells, by means of secret agents working behind the front lines, and by contamination of territory before a strategic retreat.

The Hungarians developed and field-tested a number of technologies to be applied in such situations. Glass bombs ranging from 1 to 50 kilograms were used to produce wet and dry aerosols. Studies were focused on the influence of meteorological conditions and on the effective number of germs per unit area. *Bacillus anthracis* (anthrax), *Clostridium perfringens* (gangrene), *Salmonella paratyphi* (diarrheal disease), and *Shigella dysenteriae* (dysentery) were cultured as biological agents.

They also tested the viability of pathogens transmitted by infected projectiles from pistols and guns. Pathogens could survive the heat and mechanical shock of firing, causing infections through bullet wounds. Attempts were made to increase the virulence of *Salmonella paratyphi* by serial passage in laboratory animals.

They also invented a remarkably simple method to store inactive bacteria under field conditions without any loss of viability or virulence. This breakthrough enabled them to develop ways of manufacturing chocolate, toothpaste, and other domestic items contaminated with virulent pathogens. Although most research was intended to produce an offensive arsenal, elaborate plans to develop biological defense capabilities were also formulated.

The Hungarians attempted to establish relations with German and Italian military medical institutes working in similar fields. However, they never succeeded in finding an appropriate German connection, apparently because such an organization did not exist in Germany.[2] An important consequence of these efforts was that Professor Kliewe, a German microbiologist who visited Budapest in 1944, could give detailed information about the work carried out there when he was later interrogated as a prisoner of war by US officials.[3]

The Italian connection was more fruitful. A similar but larger research program was led by Lieutenant Colonel Professor Raitano at a military hospital in Rome. Scientists of the two institutes paid visits to each other and exchanged information. This is the only available information about the pre-1945 Italian biowarfare program; otherwise that project still remains unknown to history.

The Health Control Station was destroyed on 4 April 1944, when the Allied forces bombed Budapest heavily. The offensive biowarfare R&D project had reached the highest scientific standards of those days, but it was eliminated before it could reach the large-scale manufacturing phase.

At the end of the war many soldiers and civilians were simply caught in the streets and transported to the Soviet Union for a "little work," that is, for forced labor lasting several years. Colonel Bartos, who was among them, probably kept his BW career secret during his stay in the Gulag. However, soon after getting home he wrote a report on the BW program to the Hungarian minister of defense.[4] His report provided detailed information about the activities carried out at the former Health Control Station. He suggested that in the Cold War it would be most advisable for the Communist leadership to revitalize the project.

However, his proposal was apparently not accepted, and Bartos practiced medicine in Budapest until his death in the 1970s. We do not know who made this decision and what the reasons were behind it. Probably the Soviet Union did not permit any major WMD program to be carried out in a satellite state, and it seems likely that the Hungarian Communist leadership did not even try to obtain permission. Whether the Hungarians shared their scientific-technological knowledge with the USSR remains unknown.

Thus, there was apparently no offensive BW research in Hungary after 1944, perhaps in part because in the nuclear era BW were viewed as inferior.[5] The former BW site was developed into a gun factory, and the military medical personnel were trained for defensive activities such as diagnosis and medication.

Although Hungary—and perhaps most if not all non-Soviet WP countries—had no major offensive BW programs, they might have contributed indirectly to some Third World programs as sources of know-how. A number of developing countries and the Palestine Liberation Organization sent thousands of students to WP universities during the Cold War era. Thus North Korean (in the 1950s), Libyan, Syrian, Iraqi, Palestinian, Vietnamese (in the 1970s and 1980s), Afghan (in the 1980s) and other students were abundant at campuses all over WP countries. Of course, some of these states could have had their students educated both in the East and in the West, but others were dependent on WP higher education exclusively. We do not know whether some of these nations utilized this possibility to build up an academic staff needed in a potential BW research program. Hints indicate that students of some Third World nations have been particularly interested in military medical sciences; for example, some 30 Libyan soldiers were attending the Semmelweis Medical University of Budapest in the 1970s—in uniform, and strictly separated from other students.[6]

In the 1970s the military medical staff of non-Soviet WP countries paid reciprocal visits to each other's main institutes. Although these visits were more or less formal, they mutually convinced participants that their countries were not involved in offensive BW programs. This was not surprising, since all military activities in these countries were strictly controlled by the USSR. As John Hemsley summarized: "Unlike CW [chemical weapons], in which non-Soviet WP countries carry out research into and practice the military application of CW, the USSR has a monopoly on all research and development into offensive BW. There is every indication that this tight control will continue to be maintained."[7] Similarly, Colonel Kanatjan Alibekov, deputy director of Biopreparat before his defection in 1992, says, "to our best knowledge, none of our East-European satellite states worked on a BW program."[8]

Romania

Libya's interest in sciences relevant to biowarfare may be significant in the light of a possible Romanian-Libyan cooperation to develop BW agents from about 1980, as mentioned by J. D. Douglass.[9] The situation in Romania was always different from that in other WP countries. It switched sides from Germany to the USSR as soon as the front between retreating German and advancing Soviet troops reached Romanian territory. Consequently, the Red Army left this country a few years after the war. During the Ceausescu era (1967–1989), the "supreme leader" used nationalism to gain independence from the USSR within the context of Communism. He built closer relations with China and with the West. Thus, one cannot exclude the possibility that the Soviets' BW monopoly did not apply to Romania. Theoretically, Romanian-Libyan cooperation could have been mutually beneficial, since Romania had an advanced chemical industry, while Libya could provide areas for field-testing. However, we have found no information in the public domain bearing on the existence of such a program.

Czechoslovakia

Czechoslovakia—a country under strict Soviet control, especially after 1968—was also interested in biowarfare. It possessed a collection of viral and bacterial strains that might have been established for potential biowarfare purposes. In 1994 the newspaper *Cesky Denik* reported that viral and bacterial strains remaining from the WP era were still stored at the Immunology and Microbiology Institute of the Military Medical Academy in Technonin, Bohemia, and in the Central Military Hospital in Prague-Stresovice.[10] They included strains of bacteria causing plague, cholera, tularemia, meningitis, and psittacosis, and the smallpox virus. Former and current directors of the Technonin institute denied that any offensive weapon research had been conducted there and noted that the institute had been opened to international inspection in 1990. Officials said that the strains served scientific research that was not military in nature, and did not violate any international agreement. Moreover, neither storage site had large-scale production facilities. Defense Minister Antonin Baudys ordered the destruction of pathogens soon after he learned of their existence, because they "were no longer useful." He said that the majority of the strains represented banal pathogens like influenza, but about 20 percent of them were "especially dangerous and exotic."[11]

The Czech Republic apparently did not inherit facilities needed for large-scale production and weaponization of pathogens. It seems unlikely that the collapsing Communist regime would carefully dismantle and eliminate such equipment while forgetting to destroy the strains themselves. It is much more likely that such facilities did not exist in the Cold War era. However, the possession of smallpox is notable. The possession of the smallpox virus outside two approved repositories is prohibited by international agreements since its global eradication in 1980.

We do not know whether the Slovak Republic also inherited anything related to BWs from Czechoslovakia.

Czechoslovak scientists appear to have been involved in the research, development, and application of psychoactive drugs in close cooperation with their Soviet colleagues. In the early 1950s, public testimony by Cardinal Mindszenty of Hungary and by American POWs in Korea asserting that America was evil and Communism superior shocked local people and Western experts alike.[12] These were the first signs to indicate that the USSR had developed "mind control" and behavior-modifying drugs. The extent of this program is hard to assess. The view stressed by Douglass is that it was a major and serious program starting as early as 1949, when neuropharmacology appeared as a new branch of science.[13] The scientists assigned to this KGB program included researchers from Czechoslovakia and the GDR. The first known operational use of Soviet mind-control drugs was the use of "confession drugs" in the cases mentioned above. Another family of drugs, named "friendship drugs," was used to manipulate, for example, the president of Finland and the majority of the Indonesian cabinet in the 1950s. Drugs were also used to eliminate religion from Communist societies. Friendship drugs made many bishops and priests "red," while others were driven to suicide or to insanity by drugs.

Of course, no pill can in itself turn someone into a Communist. Drugs were only one aspect of a complex psychological operation. The friendship drugs were administered over several days to suppress the target's natural inclinations or mental defenses. Thereafter, agents made friendly overtures and repeatedly advanced the desired point of view. If all went well, the target slowly adopted the new perspective. Thus the drugs were necessary but not sufficient; sustained operational efforts by intelligence agents were also needed.

This program paralleled another one to distribute classical drugs to particular target groups of certain countries, such as the army, the youth, and the "bourgeois leadership" of the US. The origin of this idea dates back to the Korean War, when the Russians observed the Chinese using drugs to soften morale in the US Army. Early Russian activities were intensified when Nikita Khrushchev formally extended the project to non-Soviet WP countries under the codename Peoples' Friendship.[14] Czechoslovakia's contribution appears to be most relevant to this project, while Cuba, Vietnam, Bulgaria, Poland, the GDR, and Hungary were also involved.

Most of our information about these programs comes from a Czech defector, General Jan Sejna, head of the Defense Council Secretariat and chief of staff to the minister of defense.[15] He was personally involved in planning and monitoring Czechoslovakia's participation in these programs from 1956 until his defection in 1968. A parallel source is Colonel Alibekov, who mentions a Soviet program codenamed Flute to develop psychotropic and behavior-modifying drugs. According to him, this was one of the two most secret projects (the BW program being the other) supervised by the Central Committee and the KGB. The program controlled many organizations such as pharmacology institutes and psychiatry clinics.[16]

Independent reports of former targets seem to verify the widespread misuse of psychoactive drugs. However, accounts by persons like Jan Svankmajer, who was a volunteer in drug experiments with his wife in Prague in 1972,[17] or Mihail Semjakin, who was subjected to forced psychiatric treatments in Moscow in 1968 to "medicate" his interest in religious arts,[18] appear to contradict the picture of a highly sophisticated neuropharmacological approach. Rather, they seem to have suffered from a forced use of brutal drugs.

Poland

Extensive military research concerning BW was also carried out in Poland. The Central Sanitary Epidemiological Laboratory and Ninth Sanitary Epidemiological Laboratory of the Front were established on 8 January 1945, just before the end of World War II. Captain Professor Edmund Mikulaszek was appointed the first head of this laboratory. He later became head of the Department of Microbiology at the Medical Faculty in Warsaw, and a member of the Polish Academy of Sciences. The laboratory was transferred on 10 February 1960 to the General Karol Kaczkowski Military Institute of Hygiene and Epidemiology (MIHE). The academic character of the MIHE is reflected in the fact that it was soon authorized to grant Ph.D. degrees in medicine and later also to supervise postdoctoral studies. Early research activities were apparently focused on military hygiene, ecology, epidemiology, microbiology, toxicology, and pharmacology. Interestingly, biological warfare against animals and plants was assessed as being particularly relevant. Once again, there is no evidence in the open literature even hinting at the presence of an offensive BW program at the MIHE. Today it is located in Warsaw and Pulawy (the Veterinary Research Center) and is involved in biological defense R&D.[19]

German Democratic Republic

Apparently the Soviet Union kept its extensive BW program secret even from the GDR, economically one of the most powerful members of the WP, despite the fact that some fermenter and dryer equipment was produced in the GDR.[20] However, whether the GDR did actually build up an offensive BW program remains a question that cannot be answered at present because of the inaccessibility of documentation pertaining to possible activities after World War II.[21]

Most of the open literature on GDR concerns about BW was written by scientists who were also advisors to the delegations representing the GDR at negotiations over the BWC and the Chemical Weapons Convention (CWC). These reports warned about possible threats due to new types of CW (such as bioregulators) as well as developments in the area of biotechnology; they urged strengthening the BWC and negotiating the CWC to a successful conclusion.[22]

Bulgaria

One of the most dramatic incidents involving the reported use of a biological weapon, which occurred in 1978 and can only be described as bizarre, was the Markov case. Allegedly, the Bulgarian government of that time was supported by the Soviet KGB in planning and executing the murder of Georgi Markov, a Bulgarian dissident working in London. The incident gained wide coverage in the press and seemed to be clear-cut. In fact the evidence was largely circumstantial, and the involvement of Bulgaria and the KGB was inferred mainly through the reports of dissidents.

Markov was a popular writer and TV commentator in Bulgaria during the 1960s, and was a personal protégé of Todor Zhivkov, the former chief of state. He had special privileges that included access to Communist archives. While looking through this material he apparently became disillusioned with the regime. He had also written a controversial play. In 1969 he fled to Italy with his brother. In 1971 he was in London, working for BBC Radio's Bulgarian service as well as the Deutsche Welle and Radio Free Europe, using these forums to criticize the Bulgarian regime.[23] Zhivkov tolerated these broadcasts for some time, until in early 1978 Markov started reading from his memoirs, in which his criticism was not only especially hard, but also poked fun at the dictator. Soon thereafter Markov started receiving death threats: unless he stopped writing for Radio Free Europe, he would be executed in "a refined way, something out of the ordinary."[24]

On 7 September 1978, Markov was waiting at a bus station in London when he felt a blow to the back of his right thigh. He turned to see a man bending down to pick up an umbrella. The man apologized in a foreign accent and then went off. Markov went on by bus to his office. He complained to a co-worker of pain in the back of his right thigh, which showed an "angry red spot." A few hours later he left to go home. Soon he became obviously ill, and went to the hospital the next day with a high temperature, swollen lymph glands, and vomiting. X rays of his right thigh did not reveal any foreign body, and blood cultures were negative for bacteria. Nevertheless he had a highly elevated white blood cell count. He received antibiotics but soon went into kidney and cardiac failure, and died on 11 September.

An autopsy revealed pulmonary edema due to heart failure, fatty change of the liver pointing to toxemia, and hemorrhagic necrosis of the small intestines and the lymph glands in the right groin. Microscopic examination showed small hemorrhages throughout the heart muscle. Because of suspected toxemia, tissue samples of the right thigh area affected, along with the matching piece of tissue from the rear part of the left thigh, were sent to the Chemical Defense Establishment at Porton Down to be examined in an attempt to isolate and identify any toxin. During the histological examination a pinhead-like metallic object was found. This seemed to be a small metallic bead with two holes drilled in it at right angles. It was made of rare metals (90 percent platinum and 10 percent iridium) and measured 1.53 millimeters across with holes 0.34 millimeters in diameter. The holes could possibly have contained about 500 micrograms of toxin covered with a wax or a sugar coating that would melt as a result of body heat.

No poison was ever detected in this pellet or in the tissues examined. However, on the basis of Markov's symptoms, the Porton Down analysts speculated that the poison was most likely ricin, a highly toxic molecule extracted from the castor bean plant *Ricinus communis*. Almost everything else was ruled out in accord with the degree of toxicity that had to be achieved and the symptoms that were observed. However, no antibodies to ricin could be found in blood samples taken at the time of autopsy (not surprising, given the short interval between the presumed exposure and death).

The presumptions seemed to be supported by another incident that occurred in Paris some two weeks before the Markov case and involved Vladimir Kostov, a Bulgarian State Radio and Television correspondent who had defected to Paris in June 1978. On 26 August he was on the Metro when he heard a sound like an air gun being discharged behind him and felt a blow to his back, which later showed a small red spot. He became ill and was in the hospital for 12 days with a fever, from which he did, however, recover. Because of the Markov case, Kostov was reexamined sometime later, and X rays showed a foreign body in the region of the wound on his back. A pellet identical with that found in Markov's thigh was removed from his back on 26 September. Furthermore, antibodies to ricin were found in Kostov's serum, suggesting that in contrast to Markov, Kostov's recovery provided the time needed for antibodies directed against the ricin in his system to develop. Kostov's recovery was attributed to the possibility that he might have received a smaller dose than Markov.

In the end, there was strong circumstantial evidence that Markov had been killed by ricin, but the examiners admitted that they could claim no more than that. Support for the claim came from another source. Oleg Kalugin, a former KGB major general, who was in charge of counterintelligence in the Soviet Foreign Ministry from 1970 through 1979, was stripped of his rank, his KGB decorations, and his pension after he broke with the organization in 1990. In an interview in London, he denied having anything directly to do with the Markov case, but implicated the KGB in the affair. He said that Dimitur Stoyanov, Zhivkov's interior minister, had asked the assistance of the KGB in the assassination of Markov. Kalugin personally sent two KGB operatives to Sofia in 1978 to provide Bulgarian secret service agents with dissolving poison pellets, concealed in the tip of an umbrella that was configured to inject them.[25] He went on to say that Stoyanov informed KGB officials that Zhivkov had ordered Markov's murder, and that Yuri Andropov, the late Soviet leader, who was the KGB director at that time, had approved the order. Oleg Gordievsky, a former KGB station chief in London, confirmed that the KGB had provided the poison pellet, which was manufactured in Moscow, and the umbrella, which was modified by KGB technicians.[26] The assassin (whose identity is still unknown) was supposedly supplied by the Bulgarians.[27]

The inquest into Georgi Markov's death returned a verdict of unlawful killing. With nothing more in hand at the time, the Markov case was closed. After the collapse of the Communist regime in 1989, it was reopened by the new Bulgarian government, which brought charges against Zhivkov and Stoyanov. However, the trial was in trouble from the beginning. All pertinent Bulgarian documents concerning the case had suddenly disappeared. The day before he was to give key testimony in the trial, General Stojan Savov, who had been the deputy interior minister under Zhivkov, committed suicide or was killed. Zhivkov, who was 80 years old at the time of the investigation, was said to ramble on in public and show signs of being close to senility, so that nothing much could be expected from his testimony.[28] The only documented evidence against Zhivkov and Stoyanov is a Politburo decree from July 1977 signed by both, stating that "all measures can be used to neutralize enemy émigrés."[29] As late as 1997, the British government was appealing to Bulgaria to provide a definitive account of the incident so that the case could be closed. Sources in Bulgaria reported that there was no new evidence, and until this was found, nothing further would happen.[30]

Conclusion

All prerequisites upon which potential BW programs could build were present in the non-Soviet WP countries after World War II. At least one of them, Hungary, already had a staff experienced in BW R&D and motivated to continue the project. Another one, the GDR, had an advanced chemical industry; even the Biopreparat ordered some of its large equipment there. Czechoslovakia, possessing significant military-medical traditions, illegally collected pathogen strains relevant to potential weaponization. Bulgaria apparently felt a need for covert biotoxin capability, at least when it ordered the Markov and Kostov assassinations from the KGB. The scientific community provided a high level of research and education in biology and medical sciences, providing a suitable background for BW programs—a potential that might have been realized by some Third World allies. However, the USSR's determination to maintain a BW monopoly prohibited most—if not all—of the non-Soviet WP armies from conducting offensive BW programs. On the other hand, evidence suggests that East European secret services applied classical drugs, modern psychoactive drugs, and even biotoxins over several decades to destroy their enemies at home and abroad. The scale of this activity is still not well understood.

CHAPTER 8

The Iraqi Biological
Weapons Program

GRAHAM S. PEARSON

Iraq acceded to the 1925 Geneva Protocol on 7 April 1931 with a reservation: "On condition that the Iraq government shall be bound by the provisions of the Protocol only towards those States which have both signed and ratified it or have acceded thereto, and that it shall not be bound by the Protocol towards any State at enmity with Iraq whose armed forces, or the forces of whose allies, do not respect the provisions of the Protocol."[1]

Iraq signed the Biological Weapons Convention (BWC) in 1972 and ratified it in 1991 as part of the implementation of UN Security Council Resolution 687 (1991), the ceasefire resolution following Iraq's invasion of Kuwait.

There is little independently available information about Iraq and its biological weapons (BW) program apart from that obtained by the United Nations Special Commission (UNSCOM) on Iraq, by its successor, the United Nations Monitoring, Verification and Inspection Commission (UNMOVIC), and by the Iraq Survey Group (ISG). While these data are based largely on information provided by Iraq, it has become clear that Iraq has gone to considerable lengths to conceal its WMD programs, an effort that has included the fabrication of false documents. UNMOVIC, in describing the Iraqi BW program, has stated that "of all its proscribed weapons programs, Iraq's biological warfare (BW) program was perhaps the most secretive. Iraq has stated that knowledge of the program was kept to a select few officials and that, to maintain secrecy, special measures were taken."[2] Consequently it was not surprising that Iraq's efforts to conceal the program complicated UNMOVIC's task of piecing together a coherent and accurate account of it.

The following appreciation of Iraq's BW program is drawn from the information collected and presented by UNSCOM and UNMOVIC.[3] In September 2004 the Iraq Survey Group (ISG) issued its "Comprehensive Report,"[4] and the information contained therein has been used to update the information provided by UNSCOM and UNMOVIC. Numerous uncertainties remain: the phrases such as "Iraq said," "Iraq has stated," and "according to Iraq" that pepper the following account indicate that the Iraqi statements are unsubstantiated and unverified. Moreover, Iraqi accounts of "research" should frequently be construed instead as "activities" rather than as mere research or development. Iraq may well have chosen to use the word *research* whenever possible to describe a range of activities, since research is not prohibited by the BWC.

Initiation of the Program

The state of Iraq was created in 1921 as a mandate under the League of Nations and was administered by the United Kingdom until gaining independence in 1932. In 1945 Iraq became one of the original members of the United Nations. The monarchy was overthrown in 1958 and a republic was established.

In 1968 the Arab Ba'ath Socialist Party seized power in Iraq and embarked on modernizing the country, including its industrial sector. Saddam Hussein become chief of the Revolutionary Command Council, as well as head of the government, the party, and the armed forces. The development of WMD began in the late 1960s or early 1970s at about the same time as the change in the Iraqi leadership. According to Iraq, chemical weapons (CW) were first studied in 1971, when small quantities of chemical agents were synthesized.

Iraq has stated that in 1974, in response to a rising threat from Israel and Iran, a government decree was issued to study "scientific, academic and applied researches in the fields of chemistry, physics and micro-organisms."[5] However, no copy of this decree was provided to UNSCOM or to UNMOVIC. Nor has there been any elaboration of what type of work was envisaged by the word *researches*. Although Iraq has stated that this work was motivated by the threat from its enemies, the fact remains that

Iraq signed the BWC on 11 May 1972, well before its entry into force on 26 March 1975. The very existence of the BWC provided an external reason for Iraq to be aware of the international movement to prohibit BW and of the associated moves in Article IX to prohibit CW, and thus to consider its national position in relation to such weapons.

The BW Program

The Ibn Sina Center

Iraq has stated that the response to the 1974 decree was to construct a facility, the Ibn Sina Center, at a site on the Al Salman peninsula, 30 kilometers south of Baghdad. The center was made part of a newly created organization, the Al Hazen Ibn Al Haithem Institute. Although Iraq declared that construction began in 1974, the relatively unsophisticated design of the center makes it likely that planning for it began in 1973 or perhaps even earlier.

The Al Hazen Institute was ostensibly part of the Ministry of Higher Education and Scientific Research, but Iraq has acknowledged that in reality it operated under the influence of the State Security Apparatus. Iraq stated that the work carried out at the center was basic and that little was actually achieved, because of poor direction and management. It is not certain that this was the case, and in fact the testimony of some of the scientists at the center indicates that more was achieved than has been declared.

Iraq has declared that activities at the center were terminated in 1978, when its director and some of the leading scientists were jailed for both scientific and perhaps financial fraud; Iraq provided documentary evidence to support this. However, it appears that activities at the center, stated by Iraq to be unrelated to BW, continued for some time into the 1980s. Although UNMOVIC interviewed Iraqis who worked in the program, it reports that the actual nature of the continued work is not clear.

The dates of the opening and closure of the Ibn Sina Center may not be particularly important, but its objectives and achievements are. The achievements, particularly those that were more military related, could have provided a sound basis for later developments, and hence shortened the lead time for production of BW. In this regard, UNMOVIC observed that at least five of the workers at the center went on to make contributions later in the BW program. The ISG notes that the generation of scientists trained and employed at Al Hazen formed the backbone of Iraq's later CW and BW programs.

The Al Muthanna State Establishment

War with Iran broke out in September 1980, and in the 1980s Iraq restarted its BW program. According to Iraq, after the closure of the Ibn Sina Center no practical work on BW was conducted until 1985. The account provided by Iraq runs as follows: In 1983, in the preamble of his annual report to the minister of defense, the director general of Iraq's CW

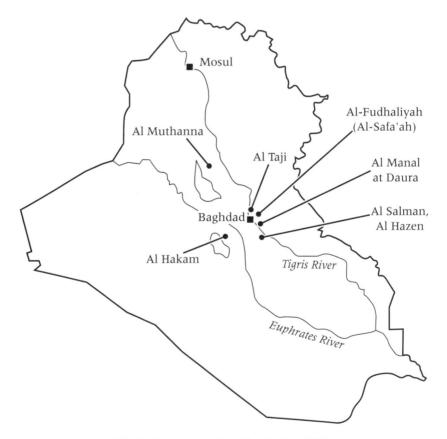

Figure 8.1 Major sites involved in the Iraqi BW program.

facility at the Al Muthanna State Establishment (MSE) stated that his mandate covered both chemical and biological agents. Since the minister did not dispute this assertion, the director general assumed that he had authorization to conduct work on BW. By Iraq's account, the director general's report also coincided with a 1983 letter to the Ba'ath Party from a senior Iraqi microbiologist, suggesting that Iraq could defend itself from Iran by the development and use of BW. However, no copy of the 1983 annual report or of the letter was provided to UNSCOM or to UNMOVIC. The ISG states that a militarily relevant BW program restarted at Al Muthanna in 1983 and that in 1986 a five-year plan was drawn up that would lead to biological agent weaponization. However, Iraq said that no action was taken until early 1985, when the first biologists were recruited for which documentary evidence has been provided. The former director general of the MSE told UNSCOM that he informed the minister of defense in 1985 that the first BW would be produced within five years.

UNMOVIC reported that there is documentation to show that bacterial strains and basic laboratory supplies were obtained in late 1985 and early 1986. According to Iraq, two agents, botulinum toxin and anthrax, were selected as candidate BW agents; Iraq said that they were selected because other countries had produced them for BW purposes and that they were relatively easy to produce. In 1986, according to Iraq, work on these agents was restricted to pathogenicity and toxicity studies, their characteristics, and methods of production at the laboratory scale.

Al Salman

In a separate stream from BW activities at the MSE, it appears that BW activities were also being conducted under the auspices of the State Security Apparatus at Al Salman. UNMOVIC has no clear understanding of what the stimulus was for the initiation of this work. According to Iraq's statements, this work began in 1984 with the investigation of wheat smut, initially to prevent crop infection; by 1987, however, research had shifted to use of the disease as an economic weapon. No other BW activities are acknowledged by Iraq during the period 1984–1986, although UNMOVIC noted that an inhalation chamber was installed at Al Salman, probably in 1984, and was later used in the BW program.

Iraq has stated that toward the end of 1986 the MSE put forward a proposal to increase the production of botulinum toxin from laboratory to pilot scale. To this end, the takeover of a facility at Al Taji was sought. However, before this could occur, a new director general of the MSE was appointed in early 1987, and, according to Iraq, he considered the expansion of BW activities at the MSE to be incompatible with the site and wanted the BW group to move from his establishment. Documentary evidence shows that there was indeed a transfer of the BW personnel from the MSE to the Forensic Research Department at Al Salman in the first half of 1987. At Al Salman the research came under the control of the Technical Research Center (TRC), which at the time was part of the State Security Apparatus.

Senior Iraqi officials have stated that in early 1987 there was dissatisfaction with progress on BW: in its two years of operation, the program was still involved with basic laboratory research, and the promise of weapons in five years seemed unrealistic. Accordingly, following the transfer of staff to the TRC, the program was accelerated. Additional laboratory supplies and equipment were acquired in 1987 and 1988, and new staff were recruited. Iraq has declared that a new building to house a pilot-scale fermenter and to allow other expansion was designed at the end of 1987, and that construction of the new building at Al Salman began in 1988. The remains of such a building have been inspected by UNSCOM.

Also sometime in 1987 (Iraq has declared from mid-1987), further development work on botulinum toxin and anthrax began. This work involved production of these agents in bench-top fermenters and experimentation on a range of animals, including sheep, donkeys, and monkeys, to study inhalation and other effects.

Another bacterial agent, *Clostridium perfringens* (gas gangrene), was added to the program, probably in 1987. There is evidence to show that work on certain fungal toxins (trichothecene mycotoxins) also began at the end of 1987. And in May 1988 a mycologist was recruited for the development of other fungal toxins, in particular aflatoxin.

Al Taji

According to UNMOVIC, the Al Taji facility appears to have come under the control of the TRC in 1987. Iraq has declared that the fermenter at that facility was refurbished and botulinum toxin produced there by the

TRC from January to October 1988. Although the dates cannot be confirmed, there is some evidence to suggest that botulinum toxin was indeed produced at the Al Taji facility, though possibly earlier than declared by Iraq.

Weapons Development

Iraq has stated that the first field trial of a crude BW dissemination device commenced in February 1988, followed by more sophisticated trials of BW bombs (the LD-250) filled with biological agent/simulant in April–May 1988. UNMOVIC has supporting evidence for at least some of these trials, which involved the expertise of the weapons engineers from the MSE. In addition to these weapons, according to Iraq an aerosol spray device was developed in 1987 and 1988 at Al Salman. The device, a modification of an agricultural crop duster, was intended for the spraying of bacteria and was tested with an anthrax simulant.

Enhancement of the Program

Iraq has stated that toward the end of 1987 the head of the TRC submitted a report on the progress of the BW work at Al Salman, and that as a consequence of this report, Lieutenant General Hussein Kamal (then the new head of the Military Industrialization Commission) instructed that the BW program should proceed toward the production of BW agent. Although Iraq stated that it was the success of the work in 1987 that stimulated the decision for production, UNMOVIC questions whether very much could have been achieved in the few months the BW team was working at Al Salman. On the other hand, there is evidence that initial preparations for large-scale production did begin in late 1987.

At the end of 1987 Iraq placed the first of a series of orders to purchase large quantities of bacterial growth media, which eventually totaled over 40 tons. In April 1988 an equipment requirement list was prepared, apparently in response to a directive from Lieutenant General Hussein Kamal. The list included three sizable (5,000-liter) fermenters, two for botulinum toxin and one for anthrax production. Dryers and other processing equipment for anthrax were also specified.

Al Hakam

At about the same time that the inquiries about materials and equipment were being made, Iraq conducted a search for a suitable production site. Iraq has stated that the Al Salman site was considered unsuitable for safety reasons, because of its proximity to Baghdad. A search for an alternative site was therefore made in early 1988. According to an account given by a senior Iraqi official, one of the early considerations for production was a facility that could be moved from site to site, but a mobile facility was rejected as being impractical. In March 1988 a production site was selected at a remote desert location about 55 kilometers southwest of Baghdad, and its acquisition is documented by Iraq. Iraqi officials have said that, in commemoration of the date of founding of the site (24 March), it was initially named Project 324. The site was later known as Al Hakam and became Iraq's main BW production facility.

Al Hakam was constructed rapidly and in secrecy. Priority was given to the production buildings, and by the end of 1988 these were said to have been completed and equipment installed. Iraq stated that transfer of staff and functions began toward the end of 1988 and finished by late 1990.

The acquisition of fermenters was crucial to the production of agent. UNMOVIC states that in a submission to senior officials, the options available to Iraq for their acquisition were considered. The submission argued that manufacture in Iraq was, at that time, not considered to be technically feasible and that in any case gearing up to domestic manufacture would take a long time. The recommended action was to purchase fermenters from overseas, and several foreign companies were contacted in early 1988. A contract with a supplier for three 5,000-liter fermenters was signed in July 1988, and at Iraq's insistence the first unit was scheduled for delivery later that year. However, in the end, the supplier could not meet the schedule, and delivery was postponed to 1989.

Since no foreign supplier was permitted to visit Al Hakam, Iraq modified another facility at Al Latifyah, 50 kilometers west of Baghdad, and presented this to the manufacturer as the plant where the fermenters were to be installed. Iraq also falsified the end-user certificate to indicate that the fermenters were for civilian use. Iraqi officials have stated that the plan was that, after the fermenters had been installed and commissioned by the company, they would be relocated to Al Hakam. In the end, the supplier could not obtain an export license, and the contract was finally cancelled in late 1989.

However, the priority on production in 1988 led Iraq to pursue another option for fermenters. The TRC became aware that a line of fermenters was available at the Veterinary Research Laboratories at Al Kindi. Consideration was given to producing botulinum toxin in these fermenters at Al Kindi. Iraq stated that this option was not adopted; instead, Al Kindi fermenters were requisitioned and transferred to Al Hakam in late 1988. UNMOVIC can confirm the acquisition but not the exact date of transfer.

According to Iraq, the fermenter from Al Taji was also transferred to Al Hakam at the end of 1988. Thus, by the end of 1988, Al Hakam, after a rushed construction and acquisition program, was ready to produce agent. The war with Iran ended in August 1988; therefore, it might have been expected that the pace of production would have slowed. However, the evidence provided to UNMOVIC indicates the contrary: for two years after the end of the war the momentum continued, because no one was sure that the war would not restart. Activities included BW agent production, new research projects, and weapons testing and development.

Botulinum Toxin Production

Iraq has stated that production of botulinum toxin began at Al Hakam, first in the fermenter from Al Taji in January 1989, and then in the veterinary fermenter line in February 1989. Iraq has also declared that, from the start of production at Al Hakam in January 1989 to August 1990, 13,600 liters of concentrated botulinum toxin were produced; in addition, 5,000 liters of botulinum toxin were produced at Al Manal, at Daura, in late 1990.

Anthrax Spore Production

Iraq has declared that in early 1989 Al Hakam did not produce anthrax spores; however, pilot-scale production of anthrax was started at Al Salman in March 1989 and continued there for four months. Anthrax spore production at Al Hakam was said to have started in June 1990. Up to August 1990, total production of (concentrated) anthrax spores was stated to be 170 liters, a relatively small amount compared with the amount of botulinum toxin produced up to this time. In addition, UNSCOM acquired evidence that anthrax spores were produced at Al Manal.

Aflatoxin Production

According to Iraq, aflatoxin production began at Al Salman in January 1989 and at Al Fudhaliyah (Al Safa'ah) in January 1990, and continued until July 1990, during which time about 400 liters of aflatoxin were produced. The Al Hakam report for 1990 says that 2,200 liters were made but does not specify the location. UNMOVIC has been unable to confirm these locations and quantities but accepts that aflatoxin was produced at both locations, with more being produced at Al Fudhaliyah.

Other Agents

Work on previously selected agents continued, although UNMOVIC is uncertain of the precise nature, timing, and scope. Iraq has declared that from 1988 to August 1990 studies on botulinum toxin, anthrax, gas gangrene, and aflatoxin continued. This work included investigation of the best growth conditions, effects on animals, and other studies, such as the determination of the optimum storage parameters.

Work on new agents also began during this period. Work on the BW agent ricin (extracted from castor oil beans) was probably commenced in the latter half of 1988. Studies included production and identification of the toxin protein and toxicity effects, including inhalation studies. Iraq has also declared that in the first half of 1990, research was conducted on *Clostridium botulinum* spores (as opposed to the toxin) as a potential infectious agent; UNMOVIC cannot confirm this statement.

A decision to study viruses as potential BW agents was probably made in May 1990 or earlier. In any event, in July 1990 a virologist was recruited to begin work on viral BW agents. Iraq has stated that the laboratories at Al Hakam were considered unsuitable for viral research and that actual BW work on viruses did not commence until December 1990.

In March 1990 a genetic engineering unit was established at Al Hakam. A senior Iraqi official said that one aim of the unit was to produce antibiotic-resistant anthrax. Another genetic engineering unit, apparently connected to the viral program, was also planned, but was said not to have been established. UNMOVIC is not entirely clear what the objectives of these BW genetic engineering projects were, but it speculates that in reality probably very little was achieved.

Other work during the period included experiments on the drying of anthrax. Spray dryers for anthrax spores were included on Iraq's equipment requirement list in April 1988. In 1989 Iraq signed a contract with a foreign company for the supply of a dryer suitable for this purpose. At the end of that year, a visit to the company by a senior Iraqi scientist was made, and a small sample of anthrax simulant was dried during a demonstration of the company's equipment. UNMOVIC has evidence that in 1990, in anticipation of receiving the dryer, Iraq conducted a series of experiments on a laboratory scale at Al Hakam to determine the best compounds to add to anthrax spores to assist in the drying process. However, in March 1990 the company withdrew from the contract because an export license for the special dryer could not be obtained. UNMOVIC has evidence that the drying experiments at Al Hakam then stopped (at least for 1990).

Field Trials of BW Weapons

UNMOVIC states that from documentation it appears that field trials of 122-millimeter rockets as a BW delivery system were conducted, possibly in November or December 1989. No test of any other BW weapon systems is declared by Iraq for 1989.

Further Enhancement of the Program

On 2 April 1990 President Saddam Hussein declared: "I say that if Israel dares to hit even one piece of steel on any industrial site, we will make the fire eat half of Israel . . . Let them hear, here and now, that we do possess binary chemical weapons which only the United States and Soviet Union have."[6]

This statement stimulated a number of WMD developments. Iraq has stated that, along with other military establishments, Al Hakam was required to respond to this new perceived threat. Consequently, a series of hurriedly organized dynamic tests of 122-millimeter BW rockets was conducted in May 1990. UNMOVIC has evidence that dynamic firings of such rockets containing botulinum toxin, anthrax simulant, and aflatoxin took place at some time in 1990. UNMOVIC has not been able to identify any other BW activity that might have constituted a specific response to the presidential statement. The ISG states that following the president's statement, the BW program was ordered to go all out for weaponization.

Iraq's invasion of Kuwait on 2 August 1990 also accelerated and changed the direction of its WMD programs. The emphasis was now on production and weaponization for the coming Gulf War. Projects that had direct relevance to the war effort had priority, and longer-term efforts were put on hold.

Decisions were also made about what munitions would be deployed during the war. For example, it was decided that the R-400 bomb, which had been developed for CW purposes, would also be deployed as a BW bomb, and that BW agent would also be deployed in Al-Hussein warheads. At the time of these decisions neither of the munitions had been tested with BW agent. According to one senior Iraqi general, the BW program at this time took a "hasty, unplanned and badly conceived course." However, it was inevitable that the coming war would have had a profound effect on the direction and nature of Iraq's WMD programs. The ISG notes that the BW program moved into high gear with the aim of fielding filled weapons as quickly as possible.

Iraq has declared that weaponization of BW agents took place in December 1990 and January 1991. However, the ordering and timing of weapons filling and deployment remains unclear, as do precisely which weapon systems were filled. Tests on R-400 bomb and Al-Hussein missile warhead remnants confirm that BW agents were filled into these weapons.

Iraq has also declared that weapons filled with biological agents were deployed from January to July 1991, although the numbers and locations of the agents deployed remain uncertain. Iraq declared to UNSCOM in 1995 that authority to launch chemical and biological warheads had been predelegated in the event that Baghdad was hit by nuclear weapons during the Gulf War. UNSCOM pointed out that this predelegation did not exclude the alternative use of such a capability and therefore did not constitute proof of intentions concerning second use. UNSCOM emphasized that it must have a complete understanding, for all proscribed weapon systems, of their intended and actual deployment plans.

Table 8.1   Biological agents declared by Iraq

Code	Agent	Summary
Agent A	*Clostridium botulinum* toxin	19,000 liters concentrated toxin (10,000 liters into munitions)
Agent B	*Bacillus anthracis* spores	8,500 liters (5,000 liters into munitions)
Agent C	Aflatoxin	2,200 liters concentrated (1,120 liters into munitions)
Agent D	Wheat cover smut	Considerable quantities
Agent G	*Clostridium perfringens* toxin	340 liters concentrated
	Ricin	10 liters Trials failed
	Trichothecene mycotoxins	Research
	Hemorrhagic conjunctivitis	Research
	Rotavirus	Research
	Camel pox virus	Research

The biological agents declared by Iraq are summarized in table 8.1, although the quantities cannot be verified. Facilities declared by Iraq as being engaged in its BW program are summarized in table 8.2, although the stated capabilities are also unconfirmed. BW delivery means declared by Iraq, though unverified, are summarized in table 8.3.

The Program after 1991

UN Security Council Resolution 687 (1991) of 3 April 1991 established UNSCOM and required Iraq to "unconditionally accept the destruction, removal, or rendering harmless, under international supervision," of its WMD, ballistic missiles exceeding a range of 150 kilometers, and all associated facilities, equipment, and materials. The same resolution required that Iraq ratify the Biological Weapons Convention, and the instrument of ratification was deposited in June 1991. It is now well known that Iraq did not cooperate with UNSCOM or later with UNMOVIC, taking steps to conceal information and capabilities relating to its WMD programs.

Indeed, Iraq has acknowledged in statements to UNSCOM that, in the biological field, its approach to ending the program was different from that in the chemical, missiles, and nuclear fields. It has been stated by Iraq

Table 8.2   Iraqi biological warfare facilities, 1975–1991

Facility	Description
Al Salman	1975 R&D program at Al Hazen Institute
	May 1987 BW program under TRC
	7-liter and 14-liter laboratory fermenters
	150-liter fermenter from Al Muthanna
	(Aug. 1990 to Al Hakam)
Al Hakam	March 1988 Research and BW production
	Nov. 1988 2 × 1850-liter fermenters from VRL
	7 × 1480-liter fermenters from VRL
Al Muthanna	1985 Research and production
	150-liter fermenter
	Dec. 1990 Weaponization on large scale
Al Manal	Nov. 1990–Jan. 1991 Botulinum toxin production
	Also evidence of anthrax spores
Al Safa'ah	1989 Aflatoxin production
(Fudhaliyah)	
Al Taji	Single-cell protein plant
	Production of botulinum toxin
	1 × 450 liter fermenter (to Al Hakam Oct. 1988)
Near Mosul	1987–1988 Wheat cover smut production under TRC

that although its BW munitions and agents were unilaterally destroyed in the summer of 1991, a decision was taken to conceal other aspects of its BW program from UNSCOM. Thus, for example, its main BW production facility at Al Hakam was converted to a civilian plant to disguise its true nature.

UNMOVIC in its 6 March 2003 paper on unresolved issues stated that, "Of all its proscribed weapons programs, Iraq's biological warfare (BW) program was perhaps the most secretive. Iraq has stated that knowledge of the program was kept to a select few officials and that, to maintain secrecy, special measures were taken. This secrecy was maintained after the Gulf War when Iraq went to considerable lengths, including the destruction of documents and the forging of other documents, to conceal its BW efforts from UNSCOM."[7]

The UNSCOM document of 1999, summarizing its work on the Iraqi BW program, points out many still-unresolved discrepancies in the Iraqi

Table 8.3   Iraqi biological warfare weapons

Delivery means	Comment
Al-Hussein warheads	25 BW warheads declared by Iraq: 16 filled with botulinum toxin; 5 filled with anthrax; 4 filled with aflatoxin. Numbers later adjusted by Iraq to 5, 16, and 4, respectively
R-400 aerial bombs	Iraq declared 200 produced: 100 filled with botulinum toxin; 50 filled with anthrax; 7 filled with aflatoxin. Numbers impossible to verify
Aircraft drop tanks: Mirage F-1	Plan for 12 modified drop tanks; Iraq claimed for use with anthrax and then said with botulinum toxin
Pilotless aircraft project: MiG-21 with drop tank	FFCD account very brief
Aerosol generators/helicopter spray system	12 devices produced; Successful field testing of *Bacillus subtilis* in Aug. 1988
Aerosol generator system for drones	Possibility that such a system was made
122-millimeter rocket warheads	Successful trials and recommended for agents A, B, C, and D
155-millimeter artillery shell	Four used in trial with ricin
LD-250 aerial bomb	Static trials of agent dispersion conducted in 1988
Fragmentation weapons	Said to have been trialed with *Clostridium perfringens*

declarations. Of particular relevance is the absence of an adequate account of the military involvement in the program and thus of the military objectives and requirements for such weapons. UNSCOM points out that the contention by Iraq that its Ministry of Defense remained wholly unaware of Iraq's BW program is implausible, since Iraq acknowledges that the MOD was involved from 1983 to 1987. It is unlikely that the program was not visible to senior MOD personnel after 1987, as programming for the adoption of BW would have been part of the strategic planning of Iraq's military development along with its nuclear and chemical capabilities. Indeed, UNSCOM has argued that such strategic planning would have developed the requirement for a militarily significant capability of BW in 1987–88 or possibly earlier. This planning would in turn have defined the technical scope of the program and the necessary funding. Furthermore, UNSCOM has pointed out that BW would have been integrated into Iraq's strategic capability and that the military objectives, concepts of use, and mechanisms for releasing these weapons must have been defined. This effort would have required extensive planning, which Iraq denies having done.

In March 2003 UNMOVIC identified the issues of greatest importance in regard to the Iraqi BW program as relating to agent production, weaponization, and unilateral destruction. Among the unresolved disarmament issues, there are 6 clusters relating to delivery means and 11 relating to biological agents.

The Iraq Survey Group, established after the military action against Iraq in 2003, has made several reports. David Kay, in an interim report submitted to the US Congress on 2 October 2003, said that the ISG has "discovered dozens of WMD-related program activities and significant amounts of equipment that Iraq concealed from the United Nations during the inspections that began in 2002" and went on, in regard to BW, to say that the ISG has "begun to unravel a clandestine network of laboratories and facilities within the security service apparatus. This network was never declared to the UN and was previously unknown . . . this clandestine capability was suitable for preserving BW expertise, BW capable facilities and continuing R&D—all key elements for maintaining a capability for resuming BW production."[8]

Only very recently have there been clear indications that the Governing Council of Iraq indeed abandoned its WMD. On 10 November 2003, at the Meeting of States Parties of the BWC in Geneva, Iraq made the following statement:

Iraq signed this Convention in 1972. We ratified it in 1991 but circumstances have not allowed us to implement the Convention sufficiently effectively nor to work to ensure the success of principles to rid ourselves of this most severe of weapons of mass destruction, biological weapons . . . Last year Iraq drafted a first set of legislative norms at international level prohibiting the development of weapons of mass destruction. However, circumstances have not allowed Iraq to complete all measures related to this endeavour. The Governing Council is engaged in ensuring respect for all international treaties and conventions signed by Iraq and as soon as circumstances allow the Council will seriously consider further development of national legislation to prohibit all forms of production of weapons of mass destruction.[9]

Further insight became available in early 2004 through published interviews with David Kay,[10] the initial leader of the ISG, and a statement by Charles Duelfer, who replaced Kay as leader of the ISG; and in September 2004 with the release of the ISG's "Comprehensive Report." The interviews with Kay indicated that the ISG had gained greater insight into Iraqi perceptions of the UNSCOM and UNMOVIC inspections and the deception and denial program carried out by Iraq. It is becoming clearer that Iraqi officials feared the UNSCOM and UNMOVIC inspections because of their effectiveness, despite the Iraqi attempts to deceive the inspectors and deny them information. Furthermore, it is becoming clear that Iraq destroyed most of its stockpiles in the summer of 1991. What is not yet clear is whether Iraq sought to retain the capability to produce CBW. The recent interviews convey the impression that senior Iraqi personnel believed that the CBW programs could be restarted quickly. According to Charles Duelfer in his March 2004 statement to the US Congress, the ISG found that the Iraqi deception efforts continued until the onset of military operations in March 2003. Duelfer also said that the ISG had developed new information regarding Iraq's dual-use facilities and ongoing work suitable for a capability to produce biological or chemical agents on short notice.

It is, however, far from clear why, if Iraq had indeed destroyed all stockpiles of CBW in 1991, it chose to hinder the work of UNSCOM and UNMOVIC and thus prolong the maintenance of sanctions. David Kay, on the assumption that Saddam Hussein's main fear was of a coup d'état or a revolution, believes that CBW were regarded principally as instruments for domestic use. Hans Blix speculates that Iraq did not mind the promulgation of a deliberate ambiguity, and that the issue involved national pride.[11]

In September 2004 the ISG said that Saddam Hussein so dominated the Iraqi regime that its strategic intent was his alone and that he wanted to end sanctions while preserving the capability to reconstitute his WMD when sanctions were lifted. The ISG reported that "depending on its scale, Iraq could have re-established an elementary BW program within a few weeks to a few months of a decision to do so, but ISG discovered no indications that the regime was pursuing such a course."[12] In seeking to explain why Iraq continued to provide a confused message regarding its WMD capabilities, the ISG presented the following conclusion:

> From the evidence available through the actions and statements of a range of Iraqis, it seems clear that the guiding theme for WMD was to sustain the intellectual capacity achieved over so many years at such a great cost and to be in a position to produce again with as short a lead time as possible—within the vital constraint that no action should threaten the prime objective of ending international sanctions and constraints. Saddam continued to see the utility of WMD. He explained that he purposely gave an ambiguous impression about possession as a deterrent to Iran. He gave explicit direction to maintain the intellectual capabilities. As UN sanctions eroded there was a concomitant expansion of activities that could support full WMD reactivation. He directed that ballistic missile work continue that would support long-range missile development. Virtually no senior Iraqi believed that Saddam had forsaken WMD forever. Evidence suggests that, as resources became available and the constraints of sanctions decayed, there was a direct expansion of activity that would have the effect of supporting future WMD reconstitution.[13]

Despite all these indications of long-term intentions to maintain the capability to restart a BW program, it seems clear that in reality, after the mid-1990s Iraq had neither stockpiles of BW nor an ongoing program to directly develop BW. As the ISG "Comprehensive Report" put it, "with the destruction of the Al Hakam facility, Iraq abandoned its ambitions to obtain advanced BW weapons quickly. ISG found no direct evidence that Iraq, after 1996, had plans for a new BW program or was conducting BW-specific work for military purposes. Indeed, from the mid-1990s, despite evidence of continuing interest in nuclear and chemical weapons, there appears to be a complete absence of discussion or even interest in BW at the Presidential level."[14]

In July 2004 reports were issued in both the US and the UK evaluating the intelligence regarding Iraq's WMD.[15] The US Senate report made it clear that the major judgments in the intelligence community's October 2002 National Intelligence Estimate, "Iraq's Continuing Programs for Weapons of Mass Destruction," were at best overstated:

> particularly that Iraq "has chemical and biological weapons," was developing an unmanned aerial vehicle (UAV) "probably intended to deliver biological warfare agents," and that "all key aspects—research and development (R&D), production and weaponization—of Iraq's offensive biological weapons (BW) program are active and that most elements are larger and more advanced than they were before the Gulf War," either overstated, or were not supported by, the underlying intelligence reporting provided to the Committee.[16]

The UK drew similar conclusions, though in different ways. It prudently pointed out that "the most important limitation on intelligence is its incompleteness . . . In fact, it is often, when first acquired, sporadic and patchy, and even after analysis may still be at best inferential."[17] It also observed that

> assessments in the chemical and biological weapons fields are intrinsically more difficult, and that analysis draws on different intelligence techniques. We are conscious in particular that, because chemical and biological weapons programmes can draw heavily on "dual use" materials, it is easier for a proliferating state to keep its programmes covert . . . Our impression is that they [Joint Intelligence Committee assessments] were less complete . . . and hence inclined towards over-cautious or worst case estimates, carrying with them a greater sense of suspicion and an accompanying propensity to disbelieve.[18]

The UK report concludes:

> Even now it would be premature to reach conclusions about Iraq's prohibited weapons. Much potential evidence may have been destroyed in the looting and disorder that followed the cessation of hostilities. Other material may be hidden in the sand, including stocks of agent and weapons. We believe that it would be a rash person who asserted at this stage that evidence of Iraqi possession of biological or chemical agents, or even of banned missiles, does not exist or will never be found. But as a result of our Review, and taking into account the evidence that has been found

by the ISG and debriefing of Iraqi personnel, we have reached the conclusion that prior to the war the Iraqi regime:

a. Had the strategic intention of resuming the pursuit of prohibited weapons programmes, including if possible its nuclear weapons programme, when United Nations inspection regimes were relaxed and sanctions were eroded or lifted.

b. In support of that goal, was carrying out illicit research and development, and procurement, activities, to seek to sustain its indigenous capabilities.

c. Was developing ballistic missiles with a range longer than permitted under relevant United Nations Security Council resolutions; but did not have significant—if any—stocks of chemical or biological weapons in a state fit for deployment, or developed plans for using them.[19]

Analysis

Iraq has gone to considerable lengths to prevent the UN and others from gaining an accurate or comprehensive appreciation of its WMD programs, with particular efforts being made to conceal its BW programs. There is very little, if any, reliable information available about the strategy or the objectives of the Iraqi BW program. Indeed, Iraq's strategy in relation to UNSCOM and UNMOVIC seems to have been to limit disclosures to what Iraq judged that UNSCOM and UNMOVIC were likely to know already from other sources. UNSCOM considered the successive "full, final, and complete declarations" (FFCD) made by Iraq incomplete and technically inconsistent.

A technical evaluation meeting (TEM) dealing with all aspects of Iraq's BW program was held in Vienna on 20–27 March 1998 and involved 18 experts from 15 countries. Considerable information about the Iraq BW program is in the TEM report, which highlights the deficiencies in the Iraqi FFCD. The TEM concluded that "Iraq's FFCD is judged to be incomplete and inadequate . . . The construction of a material balance, based primarily on recollection, provides no confidence that resources such as weapons, bulk agents, bulk media and seed stocks have been eliminated. The organizational aspects of the BW program are not clear and there is little confidence that the full scope of the BW program is revealed. Additional aspects, such as the existence of dormant or additional BW programs, remain unresolved."[20] These conclusions are further echoed by UNSCOM's January 1999 report on the disarmament of Iraq's proscribed weapons and by the March 2003 UNMOVIC Working Document on unresolved disarmament issues regarding Iraq's proscribed weapons programs.

As a result of the many persisting gaps in information, it is difficult yet to assess with any certainty why Iraq chose to develop BW, what its objectives were in developing such weapons, and why it took so much effort to conceal its work on them. It is, however, possible to construct a plausible explanation. The initial work on BW was probably seen as an extension of work on CW, since BW such as toxins and living microorganisms require significantly smaller quantities—kilograms rather than tons—to be effective. There are also some indications that Iraq regarded toxins as being CW rather than BW, although toxins are prohibited under both the 1972 BWC and the 1993 Chemical Weapons Convention.

The initial impetus for work on CBW was probably as declared by Iraq: to provide a capability to respond to regional threats from Iran and Israel. Iraq would have been aware that Israel, unlike Iraq, has not signed or acceded to the BWC. Iraq used CW extensively against Iran during the 1980s, and it has become apparent that Iraq believed that its use of CW significantly affected the outcome of the war. The ISG has reported that the Iran-Iraq war was the catalyst for Iraq's reactivation of BW efforts and that one Iraqi asserted that if the war had lasted beyond 1988, Saddam Hussein would have used BW against Iran. Iraq had produced large quantities of chemical agents for immediate use against Iran—and thus that long-term storage was not a requirement. This is a significant difference in approach from the traditional Western approach, followed in both the UK and the US, of seeking agents and weapons with long storage lives that could be used to retaliate in kind when required.

Much of Iraq's CW capability—geared to the production of thousands of tons of agent—was destroyed in the 1990 Gulf War by bombing. Iraq's decision to conceal its BW program may have reflected an appreciation that the program had been kept secret, was largely unknown outside Iraq, and, as a consequence, had probably been little damaged as a result of bombing and could therefore be successfully hidden. BW production also involved smaller plant sizes and smaller quantities of materials than the CW program. It is also possible that if bulk agents and weapons were destroyed, there would be little trace of the program left. Successful concealment of the BW program could have been seen by Iraq as simplifying and shortening the inspection process—something that could not have been done for CW or for missiles. There are also indications that Iraq monitored UNSCOM and UNMOVIC activities and utilized these observations to further conceal its program, perhaps by using mobile facilities and a network of laboratories.

What is clear from the Iraq experience is the importance of achieving universal participation in the BWC, and of strengthening the norm embodied in that Convention through enhancing its effectiveness and improving its implementation. As David Kay has said, an "international inspection regime is even more important now than it ever was. If there is effective inspection, then the need for unilateral preemptive action becomes much less critical."[21] It is also evident from the experience of UNSCOM and UNMOVIC in Iraq that the ongoing monitoring and verification program for biological materials and equipment was effective, as were the analogous programs for chemicals and missiles. In other words, the Iraq experience provides no evidence that BW capabilities could not be effectively monitored by a regime like that used for CW.

CHAPTER 9

The South African
Biological Weapons Program

CHANDRÉ GOULD

ALASTAIR HAY

One of the earliest public references to South Africa's interest in chemical and biological weapons (CBW) was in 1989, when SIPRI reported incidents suggesting a possible military capability in the area. Falling short of stating that a CBW capacity actually existed, SIPRI analyzed the South African evidence and concluded: "Although fiction heavily outweighs the facts of the case . . . in the psychological climate in southern Africa, reflecting a growing polarization between black and white, there is apparently no limit as to what the South African regime is expected to do in order to preserve white supremacy."[1]

By the time SIPRI reported, the CBW program in South Africa had been in existence for six years. Accurate about the program, SIPRI also identified the motivation underpinning it—anything that would prop up the South African government of the day.

This chapter reviews what is known about the South African BW program. It draws on original documents describing the program by those who established it, together with evidentiary material presented both to the Truth and Reconciliation Commission (TRC) and at the criminal trial of the former director of the program, Dr. Wouter Basson,[2] as well as the personal recollections of participating scientists. We outline when, how, and why South Africa became interested in BW. What we have been unable to do is access any documents of the State Security Council (a Cabinet-level committee) that refer to the BW program specifically. None of the indexes and titles of documents in the archives names the program. Any reference to the program at the State Security Council level may, therefore, be buried in documents that analyze the internal and external threats to South Africa. The secretive nature of the program and its use of "front" companies as a disguise may mean that official minuted discussions of senior government ministers do not exist. The minister of defense knew about the BW work, but we can find no official record of the extent to which he may have discussed its activities with others of high rank.

Most South Africans first learned about the existence of a chemical and biological warfare program in 1992 through press reports. Not front-page news, these reports simply described hearings on the privatization of the front companies. It was only in 1996, when South Africa's Joint Committee on Public Accounts began to probe allegations of fraud associated with the closure of the top-secret project, that further details of the program, codenamed Project Coast, emerged.

In 1998, on receipt of an application for amnesty by one of the program scientists, the TRC held a public hearing on the apartheid-era chemical and biological warfare program. The scientists and military personnel involved appeared before the commission to answer questions about both the nature and extent of Project Coast. Much of what is now known about Project Coast came through the TRC hearing. What has been revealed suggests a small-scale chemical and biological warfare program that sought to develop, produce, and weaponize novel and questionable crowd-control agents and assassination weapons. The program had a defensive element that sought to develop protective clothing and detection equipment. These defensive objectives are beyond the scope of this chapter.

Starting Up

There is no evidence to suggest that South Africa was involved with BW before Project Coast. During World War II two facilities (at Chloorkop, near Johannesburg, and at Firgrove, in the Cape) were built to produce mustard gas. They were closed down in 1945.[3]

More than 35 years elapsed before the minister of defense, General Magnus Malan, authorized, in 1981, the establishment of a secret CBW program for the South African Defence Force.[4] Unlike the nuclear program, established some eight years earlier and run under the auspices of Armscor, the parastatal arms procurement company, the CBW program was to be a military operation, under the nominal guidance of the surgeon general via the South African Medical Service (the medical wing of the military).

Very few military documents exist in the public domain dating back to the initiation of the CBW program, and none of these provides a contemporaneous explanation of what motivated those who set it up. However, such documentation as is available, together with testimony from those involved in the decision-making process, leaves little doubt that the principal motivation was the need to provide South African Defence Force troops fighting in Angola with protection against chemical weapons (CW). A subsidiary goal was the provision of novel crowd-control agents to the South African police.[5] Neither of these aims, however, provides a persuasive reason for establishing the biological component of the program.

Documents assessing the threat to South Africa in the late 1970s specifically note that there was no immediate or envisaged threat from biological warfare.[6] The BW program itself appears to have had only a limited defensive component; most of the evidence suggests that the aim was

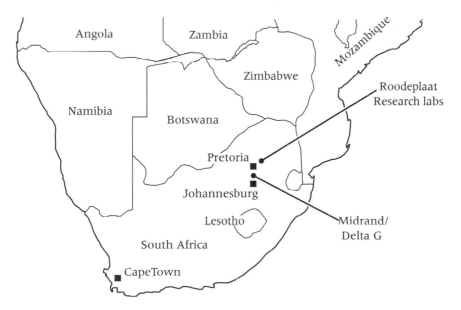

Figure 9.1 Major sites involved in the South African BW program.

offense. Some military officials have argued that the primary reason behind the development of the BW facility at Roodeplaat Research Laboratories (RRL) was to provide an animal-testing facility for chemical agents
developed at the sister company, Delta G Scientific.

Both RRL and Delta G Scientific were military front companies, established to conduct research and to develop and produce products for Project Coast. In the event of detection, the front companies were meant to
shield the CBW program and disguise its military connections. They also
made it easier to import dual-use equipment and other items that might
have raised suspicion had it been known that they were destined for a
military organization.[7]

There is certainly documentary evidence that chemical agents were
tested on animals at RRL.[8] Scientists involved in the BW program, however, claim that it was intended to supply the military and police with
covert assassination weapons for use against individuals regarded as a
threat to the apartheid government.[9]

Documents from RRL confirm claims made at the TRC hearings that
work was done to find substances that were odorless, colorless, and not
traceable postmortem.[10] It appears that either chemical or biological
agents would do. The join between the chemical and biological components of the program was seamless. Research at RRL included synthesis
and development of biological and chemical agents (particularly organophosphates),[11] while scientists at Delta G Scientific occasionally assisted
RRL researchers with projects involving biological agents.[12]

No publicly available document about Project Coast provides a clear indication of either the extent or nature of the biological program. The emphasis is rather on the perceived chemical threat to South African forces
fighting in Angola and the need to defend them, as well as on developing
agents to control internal opposition to apartheid.[13]

All the documents that do reveal the motivation behind the BW program and its development are retrospective. These documents were prepared for the minister of defense, Eugene Louw, and President F. W. De
Klerk in the early 1990s. Both men needed to be seen to be making a
break with the past,[14] and it is likely that the briefings they received deliberately obscured aspects of the program that might have caused discomfort. Unlike his predecessor, P. W. Botha, De Klerk was not a militarist;
soon after becoming president he replaced General Magnus Malan with a

civilian minister of defense. Many members of the Defence Force felt
threatened by the changes, and it is unlikely that De Klerk would have
been told about the more sinister aspects of the CBW program, particularly since the development of BW assassination weapons, which were
part of it, put South Africa in violation of its commitments under the Biological Weapons Convention (BWC).

De Klerk's 1990 briefing paper referred to the BW program in two
short and vague paragraphs, stating in part: "It is not possible to describe
the current biological threat to the world because of the speedy development of techniques to produce new bacteria as well as other organisms.
Our biological capacity is focused on staying up to date with the changing
threat. To do this we are constantly producing new organisms in order to
develop a preventative capacity as well as treatment."[15]

On the objectives of the CBW program, the briefing was also obscure:
"The aim of Project Coast is that of covert research and development of
CBW and the establishment of production technology in the sensitive
and critical areas of chemical and biological warfare to provide the South
African security forces with a CBW capacity following the CBW philosophy and strategy." Neither the so-called CBW philosophy nor the strategy
was explained. With regard to BW the objective of Project Coast was to
"establish a research, production and development capacity with regard
to biological warfare."[16] The vague language of the documents describing
Project Coast was undoubtedly intended to ensure plausible deniability if
the documents ever became public.

In 1990 Basson presented a document to the Reduced Defence Command Council, outlining the proposed philosophy with regard to chemical warfare; no corresponding document on biological warfare has been
made public. It is far from clear whether a philosophy was ever outlined
on biological warfare, and it is unlikely that a document describing it exists. Basson provided some insight into a possible reason for this silence,
noting: "This philosophy does not cover any aspects of Biological warfare.
Because of the more controlled nature of Biological Warfare there are
many more international control measures. The production of Biological
weapons is not allowed anywhere in the world."[17] This account suggests
that because of the ban on BW, the policy remained unwritten. This official silence is consistent with the way in which the South African security forces operated under apartheid. The TRC, for example, found very

few documented orders for assassinations, and police officers testifying before the TRC spoke of illegal orders being given verbally and using ambiguous terminology.[18]

While the BWC and Geneva Protocol did not appear to deter the military from seeking to develop small-scale, covert BW, awareness of the Convention may have been one factor that led to the program's being run through front companies so that the military link was hidden.

Structure

The total cost of RRL to the Defence Force, as audited, amounted to R 98,432,657. This figure includes the cost of building the facility, total operational costs, and the payment made by the South African Defence Force (SADF) when the company was privatized. The only annual figures available show the operating costs of the company for the financial years 1987–88 and 1988–89. In 1987–88 roughly R3 million was spent; the following year the costs more than tripled, to R11 million.[19] Funds were channeled from the Secret Defence Account to RRL and Delta G through a third company, Infladel, established for this purpose. The forensic auditor put the total cost of Delta G at R127,467,406, of which some R40 million went into the fixed assets of the company and R50,467,406 into operations. The cost of privatization was R37 million.[20]

The structure of the CBW program hid the relationship between the research (RRL) and production facilities (Delta G Scientific) and the military. Because the two front companies did not come under the traditional military command, higher salaries and more benefits could be offered to the scientists.[21] During interviews with 12 scientists who worked at RRL and Delta G, 11 claimed that the salaries and generous benefits offered (including housing subsidies and car allowances) were the principal factors motivating them to join these front companies.[22]

In order to maintain their covers as private companies, Delta G and RRL undertook commercial work. At RRL this commercial work represented only 15 percent of the company's income,[23] the rest coming from military contracts. According to a former RRL director and head of research, Dr. Schalk Van Rensburg, commercial projects represented 5 percent of the work done during the early stages of the company's development and later grew to about 30 percent.

Reviewing the documents from RRL during the trial of Wouter Basson, Dr. Daan Goosen, the first managing director of the company, said that of the 203 project files found in trunks after Basson's arrest in 1997, 177 dealt with research into and/or the development of chemical and biological agents, while the other 26 represented commercial projects. Of the 177, 34 dealt with antidotes and treatment for biological agents, and of these only 3 were final reports. Of the 34, 7 projects predated 1988; the rest were dated from 1988 to the early 1990s.[24] It is possible that there are other research reports that were never turned in, and some may have been retained by the scientists themselves.

Research reports from RRL indicate an intense interest in the development of organophosphates for covert use, particularly paraoxon (extracted from commercially available pesticides),[25] and that small quantities of the lethal chemical agents VX and Tabun were synthesized and tested.[26]

Roodeplaat Research Laboratories

It would appear that the impetus to establish RRL came from the relationship that developed between veterinarian Dr. Daan Goosen and Wouter Basson. During the early 1980s Goosen was head of the H. A. Grové Institute at the Hendrick Verwoerd Hospital (now Pretoria Academic), an academic research facility that had at its disposal a range of experimental animals. According to Goosen's testimony during Basson's trial, he and Basson spent many hours in discussion about CBW. Their shared political ideology and common scientific interests led Basson to recruit Goosen in 1983 to launch the biological warfare facility.

The facility began as Interlab, with a few offices in a shopping center in a suburb of Pretoria.[27] Goosen was the first managing director. In early 1984 Basson recruited the administrative director, David Spamer, as well as Dr. André Immelman, a veterinarian and toxicologist from the University of Pretoria. Veterinarian Dr. Schalk Van Rensburg was recruited in August 1984.[28]

Goosen claims that the tone for RRL was set in 1983, when he supplied Basson with a live black mamba snake and a vial of its deadly venom. Goosen believed the venom was to be used to murder a conscript allegedly passing military information to the African National Congress

(ANC). Basson later denied this charge in court.[29] But Goosen understood that he and his colleagues would be required to develop biological assassination weapons at RRL.[30]

RRL was built in 1986 on a 70-hectare site bordering the Roodeplaat dam, north of Pretoria. Basson later explained the need for the sophisticated facilities: "The only existing sophisticated high grade isolation units that worked with [biosafety level 4] biological material [in 1984] (e.g. the National Institute for Virology at Rietfontein) were internationally supervised and therefore not suitable for sensitive military work."[31] RRL was carefully sited close enough to other agricultural research facilities to avoid attracting attention[32] but just far enough away from these other organizations to prevent any accident or leakage from affecting them. A temperate climate for the primates used during experiments was also needed so that they would not require specialized housing. The actual operating period of the purpose-built BW facility was extremely short: the final buildings and laboratories were completed only in 1988, six years before the privatization of the company and the end of any BW work.

André Immelman was responsible for carrying out military contracts at RRL. His reasons for joining were set out in an affidavit before the TRC in 1998:

> I was interested because I had knowledge about substances that are used in chemical and biological warfare from my toxicological background. I realized the importance that these substances could have in the war situation that existed in the 1980s. I thought that with my knowledge I could make a contribution towards the protection of the South African population . . . As a scientist the opportunity to do fulltime research in my field would have been very fulfilling. I did not see this employment opportunity as an order to research offensive weapons which would be used against individuals or groups of people."[33]

Immelman was the single point of contact between the military and RRL. He was told from the outset that the facility was a military front company, but over time he became concerned about the legitimacy of what he was doing. Basson reassured him, saying that he would not be involved if the poisons he was making available were misused.[34]

Toward the end of 1988 Basson introduced Immelman to three men.

All were white, and only first names were used; Immelman had a false one. Basson instructed Immelman to work with the three and to see to their requirements. He met them three or four times in public places in Pretoria, where they discussed substances that could be used to "eliminate" people, how to administer these, and what symptoms to expect.[35] Immelman kept a list of the substances handed to the three men—what has come to be known as the "Verkope [sales] list." Included on the list were beer contaminated with botulinum toxin, sugar contaminated with *Salmonella,* and chocolates laced with anthrax and botulinum toxin.[36] During Basson's trial it emerged that the three were police officers.[37] In addition, Immelman acknowledged occasionally giving Basson small quantities of paraoxon, as well as thallium-contaminated whiskey. During his trial Basson claimed that these had been provided for training purposes.[38]

Veterinarian and microbiologist Dr. Mike Odendaal testified both to the TRC and during the Basson trial that as head of the Department of Microbiology at RRL he had developed some of the biological assassination weapons. He spoke of infecting cigarettes and chocolates with anthrax spores, sugar with *Salmonella,* and chocolates with botulinum toxin. Odendaal was in no doubt that these were intended for operational use.[39]

Odendaal also oversaw what was probably the most sophisticated work done at the facility, carried out by junior scientist Adriaan Botha. The outcome was the bacterium *E. coli* genetically modified to express the epsilon toxin of *Clostridium perfringens.* Botha maintains that his research was motivated by his personal interest in developing a recombinant vaccine against enterotoxemia in sheep and claims that to get his proposal passed by the management of RRL[40] he deliberately included reference to its potential military application. Botha stressed that the work went as far as the development of a vaccine for testing purposes, but that the project was nowhere near developing a biological weapon. In an interview Botha said that he was never asked to take the work further to develop its military potential and was "almost certain I would have refused" if asked.[41] During a later discussion he elaborated, saying, "I never believed they would use the substances produced, and I tried not to do any work for André Immelman. There were two kinds of work: on the one hand, one had to do commercial tests on substances, which was not in the least challenging; but then there was the other work, which involved the enterotoxemia vaccine. When I started doing that work, I was surprised to find that I was left in peace. I could go into the lab and do the work that interested me, and I was very happy there."[42]

One aspect of RRL's work, which received much media attention during the 1998 TRC hearings, was the development of an antifertility vaccine. This was a personal project of Goosen's, who believed the vaccine was to have been administered to black women without their knowledge.[43] Dr. Schalk Van Rensburg, who oversaw the fertility project, confirmed that this was indeed its purpose. RRL documents reveal that many projects were registered on both male and female antifertility vaccines, but no vaccine was ever produced because development did not get that far.[44]

In 1987 senior management at RRL commissioned the company Foster Wheeler to produce plans for an upgraded facility that would include freeze-drying and storage capacities and a biosafety level 4 laboratory for dealing with highly virulent strains.[45] Several laboratories were planned to deal with, *inter alia*, toxins, industrial chemicals, and nerve agents. Odendaal says the intention was to produce aflatoxins, T2 toxin, anthrax, *Brucella, Salmonella*, botulinum, and tetanus toxins.[46] He was convinced that the purpose of the new facility was to allow RRL to move into large-scale BW production. "At the end of the program we were planning a multimillion-rand containment facility; the plans had been drawn up, and it was going to be built. This was going to be a state-of-the-art, large-scale production facility. We were going to make big quantities . . . that was why we wanted large volumes."[47]

What is not clear is how or where these biological agents would have been used. As far as can be established from the documentation available, no military doctrine or strategy was developed to encompass the use of BW.

Changing Gears

In 1991, two years after dropping plans to upgrade RRL facilities, Basson proposed that both RRL and Delta G be privatized. He told the minister of defense that the company had attracted a great deal of commercial interest as well as the interest of scientists because of its sophisticated facilities: "the size and nature of the facility that was established to ensure good results caused some commercial 'jealousy' . . . and . . . the relatively easy way in which RRL solved complicated biological issues without being obviously part of the commercial market, made a lot of competitors suspicious of the real position of RRL."[48]

Although Basson was vague about the BW program when he later briefed Presidents De Klerk and Mandela, he was much more forthright about the BW work done at RRL when motivating Defence Minister Malan to consider privatization. According to Basson, private-sector inquiries about RRL's financing put pressure on its managers, but they did not reveal the links between the company and the state. Basson claimed, however, that the senior staff at RRL felt that their future was uncertain and that they had little influence over events. Downscaling of research and production at RRL only added to their concerns. Referring somewhat elliptically to arms control, Basson wrote: "It is . . . often a problem for the scientists of RRL, (who are not trained to think strategically), to keep perspective in the light of the renewal of Western attempts to ban chemical and biological weapons. It appears to them that South Africa should abide by these calls. The fact that no country involved is really weighing up the possibility of moving away from biological weapons is not clear to them."[49] Whether Basson was right about his colleagues' desire to abide by international arms control measures is open to doubt.

Basson's arguments prevailed, and in 1991 the company was sold to Goosen's successor, Dr. Wynand Swanepoel. The state took a significant loss on the deal, but Swanepoel and his shareholders, who included Immelman and other select members of staff, received a share of the R18 million paid out by the military.[50]

The International Context

There is evidence suggesting that the US and the UK were aware of the South African BW program. In 1981 (the year in which the CBW program was initiated) Basson attended a conference in San Antonio, Texas, where he met Brigadier General W. S. Augerson, former assistant surgeon general for research and development in the US Department of Defense. Basson discussed CBW with Augerson, who, alerted by Basson's interest, filed a report with the US Military Medical Intelligence and Information Agency at Fort Detrick. Augerson may also have alerted the US

Defense Intelligence Agency. Thus the US probably knew of South Africa's interest in CBW before the program was actually started.[51]

Former South African deputy chief of staff for intelligence General Chris Thirion has confirmed that the US intelligence services were aware of Basson's activities. He stated that in 1985–86 a CIA officer had asked if Wouter Basson was the "main brain" in CBW. The officer was told that Basson was involved in CBW countermeasures.[52]

In a 1999 interview Goosen said that he visited the US several times in the 1980s and that these trips included visits to laboratories where primates were exposed to nerve gas. Goosen claimed that during discussions with US officials he had been frank about his intentions to gain knowledge about BW and that he had not hidden his relationship with the South African CBW program. In 1984–85 he gave the US Food and Drug Administration plans of RRL and discussed the possibility that RRL might do research into antidotes for nerve agents.[53]

These incidents suggest that the US intelligence agencies knew about South Africa's interest in chemical warfare and possibly biological warfare and had good reason to make more-detailed inquiries. Other events in the mid-1990s appear to confirm that they were monitoring the activities of Basson.

In 1993 and 1994 Basson became increasingly involved in what he claimed were business ventures in Libya. His trips to Libya did not go unnoticed by South African, British, and American intelligence agencies. Indeed, such was the concern about these trips that in April 1994 the US and UK governments sent their ambassadors to meet with De Klerk.

According to former US ambassador Princeton Lyman, who was at these meetings, the US and UK were concerned that South African CBW information was "in danger of being acquired by other states, in particular Libya," and that South African scientists could be recruited by these states. Another reason for the meeting was to persuade the South African government to be frank about the offensive aspects of Project Coast in international forums. Lyman recalled that "South African officials were adamantly opposed to making such an admission, arguing that any such offensive uses were done without proper authorization and against official policy."[54]

Confirmation of the ambassador's concerns is to be found in a briefing document for the newly elected president, Nelson Mandela, in August

1994, prepared by Surgeon General D. P. Knobel. Under a subheading "Enquiry by Ambassadors of the USA and UK" it states:

> On 11 April 94 the SP [State President] and the Minister of Defence were advised by the Ambassadors of the USA and the UK of their government's position with regard to the above [chemical and biological warfare] programme as well as the CBM [confidence-building measure] declaration submitted by the RSA [Republic of South Africa] in 1993. They stated that they were fully aware of the contents and extent of the SADF CBW programme and that they had certain reservations about the RSA's CBM declaration as well as the implications for non-proliferation.[55]

A second meeting took place on 22 April 1994 between the US and UK ambassadors and De Klerk. At this meeting, held a week before South Africa's first democratic elections, De Klerk argued that a defensive program had been justified and that the data resulting from the program were a national asset that would not be destroyed.[56]

In January 1995, in a third démarche brought by the governments of the US and the UK, Basson's travels to Libya and elsewhere remained an issue.[57] The US and UK clearly knew enough about the South African CBW program to judge that the CBM submitted by South Africa was inaccurate: Why else would they have been concerned about Basson's contact with Libya? But the larger question is why the US and UK chose to do and say nothing to prevent the South African CBW program from continuing during the 1980s and early 1990s. The answer probably lies in US policy toward South Africa during the Cold War.

US policy toward South Africa during the 1970s and 1980s was openly condemnatory but privately supportive. US government statements condemned apartheid, but the administration was acutely aware both of South Africa's strategic importance and of its significance as one of the two most important producers of platinum-group metals—together with Russia it accounts for some 90 percent of the world's supply. In a 1985 research paper, the CIA noted the risk of a diminished supply should the US be included in a trade embargo against South Africa.[58]

South Africa's significance as a key mineral supplier was just one factor influencing US policy. During the 1970s US policy toward South Africa to end apartheid vacillated according to the particular US administration.

But it was not only the US which vetoed an arms embargo and sanctions against South Africa called for by other African nations; the UK and France did, too.

In 1975 South Africa announced that it had a pilot plant for uranium enrichment at Pelindaba.[59] Ninety-seven pounds of enriched uranium—enough to make seven atomic bombs—were shipped to the plant by the US Nuclear Corporation of Oak Ridge, Tennessee, to assist it.[60] Meanwhile the Organization of African Unity denounced the US, UK, and French vetoes of a UN Security Council resolution calling for a mandatory arms embargo against South Africa.[61]

By mid-July 1975 South Africa was arming two opposition groups in Angola—the FNLA (Front National de Libération de l'Angola; National Front for the Liberation of Angola) and UNITA (Uniao Nacional para a Independencia Total de Angola; National Union for the Total Independence of Angola)—to fight against the Soviet-backed government forces of the MPLA (Movimento Popular de Libertaçao de Angola; Popular Movement for Angolan Liberation). The US provided both groups with sizable donations.[62] In October 1975 Cuban troops arrived in Angola, Soviet shipments to the MPLA increased, and South Africa sent troops to support UNITA and the FNLA.[63] In December the US began a diplomatic campaign against countries allowing Soviet use of their airspace and facilities for airlifts to Angola.[64] In January 1976 the US State Department identified the Soviet incursion into Angola as a primary reason for US covert military intervention.[65] In the same month, however, Tanzanian president Julius Nyerere intervened, saying that he could convince the MPLA government to repatriate Soviet and Cuban troops. Satisfied, South Africa began to withdraw troops, but it was not until 1988 that all South African forces left Angola. The last Cuban troops left in May 1991.

On 16 June 1976 riots broke out in the South African township of Soweto when police opened fire on a group of students protesting South Africa's education system for black pupils. Within days protests spread to other townships and many were killed. The killings prompted the UN Security Council to adopt Resolution 392 condemning the South African government's violent actions and calling for an end to apartheid and racial discrimination.[66]

In June 1976 President Gerald Ford signed a law (the Clark Amendment) prohibiting US support for military operations in Angola unless such action was in US national security interests and approved by Congress.[67] Meanwhile US arms manufacturers were sending arms to South Africa through front companies. In the latter half of 1976 US attention was focused on promoting change in Namibia and Rhodesia through talks between US Secretary of State Henry Kissinger and South African President John Vorster.[68]

Beginning in 1977, US policy toward South Africa hardened. Because of its exploitation of Namibian uranium, South Africa was removed from its permanent position in the International Atomic Energy Agency. Shortly thereafter, Soviet satellite pictures showing South Africa preparing to detonate a nuclear explosive in the Kalahari desert brought a warning from US President Jimmy Carter not to do so.[69]

On 13 September 1977, black consciousness leader Steven Biko died in police detention. Repression of dissent in South Africa increased. So did international pressure. In October the US voluntary ban on arms sales to South Africa became a formal embargo, followed by UN Security Council Resolution 418, instituting a mandatory arms embargo against South Africa. This compromise resolution followed a veto by the US, UK, and France of an earlier draft calling for sanctions as well.[70]

US–South Africa relations worsened between 1978 and 1980 and were not helped by US satellite evidence recording a light signal over South Africa that some in the US government believed indicated a nuclear detonation.[71] With the inauguration of Ronald Reagan as US president in January 1981 (the year the CBW program was approved), US support for the apartheid government increased. In March 1981 Reagan asked Congress to repeal the Clark Amendment "in order to remove an 'unnecessary restriction' on his foreign policy authority."[72] The amendment was repealed.

In April 1981 US Secretary of State Alexander Haig invited South African foreign minister Pik Botha to the US,[73] and France, the UK, and the US vetoed four UN resolutions calling for sanctions against South Africa. For the next four years South Africa could count on a more tolerant US attitude.

Further examples of the US adoption of a Nelsonian eye to apartheid in South Africa abound. In April 1982 the US Commerce Department approved the sale to South Africa of 2,500 3,500-volt shock batons designed for crowd control, in violation of section 502(b) of the Foreign Assistance

Act, prohibiting exports to the police or military in countries with consistent human rights violations.[74] In September 1982 the Armaments Corporation of South Africa announced that its large-caliber howitzers could fire nuclear rounds if necessary, but that South Africa did not intend to use such weapons.[75] In January 1983 the Reagan administration modified US limitations on exports to the South African and Namibian military and police forces. New guidelines permitted the export of certain nonstrategic industrial, chemical, petroleum, and transportation equipment without a license. This was the third relaxation of export restrictions in less than a year.[76] These are but a few of the announcements and actions that both preceded and paralleled the start of South Africa's BW program. Although US congressional attitudes to South Africa hardened with the Senate's passage of the Comprehensive Anti-Apartheid Act in August 1986,[77] the Reagan administration continued to oppose sanctions. The fact that the US and other Western nations adopted a lenient attitude toward the apartheid government on matters of strategic importance during the 1970s and 1980s provided a context that limited the constraints on South African decisionmakers in authorizing the CBW program.

International opinion eventually hardened against South Africa and the government, particularly following the election of De Klerk in 1989, who was forced to begin dismantling its racially discriminatory measures.[78]

Loose Ends: A Dangerous Legacy

In 1991 RRL was privatized in a complicated deal that involved the transfer of ownership to a company owned by RRL's managing director, Dr. Wynand Swanepoel, and a group of RRL employees.[79] The intention was for RRL to increase its private-sector contracts, while the SADF contracts would be phased out. But by 1994 the company was bankrupt and was put into voluntary liquidation. A liquidation dividend was paid out to a group of nine RRL employees for just under R 18 million, of which just over 9 million was paid to Swanepoel or to companies in which he had an interest.[80]

Three of the scientists who profited from the liquidation used their payouts to establish a smaller independent biotech company, called Biocon, a few miles from the RRL facility. Biocon took over all RRL's commercial contracts and some of its equipment.[81] Two of the RRL scientists left the company to farm,[82] and the remainder returned to positions at the Onderstepoort Vaccine Institute (a division of the Agricultural Research Council) or found work in the private sector. There was no attempt by the state to ensure the reemployment of the scientists who left the organization before 1994, and there is no publicly available documentation to suggest that the matter of the employment of the scientists was ever discussed with the African National Congress government thereafter.

During the closing-down phase of the BW program, scientists were asked to hand over their research proposals and reports to the managers of the company. Some of the technical reports were allegedly scanned and recorded on optical disk in 1991.[83] According to Odendaal, shortly before the organization was closed down a "roomful of documents" was destroyed.[84]

There is a great deal of uncertainty about the fate of the RRL culture collection during this process of closure. As head of the Department of Microbiology at RRL, Odendaal was responsible for the collection. He maintains that before he left the organization in 1993 the collection was given to Immelman, who, he believed, had plans to destroy it.[85] Goosen makes a different claim, saying that scientists at RRL had retained samples of the cultures and later gave them to him.[86] According to Goosen, there was little managerial oversight at RRL when it was being closed down, and scientists simply helped themselves to cultures that they might want to use in future research.[87]

The South African government authorities appear to have given little thought to the fate of the collection. Indeed, the South African Council for the Non-Proliferation of Weapons of Mass Destruction (NPC) appeared to show interest only in 2002, when Goosen was discovered attempting to sell the collection and other products from Project Coast to former members of the CIA. In a discussion with Gould, Goosen claimed that since the culture collection was not very different from those maintained by bona fide research and industrial institutions around the world, with most of the cultures freely available, the NPC had concluded that there was no need to ensure that the collection had been destroyed.[88]

Goosen's attempt to sell the collection was a cloak-and-dagger deal involving a former general in the South African Defence Force, Tai Minnaar. The scheme began in March 2002, when Minnaar, acting as broker, contacted a personal friend and CIA veteran, Don Mayes.

Goosen claims that before this he had visited the US vaccine production company Bioport.[89] Goosen hoped to persuade Bioport to employ both him and a select group of his former RRL colleagues to produce recombinant vaccines. Bioport was half-interested. According to written communication between Mayes and his colleague, Bob Zlockie, "Dr Lallan Giri [vice president of the Scientific and Regulatory Affairs Group] said . . . that the project is too sensitive for Bioport and that [the US] DOD would pursue the issue, Bioport would however love to have the products."[90]

Minnaar wrote to Mayes in March 2002 offering an antidote for a virulent strain of an unidentified organism (presumed to be anthrax), all the "personal notes and data compiled over the years of research" (by scientists at RRL), and "stock in hand" (the culture collection), as well as the services of a research team composed of scientists who had formerly worked at RRL.[91] The correspondence between Mayes, Zlockie, and Minnaar indicates that from late March, when it was clear that Bioport could not be involved in the deal, Mayes began negotiations with the FBI to take the deal further.[92]

In response to questions posed by Mayes, Minnaar specified what the deal would involve, including:

> One hundred eighty nine strains Anthrax Strains were grown DNA finger-printed and logged. Thirty of the above strains are virulent. Twelve are very active and virulent. Three are deadly and ideally suited for Mass Destruction in a Warfare Programme. All cultures that were grown have been kept under strict Lab. Conditions under the control of the senior scientist [Goosen]. These cultures would need to be flown to the US by a US army plane under strict safety measures and US control. Antidotes were developed for EACH of the abovementioned Strains . . . All notes on the research of all the scientists would come with the group.[93]

According to Odendaal, RRL's culture collection only ever included 45 strains of anthrax, which had been collected in the Kruger wildlife reserve;[94] only one of the strains in the collection was virulent, none was antibiotic resistant, and the antidotes were old and no longer effective.[95] Indeed, it appears that Minnaar incorrectly reported the contents of the culture collection and, therefore, what was available for sale.

According to Minnaar, the deal would cost the US government some $200 million and, as part of the deal, should include the relocation of an unspecified number of South African scientists and their families to the United States.[96]

Mayes demanded proof of the products Minnaar was offering. Goosen, meanwhile, had visited China to discuss control of foot and mouth disease with the Chinese authorities. When Minnaar informed Mayes and Zlockie about the trip, the US became extremely concerned that Goosen might intend selling the package deal to the highest bidder.[97] Despite Minnaar's protestations to the contrary, suspicions had been aroused, and tensions increased.

These setbacks aside, Mayes wrote to FBI agent Rea Bliss on 9 April 2002, stating that the transfer of the "material" from its location in South Africa to the US could be successfully undertaken "with no exposure for the USG [US government] except for funding."[98] Mayes's proposal to the FBI for removing the pathogens from South Africa included a much lower costing. The proposal, attached to his letter to Bliss, stated: "The present government of SA is aware of the Biological Research Program to a limited extent. They are unaware of the advanced development and present existence of the deadly strains or the developed antidotes. The anthrax, antidotes and laboratory R&D documentation are presently in the control of Dr Goosen and a few of his research associates."[99]

Mayes proposed a plan that would not alert the South African government to the transaction. He warned of "serious consequences"; if the South African government were to become aware of the deal, it would "ultimately deny the Anthrax, technical data and antidotes to the USG."[100]

One motive of those in the US involved in the scheme may have been to secure the items from the former South African biological warfare program. Should the deal have been exposed at this stage, in the run-up to the 2003 war with Iraq, it would have been embarrassing to the US to be seen procuring BW agents.

Mayes suggested that the pathogens be shipped from South Africa on private vessels, which would attract little attention. He presented the FBI with a detailed shipping plan, including options for ensuring plausible deniability.[101]

On 6 May 2002 Goosen provided Zlockie with a sample of "*Escherichia coli* 078:K80 (+K60 GM)" to help secure the deal. This bacterium, which can cause severe intestinal upset, had had an even more toxic gene inserted into its DNA. The inserted gene, which results in the production of the epsilon toxin (causing acute mortality in sheep and goats), had been inserted at RRL by Adriaan Botha. Goosen had kept the material. The sample was sealed in a glass cylinder and inserted into an ordinary toothpaste tube surrounded by a cooling gel.[102] Zlockie carried the tube and its contents on a commercial aircraft from South Africa to the US and handed them over to Bliss.[103]

The *Washington Post* reported that the sample was tested at the US BW Defense laboratory at Fort Detrick and found to be exactly as described by Goosen. Yet, rather than being convinced to go ahead with the deal, the US authorities decided that there was "no compelling reason for paying Goosen or excluding the government of South Africa from an operation affecting the security of biological material."[104] A few days later the FBI informed the South African police about the plan.

The South African Police Service promptly searched Goosen's laboratory in South Africa. Goosen, however, claims that he had been warned in advance, with the result that the police found nothing incriminating.[105] Minnaar, in the meantime, continued to seek potential buyers for the products Goosen claims to have had. This enabled the police to set up a sting operation with someone posing as a sheikh from Qatar wanting to buy anthrax. Goosen said he cultured a noninfective strain of anthrax used in vaccine production and gave that to the sheikh.[106] If this is what was traded, it might explain why the sting failed and why there were no arrests.

Legal Remedies

According to press reports, Goosen remained closely associated with the South African National Intelligence Agency (NIA).[107] Indeed, Goosen claims that he kept the NIA informed of his activities in the deal from the outset.[108] Goosen claims to have been motivated largely by a need for financial security, although concern about the safety of the items and a desire to "safeguard the skills and material developed by the South Africans against the threat of proliferation" contributed to his decision.[109]

Goosen's claims need to be treated with a degree of skepticism because there is no confirmatory evidence to support them. If he was indeed trying to put the biological agents out of harm's way, so to speak, this motive might have made it difficult for the South African government to prosecute him successfully under the South African nonproliferation legislation.

The Non-Proliferation of Weapons of Mass Destruction Act of 1993 includes Article I of the BWC. The relevant section states that it would be an offense for an individual to seek to transfer material in "types and in quantities that have no justification for prophylactic, protective or other peaceful purposes."[110] Since this legislation has not yet been tested in court, it is unclear how it will be interpreted. The act also regulates the export and transfer of a comprehensive list of pathogens and genetically modified organisms. If Goosen has committed any violations of the South African legislation to control export of genetically modified organisms[111] through the transfer of the modified *E. coli* to the US, his activity would require the cooperation of US authorities to provide proof of shipment. It is unlikely that criminal proceedings will be initiated against Goosen.

Conclusion

The planning of the upgraded facility for RRL in the late 1980s by Foster Wheeler draws attention to the need for an internationally accepted mechanism whereby industry can report plans for the construction of biosafety level 4 facilities.

Goosen's attempt to sell pathogens from a former illegal BW program demonstrates a remarkable degree of naïveté. Had the appropriate South African authorities known about the deal, as Goosen claimed,[112] they could have been expected to have acted more decisively to contain the culture collection.

Goosen's actions may reflect personal insecurity as a former BW scientist in South Africa. He spoke of being concerned about both his own future and that of his former colleagues, who, he claimed, were under increased pressure as a result of affirmative action policies: "affirmative action is killing off all those researchers [who left RRL for a university-based institute]."[113] Whether these threats are real or imagined, Goosen's action indicates that there may be former BW scientists who are concerned enough about their position in society to undertake dangerous deals. It is not clear how actively involved the other scientists were in the plan.

This incident points to some of the difficulties a country may face when it closes down a biological warfare program as a result of a major political transition. When RRL was closed down, South Africa was undergoing a transition to democracy, and the managers of the program were apparently not concerned with ensuring that the destruction of the documentation and culture collection was verifiably undertaken. The official South African government position was that there was no official, authorized offensive BW program; a verifiable procedure to close the program would have contradicted that position.

There is a political incentive for governments to hide small-scale offensive BW programs that are of little or no strategic military value and, therefore, to take inadequate steps to oversee their closure. The same political incentive undermines the value of statements meant to serve as confidence-building measures about past programs.

The South African authorities should have undertaken an audit of the RRL culture collection at the appropriate time. Other countries might have offered advice on this issue.

The South African program also raises the difficult question of what should be done to ensure that scientists who were formerly involved in offensive BW work do not seek to sell their knowledge or services. Scientists from the former Soviet CBW programs received financial help to keep them in gainful employment in the countries that made up the Soviet Union. While something similar might have been considered for the South African scientists, given the political conditions under which the program was closed down, it is likely that such an action would have been viewed as inappropriate and unacceptable by most South Africans.

CHAPTER 10

Anticrop Biological
Weapons Programs

SIMON M. WHITBY

This chapter gives an account of the anticrop weapons activities of various countries from 1945 to the present day. It offers an appreciation of national programs devoted to the development of plant pathogens as weapons. Additionally, as a result of the biological action of certain chemical anticrop plant growth regulators[1]—substances considered midspectrum agents, between traditional biological agents and classical chemical agents (see Chapter 12)—the development and widespread use of chemical herbicides in Southeast Asia are also briefly discussed. The chapter gives only an overview of aspects of programs related to the development of agents and munitions; consequently, it necessarily neglects areas of vital importance in developing a more complete appreciation of the history and, where it has occurred, the consequences of the use of this form of warfare.

Anticrop Biological Weapons

In national anticrop biological weapons (BW) programs, the development of chemical plant growth regulators as weapons was intrinsically linked to developments in the area of anticrop BW, with parallel R&D efforts toward the common goal of the destruction of plant life. These efforts were an integral and important component of North Atlantic collaboration between the UK and US in the postwar period. Too little is still known about the anticrop programs in the former Soviet Union and Iraq for them to feature in this discussion.

The first notable contribution to the literature on anticrop biological

213

and chemical warfare programs in the postwar period was the six-part series published by SIPRI in the 1970s, which contained valuable insights into the nature and scope of such programs.[2] Further information has been made available with the recent publication of a number of books and articles, including a more comprehensive account of anticrop biological warfare based on a broad but incomplete set of primary sources.[3] Although this subject was not dealt with separately in the SIPRI publication *Biological and Toxin Weapons Research, Development and Use from the Middle Ages to 1945,*[4] it is clear that anticrop biological warfare programs were pursued by a number of nations, notably the US and the UK, both of which emerged from the war with active BW development programs.

In the absence of legal constraints, offensive anticrop BW R&D in the US was pursued in earnest between 1945 and Nixon's announcement in 1969 that the US was unilaterally ceasing its offensive program. (For a general account of the US BW program, see Chapter 2.)

Several newly declassified primary sources provide new insight into US R&D activities in this period in collaboration with the UK.

UK and US Collaboration

British and American collaboration on problems relating to chemical warfare can be traced back to World War I, when, according to Gradon Carter and Graham Pearson, "Anglo-American liaison underpinned the US Chemical Warfare Service created in 1918." Although collaboration established clear channels of communication between the UK and the US, informal collaboration and information exchange on matters relating to anticrop biological and chemical warfare took place in the early days of World War II. The process of tripartite collaboration including Canada was formalized in 1947, with the number of participating countries expanding in 1964 to include Australia, and the quadripartite arrangement being again formalized on matters relating to chemical and biological defense under the 1980 memorandum of understanding.[5]

Political leaders on both sides of the Atlantic appear to have been intimately aware of the particulars of R&D concerning an offensive anticrop warfare capability, and that R&D into both biological and chemical agents were closely bound up. Churchill had considered the possible use of chemical agents against German agricultural targets in 1942. Discovered by Imperial Chemical Industries and codenamed 1313 and 1414, these agents were found to have deleterious effects on different types of crops. Through Sir John Anderson, an official with responsibility for the organization of Britain's wartime civilian and economic resources, details of British findings regarding these agents were communicated to Vannevar Bush, an official with close links to Roosevelt who had been intimately involved in the establishment, rapid expansion, and continued oversight of offensive BW R&D in the US. Confirmation that this information exchange took place is detailed in correspondence between Anderson and Churchill, where Anderson speculated about the possible use of these agents by the US against Japan. In the context of R&D on anticrop chemical and biological weapons (CBW) agents, the US announced that considerable progress had been made in the screening of some 800 chemical anticrop warfare agents, and that a number of plant pathogens had been identified as potential agents for use against food and cash crops. Research would continue in the US throughout its offensive anticrop CW program (some of which would see use in the conflict in Vietnam) and on fungal plant pathogens and other agents for use against rice, potatoes, and a wide range of herbaceous annuals, including tobacco, soybeans, sugar beets, sweet potatoes, and cotton, and later for their potential to destroy drug crops.

R&D into anticrop pathogens continued in the UK up to 1958, centering on basic offensive research guided by a long-term strategy to develop an effective offensive future BW capability. Biannual reports published by the UK's Crop Committee and its successor, the Agricultural Defence Advisory Committee (ADAC), noted the nature of fundamental research conducted in this component of the UK's BW program between 1948 and 1955. Investigations included the following subject areas:

- Herbicide research (1948)[6]
- Water content of cereals (1948)[7]
- Spore suspension research (1949)[8]
- Use of aircraft in agriculture (1949)[9]
- Anticrop chemical warfare (1949)[10]
- Countermeasures to anticrop and antianimal warfare (1949)[11]
- 2–4,D (1950)[12]
- Phytotoxicity (1950)[13]

- Effects of insecticides and herbicides on animals (1951)[14]
- Anticrop detection and destruction of destructive agents (1952)[15]
- Anticrop aerial spray trials (1954)[16]
- Analysis of rice blast epidemic caused by *Piricularia oryzae* (1954)[17]
- Cereal rust[18]
- Screening of *Piricularia* isolates for pathogenicity to rice (1955)[19]
- Clandestine attacks on crops and livestock (1955)[20]

With new terms of reference and a widened remit emphasizing "the potentialities of crop warfare by chemical and biological means," the ADAC continued to oversee UK R&D on matters relating to offensive anticrop warfare.[21] Work on anticrop chemical agents resulted in the testing in Tanganyika of the chemical agents 2,4-D, 2,4,5-T, and endothal (3,6-endoxohexahydrophthalic acid) as insecticides, with 2,4,5-T subsequently used in an aerial tsetse fly eradication campaign in Kenya. The 14-year Malayan conflict saw a chemical derivative of 2,4,5-T used in the first military application of chemical agents in a campaign to reduce jungle cover. In a related food denial program, chemical agents were disseminated by helicopter. This tactic was considered by the commanding officer in Malaya to have been a decisive weapon in the campaign against Malayan insurgents, and according to one source, this deployment formed the basis for future US involvement in herbicidal warfare in Vietnam.[22]

In the US, activities were characterized by a focus on the identification and bulk production of agents and on short-term projects relating to the acquisition of a retaliatory capability. In regard to the latter, according to one UK source, by 1952 the Americans had begun to exhibit a most aggressive outlook, which was thought to be linked to a desire to bring a speedy conclusion to hostilities in Korea.[23] This strategy created considerable UK reliance on subsequent US development of an offensive anticrop BW capability.

US Army Chemical Corps anticrop warfare activities were organized under the four constituent parts of the Crops Division: the Chemistry Branch, the Biology Branch, the Plant Physiology Branch, and the Operational Requirements Branch. The remit of the Chemistry Branch centered on the development of a universal anticrop chemical, and efforts were directed to developing a chemical agent that would, at very low concentrations, be effective in reducing the yields of both narrow- and broad-leaved crops. Investigations in this area included the screening of some 200 potential anticrop chemicals per month and resulted in the standardization of a number of agents that would see use on an unprecedented scale in Southeast Asia in the later 1960s.

Investigations by the Biology Branch centered on those agents that, according to one declassified source, "offered the greatest potential in the attack of [on] food crops." Investigations addressed a wide variety of problems related to the large-scale production, storage, stockpiling, and military use of three agents: the causal agent of stem rust of wheat *(Puccinia graminis)*, codenamed TX; the causal agent of rice blast *(Piricularia oryzae)*, codenamed LX; and the causal agent of late blight of potatoes *(Phytophthora infestans)*, codenamed LO. Two additional agents are known to have been produced and standardized, but their identities are not known. A 1969 summary of biological operations, antipersonnel agents, antiplant agents, and munition systems listed TX as "*Puccinia graminis* var. *tritici* Erikss. & E. Henn., race 56," with an infection dose of just 0.1 gram/ acre or 1 pound for 10 square miles, with spores remaining viable in aerosol form for several days.[24]

Other agents under review at this time for their potential as anticrop biological warfare agents included the causal agent of stripe rust of wheat, *Puccinia striiformis* West; the causal agent of Hoja Blanca of rice, Hoja Blanca virus (transmitted by the leafhopper *Sogata orizicola*); the causal agent of bacterial leaf blight of rice, *Xanthomonas oryzae* Uyeda and Ishiyama; and the causal agent of downy mildew of poppy, *Peronospora arborescens*—the last reflecting an increase in Chemical Corps interest in the 1960s in developing such agents for use against illicit narcotic crop production.

A number of these saw development and eventual assimilation into the strategic US war-fighting arsenal. In a letter to the chief chemical officer, the US Air Force first stated the military requirements of an anticrop BW munition in September 1947, and procurement action was initiated for some 4,800 E73 500-pound cluster munitions in October 1950. This munition was based on a modified propaganda-leaflet bomb and was intended for the dissemination of the causal agent of stem rust of wheat.

The munition was also referred to as the "feather bomb" because of its use of agent-impregnated turkey feathers containing, according to D. L. Miller, "enough particulate matter dusted with agent to create 100,000 foci of infection within a 50 square-mile area."[25] This munition gave the US military its first limited anticrop biological warfare capability.

Work on the E77 anticrop balloon bomb was inspired by, and based upon, a Japanese design tied to the deployment of a free-floating unmanned balloon with a biological anticrop warfare payload. In the final stages of the war in the Pacific some 9,300 Japanese balloon bombs had been propelled by the jet stream at high altitude to carry incendiary and antipersonnel agents over the Pacific to the North American mainland. Adopting the feather dissemination method devised for the E73 cluster munition, according to a US Operations Research Study Group report, the 80-pound E77 antiplant bomb was capable of carrying sufficient quantities of agent and carrier to "cause high levels of plant infection when impact [was] on target crops."[26] The weapon was designated as a strategic munition, with a group assigned for its deployment to the theater air command. However, subsequent developments in munitions saw the emergence of another cluster munition, and the balloon bomb was never deployed.

Larger in capacity but based on the same operational principles as the E73 cluster bomb, the E86 weighed in at some 750 pounds. Procurement was to be initiated in 1953 for the production of some 6,000 of these munitions by 1958, but further development of this weapon system ceased when munition requirements were reviewed in the first half of the 1950s. Spray tanks similar in design characteristics to those used for the large-scale dissemination of chemical agents during the Vietnam War superseded these munitions. Neither agents nor munitions saw use in wartime.

The anticrop BW programs can be traced to a conclusion in the UK in 1963 and in the US in 1969.

World War II adversaries, including Germany and Japan, and later Cold War adversaries, including the former Soviet Union and China, had been identified as potential targets during the anticrop BW activities, and it was estimated that such weapons were capable of bringing about the widespread destruction of staple food crops in both countries.[27]

Soviet Union

Interest in BW in the Soviet Union can be traced back to 1928, when it launched a program in the same year that it acceded to the Geneva Protocol. As Arthur Rimmington has noted, the period between 1945 and the late 1960s is not well documented "either in Soviet or more recent publications."[28] However, it is known that an extensive BW R&D infrastructure was created during this period through a network of links established between the Soviet Ministry of Defense and a number of civilian facilities. With the BW infrastructure established between 1945 and 1970 as its foundation, offensive BW R&D saw a period of rapid expansion from the early 1970s through the early 1990s (see Chapter 5).

Established in 1974, the All-Union Science-Production Association, known as Biopreparat, oversaw development and production of BW activities. This organization formed an expanded network of civilian and military facilities and institutions, including the USSR's Ministry of Agriculture, that were central to Soviet BW activities. As Rimmington has observed, key institutions in the network were headed by senior military personnel. Important institutional links between the antiagricultural components of the USSR's program were established with Soviet Ministry of Defense facilities in Sverdlovsk, Sergeev-Posad, and Biopreparat's Vector research center. Investigations relating to the development of an offensive antiagricultural capability were conducted under the auspices of the Ministry of Agriculture and were aimed at the destruction of both crops and livestock (see Chapter 11 for discussion of antianimal activities).[29]

As part of its antiagricultural activities, codenamed Ekologiya, the anticrop aspects of the program are reported to have centered on the production of agents, including the causal agents of diseases affecting corn, rice, rye, and wheat. Some 10,000 BW workers are thought to have been employed in the antiagricultural component of the Soviet BW program through a network of some six facilities and institutes with links to the Ministry of Agriculture. One such organization, the Central Asian Scientific-Research Institute of Phytopathology, created in Durmen, Tashkent, in 1958, included laboratories, greenhouses, an experimental farm, and a facility for the testing of weapons.[30] Another facility for the testing of anticrop BW existed from the beginning of the 1970s in Kazakhstan. According to Ken Alibek, a notable characteristic of the anti-agricultural capability was that, in contrast with the USSR's offensive antipersonnel BW capability, antiagricultural weapons were never produced on a regular basis and were not stockpiled.[31] Rather, the USSR maintained a number of mobilization facilities with rapid-production capacities that could be called upon to produce such agents on demand. In this regard, the Soviet program differed significantly from the US program, which accumulated large stockpiles of anticrop agents, and which gave substantial attention to improving their storage life.

Weapons for the delivery of the USSR's tactical and offensive biological warfare capability ranged from medium-range bombers fitted with spray tanks and cluster bombs, to ballistic missiles, and even cruise missiles.[32]

Iraq

The origins and institutional infrastructure of Iraq's offensive BW program have been well documented in the open literature, and in official documentation arising from the investigations conducted by the United Nations Special Commission (UNSCOM) on Iraq. By the early 1990s Iraq had progressed to weaponization and emergency deployment of anti-personnel BW (see Chapter 8).

The 1995 UNSCOM report noted that within its overall BW R&D program Iraq had a component related to anticrop warfare. Details of this component remain extremely limited. However, it appears that Iraq had conducted research into the crop-destructive capabilities of a fungal plant pathogen that is the causal agent of the plant disease variously called cover smut, stinking smut, or bunt of wheat. It is thought that weaponization of this agent progressed only to a limited extent, but it is understood that a reasonably mature program of testing had developed by the end of the 1980s. According to UNSCOM, "After small production at Al Salman, larger-scale production was carried out near Mosul in 1987 and 1988 and considerable quantities of contaminated grain were harvested. The idea was said not to have been further developed; however, it was only sometime in 1990 that the contaminated grain was destroyed by burning at the Fudaliyah site."[33]

Information relating to the intended target for this type of weapon also remains sketchy. However, according to the External Relations Unit of UNSCOM, "Iraq declared that it intended to use the smut as an economic weapon—their plan was to cause food shortages as part of a potential attrition war with Iran."[34]

Midspectrum Anticrop Agents in Vietnam

From collaborative arrangements between the UK and the US in the early 1940s emerged a devastating capability to wage warfare against plant life using midspectrum agents. The widespread use in Southeast Asia throughout the 1960s of biologically active chemical plant growth regulators has of course received more attention than other national anticrop BW programs. Great controversy persists about the environmental impact of spraying with plant growth regulators, and debate continues about the health implications of the dissemination of huge quantities of 2,4,5-T, a chemical plant growth regulator that contained significant levels of highly teratogenic dioxin.

Plant growth regulators are natural or synthetic organic compounds that modify or control specific physiological processes within plants. As part of the US Project Agile and under the auspices of the DOD Advanced Research Projects Agency, chemical defoliation and chemical anticrop plant growth regulators were developed in 1961 for counterinsurgency in Southeast Asia.[35] Under a directive issued by President Kennedy in December 1961, a joint US and Republic of Vietnam (RVN) program of chemical spray tests was conducted in Vietnam with commercially available plant growth regulators, utilizing RVN aircraft fitted with experimental spray devices. The tests confirmed the British experience of success with the deployment of such agents in Malaya and later US success in defoliation using plant growth regulators at Camp Drum, New York, in 1958,[36] demonstrating that the chemicals were effective both in the defoliation of tropical forests and in the destruction of food crops.

The first supplies of chemical plant growth regulators for dissemination from MC-1 Hourglass spray systems fitted to C-123 US Air Force aircraft arrived in Vietnam in January 1962. Subsequently, operations were conducted utilizing 1,000-gallon A/A45y-1 spray systems capable of producing a spray swathe 15 kilometers long and 85 meters wide.[37] Full-scale defoliation operations were accompanied by an expansion of the R&D ef-

Table 10.1   Chemical plant growth regulators used in Southeast Asia

Agent	Active component of agent	Percentage by weight
Blue	Sodium dimenthylarsinate	27.7
	dimethylarsinic acid (cacodylic acid)	4.8
Orange	n-butyl 2,4-dichlororphenoxyacetate (2,4-D)	50
	n-butyl 2,4,5-trichlororphenoxyacetate (2,4,5-T)	50
Purple	n-butyl 2,4-dichlororphenoxyacetate	50
	n-butyl 2,4,5-trichlororphenoxyacetate	30
	iso-butyl 2,4,5-trichlororphenoxyacetate	20
White	Triisopropanolammonium 2,4-dichlorophenoxyacetate	
	Triisopropanolammonium 4-amino-3,5,6-tricholoropicolinate (Picloram)	

Source: Modified from a table in SIPRI, *The Rise of CB Weapons* (Stockholm: Almqvist & Wiksell, 1971), p. 172.

fort in the same year. The program, known as Project Ranch Hand, conducted its first major operation utilizing three C-123 aircraft in September and October 1962. Crop destruction operations were also conducted from the inception of the program.

Table 10.1 identifies the chemical plant growth regulators and their intended targets. Codenames were subsequently assigned to these agents in accordance with the colors of the identification stripes painted around the barrels in which the agents were shipped.

The plant growth regulators in table 10.1 were sprayed with the intention of defoliating forest and mangrove in order to improve visibility and enemy vulnerability, to reduce the likelihood of ambush around the perimeter of military bases, and to destroy food crops upon which enemy forces were thought to rely.

Agent Purple, for use in the destruction of broad-leaved crops and forest and bush defoliation, and Agent Blue, for use in the destruction of rice crops, were first used in Vietnam in 1961. Agent Orange, which had effects similar to those of Agent Purple, subsequently replaced it, and was itself replaced by Agent White in 1966, when Agent Orange was temporarily unavailable.[38]

It is reported that military forces disseminated defoliants and anticrop agents at rates of more than an order of magnitude[39] greater than their application rates for domestic use, with early application rate recommendations being revised upward from an initial 1 gallon per acre to 3 gallons per acre. It is estimated that an area representing more than 10 percent of the land surface area (approximately 17,000 square kilometers) of South Vietnam was sprayed with such agents, covering approximately 1 million acres per annum between the beginning of the program in 1962 and its termination in the early 1970s. Approximately 84 percent of agents were disseminated for defoliation and approximately 14 percent for use in the destruction of food crops.[40]

Conclusion

From 1945 to the present day a number of nations have devoted considerable resources to the development of BW in order to bring about the destruction, for hostile purposes, of the plant resources of an adversary. A recurring feature of the programs discussed above was their reliance on agents selected for their capacity to bring about outbreaks of natural disease in some of the world's most important food crops. Fungal plant pathogens that are prone to spread rapidly to epidemic proportions within a single growing season were selected as agents of choice. It is obviously of concern that new scientific and technological developments in biotechnology may considerably enhance a nation's future ability to wage this form of warfare.

Synthetic chemical anticrop plant growth regulators, midspectrum agents considered both biological and chemical weapons agents, were used on a massive scale in Southeast Asia in the destruction of ground cover and food crops, and commentators have sounded a note of concern regarding the current and future potential of such agents in the destruction of plant life. According to Malcolm Dando, "an attack on plants using synthetic chemicals could be carried out today on the basis of a much more systematic and effective knowledge base than that which underpinned the US program in Vietnam."[41]

With these capabilities in mind it is imperative that efforts be redoubled to strengthen the international legal prohibitions relating to CBW so as to minimize the risks posed by these potentially devastating forms of warfare.

CHAPTER 11

Antianimal Biological Weapons Programs

PIERS MILLETT

This chapter discusses offensive antianimal biological weapons (BW) programs from 1945 until the entry into force of the Biological Weapons Convention (BWC) in 1975, and then from 1975 to the present. This history assesses the nature of these programs, including how such projects fitted in with broader BW programs, the duality of this form of warfare, and some motivations for targeting animals in this manner. It concludes with a brief examination of recent developments that may influence this form of warfare in the future.

1945–1975

From 1945 to 1975 there were at least three offensive antianimal BW programs. These were run by the UK, the US, and Canada. All three had conducted antianimal BW development during World War II, so their postwar programs built on these early foundations.[1]

The US Program

The US antianimal BW program featured at least one dedicated facility and the development of four agents and a number of delivery systems. Antianimal BW were viewed as strategic instruments to be used to reduce enemy food supplies or to cause economic damage (see Chapter 2).

The US was bound by federal statute banning research with certain highly infectious animal diseases within the continental United States. This limitation was initially overcome by carrying out work offshore at

Plum Island, New York, and later by coordinating antianimal activities with allies.

Antianimal R&D was initiated as a result of a study conducted by the Operations Research Office into the offensive potential of such agents. The report concluded that because of the importance of livestock in the economy of the USSR and the feasibility of developing the necessary capabilities there was a need to develop antianimal BW. This decision was followed by "a firm requirement for offensive munitions and agents for use against horses, cattle and swine" from the US Air Force. The US Army Chemical Corps responded by assessing a number of potential sites to conduct the necessary R&D and in 1952 selected Fort Terry on Plum Island, New York. Existing plans by the US Department of Agriculture (USDA) to build an animal disease research facility on the island offered the advantage of common utilities, including a steam plant, decontamination plant, and water system.[2]

Plum Island lies in Long Island Sound, Suffield County, New York, and covers about 840 acres. It is about 2.5 miles long, tapering from a width of one mile at one end to around 800 meters at the other. It possesses a natural harbor. The Chemical Corps facility covered about 3 acres and included laboratories, animal houses, administration buildings, communication buildings, a hospital and fire station, a motor pool, a dock, warehousing, staff quarters, a commissary, cafeteria, and guardhouses.[3]

Fort Terry was activated on 15 April 1952 as a permanent installation under the control of the Chemical Corps but under the custody of the First Army. Its originally scheduled opening, on 1 March 1952, was delayed by protests from local residents against the USDA's plans. At its inception, it was planned that 9 military and 10 civilian personnel would staff Fort Terry. By the time it was deactivated and transferred to the USDA in 1954, it had at least 9 military and 8 civilian personnel. Another 162 people were directly contracted to service the facility. Records and files from Fort Terry were transferred to Camp (later Fort) Detrick, which retained responsibility for antianimal BW development, and many were later destroyed.[4]

The mission of Fort Terry was "to establish and pursue a program of research and development of certain anti-animal (BW) agents." Foot and mouth disease (FMD) virus was the first candidate agent for development. Fort Terry was also charged with screening other exotic animal dis-

Table 11.1   Summary of animal pathogens and antisera housed at Fort Terry, Plum Island, New York, before its deactivation in 1954

Agent	No. of strains	No. of antisera	No. of hyperimmune sera	No. of convalescent sera
African swine fever virus	7	—	—	—
Blue tongue virus	5	—	1	—
Bovine influenza virus	1	—	—	—
Diarrhea of cattle virus	—	—	—	3
Foot and mouth disease virus	11	6	—	—
Fowl plague virus	34	—	—	—
Goat pneumonitis virus	4	—	—	—
Mycobacteria	1	—	—	—
"N" virus	1	—	—	—
Newcastle disease virus	3	—	—	—
Rift Valley fever virus	4	—	1	—
Rinderpest virus	41	4	—	—
Sheep pox virus	2	—	—	—
Teschers disease virus	11	—	—	7
Vesicular stomatitis virus	7	—	2	—

eases for potential as BW agents. It appears that development of the causative organism of rinderpest (a disease of cattle) had already been achieved in Project 1001, which took place outside the US. During its operation, personnel from Fort Terry were involved in collaborative exchanges with other facilities, including the FMD laboratory at Pirbright in the UK. The inventory taken at Fort Terry before its deactivation reveals the presence of 15 agents and antisera for many of them (table 11.1).[5]

Antianimal activities, however, were not limited to Plum Island. A 1950 overview of the operational divisions present at Fort Detrick indicated the presence of V and C Sections, representing veterinary and crop capabilities.[6]

Despite the early interest in this form of warfare, by 1952 antianimal BW had been dismissed as "relatively insignificant," and military support for R&D was withdrawn in September 1953. This move followed a report by the Joint Staff Planners in August asserting that the US military had "no capability in anti-animal BW," a claim that contradicted documentation from Fort Terry. There was a resurgence during the years that followed, however; a 1957 Air Force historical study of US BW activities indicated that two weapon systems had been successfully tested for use with antianimal agents.[7]

Contemporary studies of this program provide additional details of these antianimal activities. Agents were arranged into three schedules that presumably related to desirability and utility: schedule 1 contained foot and mouth disease and rinderpest; schedule 2 included classical swine fever (hog cholera), fowl plague, and Newcastle disease (of fowl); and schedule 3 consisted solely of wart hog disease (African swine fever). In addition, the antianimal utility of anthrax and Venezuelan equine encephalitis were noted.

Four weapon systems suitable for use with these agents were developed:

- Spray tanks, developed by a commercial corporation for use by the Navy with hog cholera and Newcastle disease
- Balloon bombs, codenamed E77, for use with a variety of agents and considered a strategic weapon
- Feather bombs, codenamed E73, which were developed from a leaflet dispersal device and contained four packets of inoculated feathers
- Particulate bombs, which were still under development in 1950[8]

As well as these four military delivery devices, antianimal agents were considered for sabotage operations. The 1950 Creasy report asserted that covert contamination of animal food should initiate an animal disease outbreak and that contamination with certain agents would also prevent the reestablishment of animals in certain areas. Such capabilities, the report declared, were easily attainable.[9]

It has proved difficult to trace the history of antianimal biological sabotage because, although there are references to sabotage capabilities, there are few details about what these were or how, why, or when they were developed, and if they were used. It is unclear when antianimal activities were abandoned in the US, or whether they were sidelined before the offensive antipersonnel program. Certainly, there are no indications that such activities were pursued after the US unilaterally rejected all BW in 1969.

The UK Program

The UK's Biological Research Advisory Board (BRAB) was founded in 1946 (see Chapter 3). In its initial meeting it recommended that future research should include "information on sabotage methods calculated to initiate at least certain animal diseases in epizootic form." During this period the focus of antianimal activities in the UK shifted from anthrax (agent of choice in the World War II program) to FMD . By 1951 the BRAB had established that FMD represented a considerable technical challenge. In order to confront such challenges, the UK BW organization became interested in the research capabilities of the Pirbright Veterinary Laboratory (since renamed the Institute of Animal Health, Pirbright). It expressed interest in closer ties between the Ministry of Agriculture and BW-related institutions and in a closer relationship between Porton Down and the Veterinary Research Laboratories (primarily the Pirbright Veterinary Laboratory and its sister facility at Weybridge, in Surrey).[10]

It is unclear whether this initial interest was related to offensive or defensive goals. In any event, by 1952 offensive research on FMD was being carried out at Pirbright, and certain parties considered this to be the primary function of the veterinary facility. The closer relationship between the BRAB and the veterinary laboratories continued into 1953, for at least one senior figure from Pirbright was present at a transatlantic meeting on 19 September.[11]

A report from the BRAB in 1955 outlined what the UK saw as desirable in antianimal BW:

> To be effective in an active war, BW attacks would need to cause very great losses in a short time and over large areas. They would need to affect crops or livestock which were essential to the war effort of the country attacked and that on a scale to impair fighting capacity either directly or through civilian production ability. Further, they would need to be of such a nature that the country attacked could not effectively combat them by established means readily available.[12]

There is no record of an order cancelling antianimal activities in the UK. It is possible that such activities spanned the entire period of BW development.

The Tripartite Alliance

The BW programs of the US, UK, and Canada were linked through the TriPartite process (see Chapter 4). This arrangement provided a conduit for information sharing and provided each of these countries (and later Australia) with access to secret documents and capabilities relating to BW.

Within this arrangement, the UK was responsible for basic research, Canada was responsible for testing that could not be carried out in either the US or the UK, and the US was responsible for production. Thus, Canada performed a crucial role in facilitating the field testing of antianimal agents.

The Canadians possessed a number of facilities involved in BW-related activities, including one on Grosse Isle, Québec. This facility conducted offensive and defensive rinderpest research for the US, UK, and Canada.[13]

Antianimal biowarfare activities ceased to fall under the remit of the Tripartite process in 1953. In September of that year a reorganization transferred jurisdiction over US animal facilities (including Plum Island) from the US Army to the USDA. It is unclear whether the USDA also assumed responsibility for antianimal BW work. This reshuffle meant that activities carried out at these sites fell outside of military jurisdiction and therefore of the TriPartite process. During the final TriPartite meeting to discuss antianimal issues, a decision was taken not to abandon such research but to adopt "a purely informal approach." This informal approach resulted in a scarcity of documentation relating to later developments.[14]

1975 to the Present

Although Russia has admitted that the former Soviet Union pursued BW development after the entry into force of the BWC (see Chapter 5), details about any antianimal components of this program have remained elusive. Testimony from Ken Alibek, a senior figure in the USSR's former antipersonnel program, has yielded some information on antianimal activities, but it has proved difficult to corroborate. According to Alibek, "anti-agricultural weapons were generally produced by more primitive method . . . For anti-livestock weapons, cultivation generally involved live-animal techniques."[15]

Alibek's knowledge of this program was probably limited, as it is believed to have been run by the Ministry of Agriculture. Codenamed Ecology, it was allegedly composed of three sections, the first developing anticrop BW (see Chapter 10), the second focusing on antianimal BW, and the last concentrating on combined antipersonnel and antianimal agents. Alibek believed it to be "one of the most successful programs." Ecology might have utilized the same mechanisms for agent and technology acquisitions, funding, and standard operating procedures as the antipersonnel programs in the Soviet Union.[16]

A number of Soviet facilities, which appear to have served as R&D centers, test sites, or production centers, have been linked to antianimal activities. There appear to have been at least two research centers. The first, the Scientific Research Agricultural Institute (NISKhI), was established in 1958 in the settlement of Gvardeyskiy, outside the city of Otar, about 180 kilometers from Almaty, in Kazakhstan. The second was the Scientific Research Institute for Animal Protection in Vladimir. It has been asserted that this facility "researched and developed antilivestock weapons: African swine fever, FMD, Rinderpest, etc."[17]

A production facility has also been connected to this program. The animal vaccine production facility at Pokrov has been described by its former director as "one of the biggest virus mills in the Soviet Union," and may have been capable of producing tens of tons of viral agents per year.[18]

Two other facilities have been identified as test sites. Alibek has stated that he believed antianimal BW were tested at a scientific institute and test site at the Otar railway station in Kazakhstan; and researchers at the Monterey Institute of International Studies have asserted that the agents produced at NISKhI were tested at the State Research Center of Virology and Biotechnology (Vector), near Novosibirsk, in Siberia.

There have been accusations that BW derived from this program were used sometime between 1982 and 1984 to target the horses of the *mujahideen* in Afghanistan. Although both the occurrence and details of these attacks are still the subject of debate, one proponent of their having taken place has said that they "successfully killed a significant percentage of Mujaheddin pack animals but because of their careless . . . management and excess numbers it had zero impact."[19]

Alibek asserts antianimal activities ceased before the collapse of the Soviet Union. Once again, it has proved difficult to corroborate this claim or to determine when such activities were abandoned.

Two other state-run programs have come to light from this period— one in South Africa, and the other in Iraq. There is no indication that either had an antianimal component.

Allegations of State Use

In 1979 and 1980, during the civil war that led to independence, Zimbabwe (then Rhodesia) suffered from a devastating outbreak of anthrax. During this period there were over 10,700 human cases (intestinal and cutaneous), resulting in 182 deaths. A number of unusual features have led experts to propose that this outbreak had unnatural origins. Dr. Meryl Nass lists these characteristics:

- Unusually large numbers of cases
- Unusual geographic distribution
- Confinement of the outbreak to former Rhodesia despite the presence of favorable meteorological conditions throughout the region
- Confinement of the outbreak to tribal trust lands
- Occurrence of the outbreak during the final stages of hostilities[20]

Potential motivations to carry out such an attack have also been proposed. These include: to reduce the food supply of guerrilla fighters, to increase animosity between these fighters (on whom the outbreak was blamed) and the local populace, and to impoverish the rural population. There have also been uncorroborated statements made by individuals claiming to have been involved in similar operations.[21]

Although proponents of a deliberate origin for the outbreak appear to agree that it was designed to target the animal population, no veterinary data or epidemiological studies have been produced to support such a supposition. Given the amount of time that has passed, controversy can be expected to continue. Nass states, however, that "human cases were secondary to an unprecedented outbreak in cattle." It has been established that before this outbreak anthrax was rare in what is now Zimbabwe. Hence, any sizable outbreak might be considered almost unprecedented.

These figures become important when the contrary position is considered—that the outbreak had natural but unusual origins. It is possible that the favorable meteorological conditions combined with a localized collapse of animal health infrastructure in the tribal trust lands led to the outbreak. During this period of southern African history, it is also possible that differences in access to resources could explain the comparative lack of infections on certain farms. Proponents of this position also cite a lack of documentary evidence for offensive BW development on a scale necessary to achieve such wide coverage.

Nonstate Antianimal Biological Sabotage

To date, nonstate BW events have included cases of bioterrorism, biocrimes, the threatened use of BWs, and allegations of deliberate instigation of disease (see Chapter 14).

A few of these events have specifically targeted animals. The Center for Nonproliferation Studies at the Monterey Institute in California compiled a database of WMD terrorism incidents. Of the 853 incidents identified by 2000, only 25 were classified as antiagricultural and utilizing biological agents. Only 4 incidents resulted in animal deaths. These numbers demonstrate the relative scarcity of their occurrence, as well as the complexities in establishing an unnatural origin for such events.

An antianimal biological attack occurred in Kenya in 1952, when the Mau Mau guerrillas used an extract from the African milk bush plant to poison 33 cattle, eventually killing 8 of them.[22]

An example of a biocrime occurred in New Zealand in August 1997, when a rabbit pathogen was smuggled into the country. A number of farmers went on record to admit that they were intentionally spreading the disease to control the rabbit population. Government inquiries into this practice established that the intentional spread of disease for such purposes was not against the law but that the original importation of the agent was illegal. The perpetrator was never identified.[23]

As recently as 2000, accusations of attacks with antianimal BW were published in the reputable publication *New Scientist,* when an article reported claims that Brazilian authorities believed that a recent outbreak of FMD in the southern state of Rio Grande do Sol had been "started deliberately."[24]

The Nature of Antianimal BW Activities

This brief overview of antianimal BW activities highlights several emergent general themes that differ considerably from modalities associated with antipersonnel or antiplant BW development and use, and together provide a partial insight into this form of warfare.

Antianimal Projects and BW Programs

Little policy-related documentation dedicated to antianimal biowarfare has been declassified. It may be that antianimal projects were subsumed into antipersonnel programs and that policy documentation related to both.

However, there appears to have been a clear dedication of resources to antianimal activities separate from other forms of BW development. All the antianimal projects discussed above included facilities that appear not to have carried out research for antipersonnel or antiplant weapons development (even if they did carry out non-weapons-related civilian research).

The concept of dual use—the utility of information or equipment for both peaceful and hostile purposes—is prominent in the field of antianimal BW programs. Much of the technology (both tangible and intangible) required to carry out the activities detailed above has both weapons-related and civilian applications. Support for this claim comes from the disease prevention work carried out at many of the facilities associated with antianimal BWs development, such as Pirbright (which is now the world reference center for FMD and a number of other animal pathogens), Plum Island (which is still the USDA animal disease research center), and Grosse Isle (which developed important animal health resources for highly infectious animal diseases, such as a vaccine for rinderpest).[25]

One clear theme that emerges from the history of antianimal activities is the relationship between military development and sabotage methodologies. There are two recognizably distinct forms of antianimal biological warfare. The first relates to developing a military capability that resembles other traditional forms of biological warfare: resource-intensive R&D facilities; large-scale production plants; tactical or strategic delivery devices such as spray tanks, bombs, and other munitions; and inclusion in military doctrine. The second form, sabotage, utilizes the innate characteristics of the agents (such as infectivity) to initiate outbreaks of disease from limited-point sources of dissemination, using the most rudimentary of technology. This form of antianimal BW development appears to have received a much higher profile than similar efforts in antipersonnel and antiplant weaponry. Such an approach does not require the dedication of resources associated with the military approach and appears to be within the capabilities of every country in the world. The history of such an approach can be traced back to the origins of modern biological warfare and the German antianimal program of World War I.[26] Although few details of similar programs carried out after 1945 have survived, there are numerous connections between references to antianimal activities and those concerning sabotage operations in the archives of the Tripartite countries.

Motivations to Pursue Antianimal Activities

It is unclear what factors influenced the routes that individual nations opted to pursue, but the history does provide some indications of the motivations that prompted them to pursue this form of warfare. Irrespective of the type of antianimal program employed, three envisioned effects appear to have prompted nations to pursue this form of warfare.

The early programs of World War I utilized animal diseases for tactical purposes: they attempted to initiate disease outbreaks among animals destined for military use (mainly transport and logistics). Although much of the military capability of the developed world is now mechanized, the same cannot be said of the developing world. As recently as the closing years of the 20th century, horses and mules were used to transport military hardware in the Kashmir dispute between India and Pakistan. In many areas of the world, antianimal biological warfare retains a tactical utility.

There are indications, such as those from Project Vegetarian, that antianimal BW can be used to target a state's food supply. Against states that must rely upon domestic animal production, these weapons could be used not only to apply pressure on military capabilities but also to target sociopolitical stability. Although many developed countries are not reliant on a single source of animal production, many regions of the world lack the capability to switch the source of supply at short notice.

Finally, the later programs acknowledged the important economic contribution of animal production (and its associated industries) to the economic health of states. Such an economic attack was probably not envisaged as a single deciding strike; more likely it was seen as part of a prolonged attritional campaign.

The potential effects of antianimal BW suggest that although they may have diminished utility in hostilities between developed countries, they may still feature in scenarios involving developed and developing countries, between developing countries, or in nonstate use for sabotage.

Recent Developments

A number of recent developments may influence this form of the potential of antianimal BW in the future. These include:

- Developments in the biological sciences
- The impact of current agricultural practices (including the use of biocontrol agents)
- The impact of international trade in animals (and the creation of a global marketplace)
- The increasing importance of the interface between domestic animals and wildlife
- The interaction of antianimal BW developments with the international BW prevention regime

Although there is no reliable method for predicting what will happen in the future, it is conceivable that this form of warfare will change to such a degree that it no longer resembles the historical programs detailed here.

CHAPTER 12

Midspectrum Incapacitant Programs

MALCOLM DANDO
MARTIN FURMANSKI

The spectrum of biochemical threats ranges from classical lethal chemical weapons (CW) agents to genetically modified biological agents.[1] Bioregulators such as neurotransmitters fall midway within this spectrum and are termed midspectrum agents.[2] These are covered by Article I of the Biological Weapons Convention (BWC) by the term "toxins."[3]

A number of states have attempted to develop agents with specific incapacitating effects on the nervous system.[4] This chapter covers in detail the British attempt to produce an incapacitating agent during the Cold War; reviews the aspect of the much larger US program related to the agent 3-quinuclidinyl benzilate (BZ), which was actually weaponized during the Cold War; and surveys current concerns about military interest in new nonlethal chemical agents, and the impact of the genomics revolution in enabling the discovery of such agents.

The UK Program

During the middle of the Cold War the UK unquestionably had an incapacitating agents program.[5] At a meeting in May 1963 the Cabinet Defence Committee agreed to a proposal that included "an increase in research and development on lethal and incapacitating chemical agents and the means of their dissemination."[6] From the official documentation it appears that the program came to a halt within about a decade.

Available Data

The center of the British work was Porton Down, in Wiltshire, then called the Chemical Defence Experimental Establishment (CDEE).[7] Work on incapacitants preceded the Cabinet decision of May 1963.[8] The collected papers from the CDEE for 1962 and 1964[9] detail three publications in the open literature.[10] Moreover, as early as 1964 there were press reports of work on lysergic acid diethylamide (LSD) at Porton,[11] and of studies of the effects of LSD on troops by 1969.[12] There was also a collection of papers from a symposium held at Porton in 1972,[13] and several books by a scientist involved in the early 1970s.[14]

The work at Porton, and any associated extramural research elsewhere, was reported to secret committees. The Chemical Defence Advisory Board (CDAB) had a series of subordinate committees such as the Chemistry Committee and, in turn, reported to the Advisory Council on Scientific Research and Technical Development.[15]

In the mid-1990s only one group of chemicals was thought likely to be used as military incapacitating agents.[16] These are anticholinergics, which block the effect of the neurotransmitter acetylcholine (ACh). One of these, BZ, was weaponized by the US during the 1960s,[17] and its mode of action is understood (see below).[18]

There have been two studies of aspects of the British search for an incapacitant.[19] The next section uses official documentation to give a year-by-year account of the development of the whole program.

Annual Developments

BEFORE 1963   Studies of LSD had commenced by 1956.[20] Studies of new agents in general were under way in 1957, when a working party was set up at the CDEE.[21] Investigations of incapacitants began on indole-alkylamines (which might interfere with 5-hydroxytryptamine neurotransmission) and atropine-like substances (which might interfere with acetylcholine neurotransmission).[22] Yet by the end of 1960 no clear lead had emerged.[23]

Dr. D. F. Downing reported on indoles to the Chemistry Committee in late 1959.[24] He also introduced a paper to the committee in mid-1960 updating the work on indoles and in addition mentioning work on analogues of tremorine (which cause tremors—as we now know through the disruption of muscarinic acetylcholine neurotransmission) among other agents.[25] The CDAB's annual review for 1961 noted ongoing work on tryptamines and indoles and on agents that might disturb the balance of the gamma-aminobutyric acid (GABA) transmitter in the brain.[26]

One problem was how to screen the agents in animal tests and to move on from tests on animals to operational agents[27]—that is, to testing on human beings. Human testing of psychotomimetic substances started in late 1961.[28] One reason for the urgency was the increased perception of the need for an incapacitant.[29]

The UK's approach was set out in a discussion paper for the 15th Tripartite Conference in 1960: "ideally, the best possible method for preparing a new agent with a given action would be to design a molecule which would have this specific type of action." However, the paper noted that "knowledge of structure-activity relationships is not sufficiently exact for this to be possible." Thus, it argued, another approach would be a wide-ranging literature survey, which might "perhaps almost by chance" throw up promising leads.[30]

In the early 1960s three papers addressed the problem of biological testing of incapacitants in animals. The first of these reviewed available testing methods;[31] the second assessed the results, in such tests, of using drugs with known effects on humans;[32] and the third considered the screening of new compounds using the available tests.[33] More-complex tests were the subject of another paper in mid-1962.[34]

A variety of potential agents were being synthesized[35] and subjected to screening.[36] Sometimes, as discussed by Catriona McLeish,[37] an industrial liaison program produced a chemical thought worthy of further intensive study.[38] A detailed report given at the Chemistry Committee's 42nd meeting in 1962 discussed "Incapacitating Compounds," noting work on indoles and tryptamines, tremorine and derivatives, substituted hydroxylamines and hydrazines (to interfere with GABA metabolism), pyrroles, and benzimidazoles.[39]

Downing also wrote a review for the ninth meeting of the Offensive Evaluation Committee in January 1962.[40] He stated that there were three lines of investigation: systematic literature review, liaison with industrial laboratories and universities, and research stimulated at the CDEE.

Downing also noted that two major leads were being followed up in the US. These were quinuclidinyl benzilate (EA2277/BZ) and tetrahydrocannabinol (THC; one of the active ingredients of marijuana).

However, as the director of Porton, E. E. Haddon, explained to the CDAB in May 1962, for the past five or six years the work at CDEE had been guided by a policy directive "to study and develop means . . . to defend ourselves against chemical warfare" (that is, a defensive program).[41]

1963   In June 1963 Haddon noted that "CDEE now had a directive for the development of incapacitating agents,"[42] and in October that the services had formulated their requirements for the period 1964–1969.[43] A December 1963 report on "new agent research" mentioned a development in regard to the work on tremoram (a chemical closely related to tremorine).[44] It stated that this compound had effects at acetylcholine receptors and discussed the structural requirements necessary for effects at the receptor. Three Porton Technical Papers investigated these issues in considerable detail.[45] BZ was also under consideration. While the search for new compounds related to BZ was being left to the US, the UK was evaluating BZ itself.

1964   At a meeting of the Advisory Council in November 1964 General Sir John Hackett stated that "it was very desirable to find a safe incapacitating agent . . . General Staff Targets had been issued."[46]

At the CDAB's meeting in June 1964 there was a report of the 48th meeting of the Chemistry Committee. This stated that "there had been five papers on new agent research relating to psychotomimetic compounds and compounds acting on the central nervous system."[47] One of these papers dealt with the search for pharmacologically active benzimidazoles. These compounds were considered for possible effects on the central nervous system, but the effects detected "were due to toxic action on the heart" and not on the central nervous system.[48]

At its meeting in October 1964 the CDAB considered the CDEE annual report for the period ending 30 June 1964.[49] The report on new agents stated that some 150 compounds had been screened during the year, with some 10–15 percent of these coming from industry or universities. More-detailed figures are also available in the fourth of the series of Porton Technical Papers on "Biological Testing of Incapacitating Agents."[50]

During the period August 1961–December 1963 240 new compounds had been received for testing as potential incapacitating agents, and several of these warranted further study.

Also in 1964, two papers appeared in the open literature.[51] As Haddon had pointed out in 1963, staff were encouraged to publish in the open literature, and "even if the application of work was classified, security difficulties could be overcome by employing suitable means of presentation."[52]

1965   The Advisory Council stressed the perceived importance of the work: "we consider that experiments on this humane type of warfare should be pressed forward with all speed."[53] In introducing the CDEE report for 1964–65 to the CDAB, Haddon said that it dealt essentially with the first year of the expanded program.[54] When the board visited the CDEE in May 1965, one of the demonstrations provided was of the new neuropharmacology laboratory by R. W. Brimblecombe.[55]

The 50th meeting of the Chemistry Committee was focused on the new agents program.[56] Dr. A. Bebbington presented an overview,[57] and R. R. Hunt discussed CDEE work on indole derivatives from 1959 to 1964. Many papers had resulted, but because of "the limited incapacitating effect, at relatively high doses, of the simple tryptamines, and the higher lethality of the more active ring-substituted tryptamines," it was recommended that no further synthetic work be carried out on simple indoles and tryptamines.[58] Another paper considered the interaction of drugs with the postganglionic cholinergic receptor and supported the idea of a three-point interaction.[59] This paper was obviously related to a series of secret technical papers, and a major paper appeared in the open literature.[60]

A summary of the CDEE annual report for 1964–65 stated that about 150 compounds, mainly potential incapacitants, had been examined during the year.[61] Work with LSD included a small-scale field trial of its effects on trained troops.[62]

1966   The main areas of investigation were set out in the CDAB's 1966 annual review. These included studying "the operational effectiveness of incapacitating agents."[63] New agent research at this time occupied about one-fifth of the establishment's research effort.[64]

In June 1966 Bebbington summarized how work had been narrowed down: "The fields chosen were those drugs affecting cholinergic systems, including the psychotomimetic catecholinergics; those affecting adrenergic systems; those, mainly morphine-like, causing depression of the central nervous system; and those, other than anticholinergics, having psychotomimetic properties."[65] There was again considerable emphasis on studies of the effects of LSD.[66]

A report presented to the 53rd meeting of the Chemistry Committee in June 1966 noted that "the current approach to the research on biologically active compounds is based as far as possible on studies of the mode of action of drugs at the molecular level."[67] Two reports in the open literature covered the work on (oxo)tremorine and the investigation of the structure of its receptor.[68] What is of most interest is the appearance of glycollates (that is, agents related to BZ) under the category of cholinergic and anticholinergic compounds, on which work was being carried out.

The drive toward molecular studies was emphasized in comments made on the CDEE annual report by Professor R. B. Fisher, the chair of the CDAB. It seemed to him that with some regularity the first lead in a new class of compound was found to be the most effective even when many modifications were synthesized.[69] Moves toward a more systematic approach can be seen in the 1966 Porton Technical Papers.[70]

1967   Studies were not done in isolation from the armed services.[71] At the 66th meeting of the CDAB, Colonel W. Nicholson noted the offensive requirement that the Combat Development Directorate had advised for "a range of incapacitating agents giving a variety of onset times and durations of effect."[72] The meaning can be gathered from a presentation on anticholinergics, which stated that "effects referred to as long duration would persist for 3 to 5 days, those of intermediate duration about one day, and short acting effects for 2 to 5 hours."[73] The ongoing review of defense expenditure was expected to result in only a small cut for the CDEE.[74]

Papers given at a meeting of the Applied Biology Committee show the changing nature of the program in 1967.[75] These consisted of a progress report on work with T3456 (LSD), a report of US experience with BZ and other benzilates and glycollates, a report on early exploratory work with BZ in the UK, and a report on future plans for UK work on glycollates. In regard to the American experience with BZ, R. J. Moylan-Jones stated that the pharmacological action of these compounds was anticholinergic. The severity and duration of the effects depended on the type of glycollate, the dose, and the route of administration. All were effective by inhalation, and those that were liquid were active percutaneously.[76]

A paper titled "Future Plans for Work in the UK" added that the US was giving priority to four glycollates. Two glycollates were selected for early study in the UK.[77] Porton Technical Paper No. 959 had concluded, from detailed study of 19 compounds, that "T3436 . . . a compound with high activity, a low ratio between peripheral and central activity and an action which is rapid in onset but short in duration might satisfy many of the requirements for a mental incapacitating agent."[78] The committee also considered a wide-ranging survey of possible sources for new ideas on incapacitants.[79]

Among other reports, "The Anticholinergic Properties of Enantiomeric Glycollates" and its associated Porton Technical Paper were of interest in exploring more deeply how drugs and receptors interacted.[80] Other, more-detailed work on receptor/drug interactions was reported in regard to morphine-like receptors,[81] and in the then secret but soon open literature in regard to cholinergic receptors.[82]

1968   The CDEE annual report for 1967–68 reviewed the ongoing work and also reported the purchase of more advanced automated apparatus for carrying out animal testing.[83]

At the 50th meeting of the Advisory Council on Scientific Research and Technical Development, the director of the CDEE said that "it was important to note that . . . CDEE had to achieve a cut of £150,000 in annual expenditure by the financial year 1970–1971." This, he said, would affect the program, and "less effort would be devoted to the search for new incapacitating agents."[84] The annual report noted that the council itself would be replaced. However, the report concluded that work on enantiomeric glycollates should be "vigorously pursued." No further trials were required on LSD, as it was "unlikely to be used as a CW agent."[85] Moreover, as Dr. D. C. Barrass explained to the Chemistry Committee, "little attention had so far been paid to elucidating the mode of action of hallucinogens such as LSD and mescaline."[86] Dr. F. W. Beswick explained that T3456 (LSD) was not a practical agent because "there were problems of dissemination, the 100% effective dose by inhalation was relatively high, and the material was expensive."[87] The CDAB heard that a third field experiment with LSD had been satisfactory, but that "work on TL2636 (the oripavine derivative) was of academic interest only."[88]

Despite such setbacks, a joint meeting of the Applied Biology Committee and the Biology Committee in late 1968 took the form of an extended seminar on "Behavioural Studies."[89] Work was clearly pressing ahead on glycollates.[90] Yet in a paper for the Offensive Evaluation Committee, Fisher stated: "On general grounds I think it unlikely that . . . a pure incapacitator agent will emerge. Any chemical agent, a small dose of which is capable of profound disturbance of bodily or mental function, is certain to be able to cause death in large dose . . . and no attack with a chemical warfare agent is likely to be designed with the primary objective of avoiding overhitting."[91] Nevertheless, work on these agents clearly continued at least into the early 1970s in the UK.[92]

The US Psychochemical Weapons Program

The entirety of the huge US effort during the Cold War to investigate nonlethal psychochemicals is beyond the scope of this chapter. However, the development and eventual weaponization of the agent BZ is of particular interest.

The US program to develop psychochemical weapons for orthodox military purposes can be dated to a 1949 US Army Chemical Corps (CmlC) report suggesting that psychochemicals might replace WMD. It proposed three groups of candidate agents: LSD and its congeners, THC, and phenethylamines. Its inspiration may well have been the efforts to use psychoactive agents in interrogation of prisoners, notably the Allied discovery of Nazi studies using mescaline. This interest in manipulating individuals with psychoactive agents for interrogation was enthusiastically embraced by the CIA, and engendered the MK Ultra and MK Niaomi mind-control projects. The CIA and Army programs investigated the same agents, often in the same institutions. Because nearly all of the CIA documentation was destroyed in 1973, and because much of the military documentation remains classified or is otherwise unavailable, it is often difficult to separate these programs from the fragmentary record. Our discussion will attempt to address only the program that sought a military munition using a psychochemical incapacitant.[93]

Army-sponsored research into psychochemicals began in 1951 with a contract with the New York State Psychiatric Institute to investigate mescaline and its derivatives in patients at that institution. Six derivatives were studied at the facility. An additional 35 mescaline derivatives were studied by the CmlC itself, though apparently not in humans. The ultimate conclusion was that no mescaline derivative was effective in a low enough dose to meet the requirements of a military agent that was to be disseminated widely in the open air. Army interest switched to LSD. From 1952 into 1956 LSD was studied under contract at several institutions, including the New York State Psychiatric Institute and Tulane University.[94]

In May 1955 the CmlC established a new long-term project, "Psychochemical Agents" (Project M-1605), for development of a military psychochemical incapacitant with the following requirements:

- Onset of action less than one hour
- No permanent effect a desirable but not essential characteristic
- As potent as nerve gases as a munition fill
- Low toxicity in handling and stable in storage
- Capable of dissemination from aircraft in all weather conditions

This project was initiated to seek an improved incapacitating agent that would avoid the slow onset of mustard gas. A study group organized under Harold G. Wolff and including General S. L. A. Marshal made recommendations on 19 November 1955. The program investigated candidate agents from the mescaline, LSD, and THC groups. A major problem was that no agent was effective in sufficiently small quantities to meet the requirements of an open-air release. By the end of 1955, 45 compounds had been received for study, and 22 had undergone animal testing.[95]

As of May 1956 none had as yet undergone human testing in a formal Army program, although such testing was anticipated if an appropriate agent should be identified. Van M. Sim assumed responsibility for CW clinical research programs, and on 24 May 1956 formal permission was granted for human studies using psychochemical agents.

Human Tests of Psychochemical Agents

The Army began studies at the Chemical Center at the Edgewood Arsenal, in Maryland, using volunteers from the Second Army, as part of a program that dated back to World War II to test chemical agents, therapies, and protective equipment. The program of contract human studies continued, and in collaboration with state governments it was expanded to studies on prisoners under civil detention. Over the 20-year period 1956–1975, at least 6,720 soldiers and approximately 1,000 civilian patients or prisoners participated in evaluation of 254 chemical agents in at least 2,000 trials of psychochemicals. Additional contract studies were established, notably in July 1956 to the University of Maryland Psychiatric Institute to study LSD.

On 1 October 1956 the psychochemicals were given the Army designation "K-agents." The primary agent under investigation was LSD. A potent THC analog designated EA1476 was shelved when it produced pathological toxic changes in animals and only mild mental-status effects in human trials. Interest in EA1476 was renewed in 1963–1966, when its ability to produce disabling hypotension was investigated.[96]

By the fall of 1958 LSD was considered effective in disorganizing small military units. A publicity campaign announcing the availability of psychochemicals was begun, with a half-hour television program in June 1959 made notable by film sequence of a cat, under the influence of a psychochemical, being terrified by a mouse. At congressional hearings held in August 1959 on psychochemicals, additional funding for psychochemical development was urged.[97]

In 1959 and 1960 additional compounds were evaluated. Notable was a crash program to standardize phencyclidine (Sernyl), which failed in 1960 because of the large doses needed. This agent, briefly used later as a veterinary anesthetic, was to become well known among law enforcement authorities as "angel dust" or PCP. Notorious for causing blind rages, it gave apparently superhuman strength to berserk recreational drug users. Other candidate drugs came from the atropine/scopolamine anticholinergic group, originally evaluated as antidotes for nerve agent intoxication.[98]

The advent of the Kennedy administration resulted in a reevaluation of the entire chemical and biological weapons (CBW) program under Project 112, resulting in an enthusiastic expansion. Urgent priority was given to developing a standardized incapacitating CW agent, and in March 1961 top priority was given to the anticholinergic agent 3-quinuclidinyl benzilate (figure 12.1), officially designated BZ, which had been under study since 1958. Its effects were initial torpor, followed by delirium, hallucinations, and unpredictable behavior, including paranoia, that lasted for several days.[99]

The archival record is fragmentary after 1960, but milestones can be identified in the psychochemical program. In the autumn of 1961 a 12-month program under Project 112 was created to provide a standardized munition for BZ, resulting in standardization on 12 March 1962 of BZ in the 750-pound M43 cluster bomb and 175-pound M44 generator cluster. These were clusters made on the basis of a ¾-pound canister, the M16 BZ generator, based on the M42 smoke pot. The 1962 BZ standardizations were designated "level B," indicating that they had important shortcomings and were considered interim weapons. Indeed, the British considered BZ and its 1962 US munitions unsatisfactory. They failed to meet the specifications in several important ways. BZ's onset time was slow and variable: the dose capable of disabling half of exposed soldiers took at least one hour to have any effect and did not reach full effect for 8 hours (figure 12.2). This dose would make a soldier incapable of combat for about 24 hours. Soldiers receiving twice this dose would be incapacitated within an hour and remain incapacitated for at least 48 hours. Much

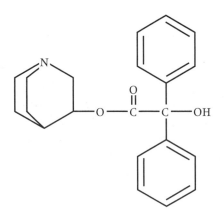

Figure 12.1 Structure of quinuclidinyl-benzylate (BZ).

wider variations in dose would be expected in any military release. BZ was a solid, and the method of disseminating it in these munitions was to generate a visible particulate cloud by thermal "BZ smoke pots." The unpredictable and irrational behavior it produced also had military disadvantages.[100]

The slow onset precluded any immediate tactical advantage through a

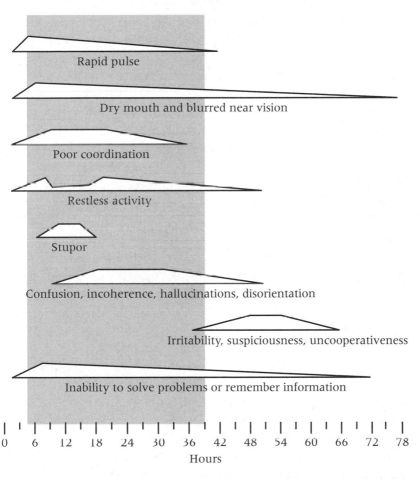

Figure 12.2 Symptoms of BZ intoxication and their duration. The shaded area indicates the period of effective incapacitation. (Source: *Joint CB Technical Data Source Book,* vol. 2: *Riot Control and Incapacitating Agents,* Part 3: *Agent BZ* [Fort Douglas, Utah: Deseret Test Center, 1972].)

surprise attack, and it negated the utility of BZ to repel attacks. Because onset would be variable, early high-dose "sentinel" victims might allow recognition of a BZ attack and reinforcement or replacement of enemy troops before the combat effectiveness of "low dose" troops was significantly reduced.

The visibility of the cloud rendered surprise or a successful clandestine attack highly unlikely. Well-equipped troops would mask immediately, and irregulars might flee the cloud or improvise mask protection. Because BZ was not absorbed through though the skin, simple mask protection was effective. Although BZ had been demonstrated to disrupt the effectiveness of small-unit infantry forces and crew-served weapons such as artillery and tanks, the persistent delirium produced by BZ was itself a problem.

One potential use of BZ was in limited warfare areas containing noncombatant civilians. Because enemy soldiers were likely to be better protected than civilians, BZ might well result in large numbers of mentally deranged civilians in an environment of active combat. Civilians might no longer seek shelter or avoid fields of fire. Even if BZ effectively neutralized organized enemy resistance, the presence of thousands of heavily armed mentally deranged enemy soldiers posed a daunting obstacle to "mop-up" operations involving disarmament and confinement. The delirium induced by BZ made such operations particularly hazardous: the initial hours of torpor were followed by days of excitation, irrationality, and paranoia. "Reviving" pockets might become irrational "berserkers" unappreciative that the battle had been lost, and even disarmed POWs would remain for days subject to episodes of violent irrationality, a body of unpredictable and uncooperative mental patients.

The second potential use of incapacitating CW was as a step in a "measured escalation" in a conflict with the Warsaw Pact. Chemical warfare advocates believed that use of nonlethal CW could overcome Soviet superiority in conventional arms by rendering combat units ineffective and disordering supply and lines of communication. This proposed strategy relied on the debatable assumption that the Soviet leaders would recognize the incapacitant CW attack as such, and interpret such use according to the uniquely American contention that "nonlethal" CW did not violate the rules of war and warrant lethal CW retaliation.[101]

However, use of BZ as the incapacitant made this scenario even more

problematic. In a NATO–Warsaw Pact confrontation, theater commanders would be assumed to have both lethal CW and tactical nuclear weapons available. Although these weapons were tightly controlled by the
normal chain of command, rendering field commanders or groups of
lower-ranking men irrational by a BZ-induced delirium was a significant
liability. Reports to central authorities could be expected to be inaccurate,
possibly filled with panic and paranoia. Moreover, unauthorized use of
lethal CW and nuclear weapons by local forces was a distinct possibility.
By virtue of their short range, some of these weapons needed to be physi-
cally located in forward areas, where BZ would be used, and they were
capable of independent use by local forces. By analogy, the US "Davy
Crocket" tactical nuclear weapon had a range of only 2,000 meters and
could be operated by as few as three men with a jeep. Moreover, land
mines and artillery shells containing lethal nerve agents would by necessity be positioned in forward areas. Warsaw Pact forces had similar chemical and tactical nuclear weapons subject to independent use in forward
areas where BZ might be used. Use of BZ in these circumstances might
well precipitate the escalation it sought to avoid.[102]

BZ was produced at the Pine Bluff Arsenal in Arkansas from 1962 to
1965. In 1970 the total stockpile was 49 tons. Because of the shortcomings of the 1962 BZ weapons, it appears that they never actually entered
the operational US CW arsenal. Despite the widespread popular and congressional publicity regarding their effectiveness, psychochemical incapacitants are not mentioned in 1963 CW manuals, even as a class of CW
agents or in characterizations of enemy threats. In 1967 the effects of CW
psychochemical incapacitants were described briefly for recognition in
defensive training, and incapacitants were treated briefly in operational
manuals. However, although the 1967 field manual for CW listed BZ as a
standardized CW agent, it did not include its munitions in the charts giving operational information for field employment of these munitions.
Since these charts included both level A and level B munitions for lethal
CW agents and for the nonlethal riot-control agents then widely in use in
combat in Vietnam, the omission of BZ munitions must be considered significant.[103]

British investigators in 1965 apparently found that BZ was more dangerous than US investigators had reported, and continued to doubt its
suitability as an incapacitating agent.[104] In 1966 BZ was tested with new

munitions in Hawaii as part of the Deseret Test Center program, but no documentation of standardization of these new munitions has appeared.[105]

Nixon's November 1969 disavowal of BW included a renunciation of first use of incapacitating CW. In the discussions leading to this decision the negative evaluation of the military value of BZ was clear. The Interdepartmental Political-Military Working Group reported to the National Security Council that BZ was "unlikely to be employed due to its wide range of variability of effects, long onset time, and inefficiency of existing munitions," and took the position that "the US currently does not have an effective operational incapacitating chemical capability."[106]

Research continued on psychochemicals after BZ was standardized and after the renunciation of first use of CW incapacitants. Several agents, including EA3167, a long-acting congener of BZ, progressed to testing on humans, in a program that concluded in 1975. None, however, was ever found to be effective enough to warrant standardization.[107]

After 1975 the perceived military threat of psychochemical incapacitants apparently waned. It has been reported that BZ was declared obsolete in 1976, and although psychochemical incapacitants continued to be described in medical texts for treatment of CW casualties, they were no longer included as a class of CW agents in the 1977 or 1983 NBC defense manuals designed for training of rank-and-file soldiers.[108]

The BZ stockpile was destroyed by incineration at the Pine Bluff Arsenal in 1988–1990, a process reportedly complicated by an accidental fire in storage facilities. This action was apparently unilateral: it began before the June 1990 US/USSR Bilateral Chemical Weapons Destruction agreement and the September 1989 Jackson Hole Memorandum of Understanding that preceded it. The political and military antecedents of this unilateral and unpublicized elective destruction of an entire class of CW warrants scholarly attention.[109] The Chemical Weapons Convention of 24 April 1997 specifically outlawed the possession of such incapacitating chemical warfare agents and the means to produce them. In 1999 the BZ filling plant at Pine Bluff was destroyed.[110]

Conclusion

On the evidence presented above, the effort in the West to find a non-lethal chemical incapacitant during the Cold War was a distinct failure.

Given the number of deaths of hostages in the attempt to use a fentanyl derivative to break the Moscow theater hostage siege in 2002,[111] it is also probable that the Soviet Union was no more successful at that time.

Nevertheless, efforts clearly continued during the 1990s to overcome the problems involved in perfecting a usable nonlethal chemical incapacitant. One US presentation at the 1995 Edgewood Annual Scientific Conference on Chemical and Biological Defense Research suggested that "depending on the specific scenario, several classes of chemicals have potential use, to include: potent analgesics/anesthetics as rapid acting immobilisers; sedatives as immobilisers; and calmatives that have the subject awake and mobile but without the will or ability to meet objectives."[112] Follow-up work along these lines is reported to be continuing.[113]

Since the end of the Cold War, the genomics revolution has allowed much greater elucidation of the receptors for the natural transmitters, and other such nonlethal chemical agents, in the brain.[114] We now know much more about the structures of the individual receptor subtypes and of the circuits in which they operate. Moreover, as the military institutions of technologically advanced nations have become more involved in operations other than war in trouble spots in the developing world, they have again become interested in obtaining usable nonlethal weapons—including chemical incapacitants.

In such circumstances it is hardly surprising that specialists have warned that "it is hard to think of any issue having as much potential for jeopardising the long-term future of the Chemical and Biological Weapons Conventions as does the interest in creating special exemptions for so-called non-lethal chemical weapons."[115] Despite the failures recorded here, it is unlikely that the search for nonlethal chemical incapacitants is over.

CHAPTER 13

Allegations of Biological
Weapons Use

MARTIN FURMANSKI

MARK WHEELIS

Only a few instances of actual use of biological weapons
(BW) by nations have been conclusively documented, all before 1945.[1]
However, there have been many accusations and speculations since. In
this chapter we analyze the four most important sets of allegations: that
the US employed BW in Korea and China in 1952; that the US used BW
against Cuba in 1962 and thereafter; that in the 1970s and 1980s the Soviet Union used toxin weapons in Afghanistan and provided them and
supervised their use by the Vietnamese and Laotian governments in Laos
and Kampuchea; and that an anthrax outbreak in 1979 in the Soviet city
of Sverdlovsk was the result of an accidental release from a bioweapons
facility. Retrospective scientific analysis has proved that such a release did
occur. The three allegations of deliberate use all deserve to be considered
seriously, but in all three cases the available evidence is insufficient to
support the allegations, and in two of them the evidence strongly suggests that the allegations are incorrect.

While we use allegations against the US and Soviet Union as case studies here, they are not the only countries alleged to have used BW since
1945. Others include South Africa,[2] Myanmar,[3] and Rhodesia.[4] We do
not address these here, because they are not supported by sufficient evidence to permit even a tentative conclusion, and they do not have the
geopolitical significance of the US and Soviet allegations. Additional discussion of the South African and Rhodesian allegations can be found in
Chapters 9 and 11.

Allegations of US BW Use in the Korean War

During the Korean War, the Chinese and North Korean governments accused the US of waging biological warfare, using techniques obtained from the World War II Japanese BW program. The US responded with unqualified denials, characterizing the allegations as total fabrications.[5] The opposing sides could not find a mutually acceptable body to investigate the charges scientifically, and they entered the realm of a worldwide propaganda campaign in which both sides engaged in deliberate deceptions.[6]

Fifty years later it is clear that a massive biological campaign did not occur,[7] and that at least some evidence was fabricated.[8] Likewise, it is clear that the US did indeed protect Japanese BW war criminals and assimilated their information into the US BW program.[9] Yet positions remain strongly polarized. The Chinese have made no official retraction of the charges, and privately maintain that their allegations were made in good faith and reflect actual biological attacks.[10] The US has not modified its position that no biological attacks occurred and that the allegations were deliberate fabrications. Scholars are nearly as polarized as the national positions.

The Allegations

The Soviet Union had been pursuing a propaganda campaign against the United States since late 1949 because the US had shielded Japanese BW war criminals.[11] This campaign continued after the beginning of the Korean War,[12] and in the spring of 1951 Soviet intelligence operatives collaborated with North Korea to make a belated claim that retreating US troops had spread smallpox during the retreat from the Yalu.[13]

Then in late February 1952, North Korea and China claimed that US aircraft had dropped insects infected with pathogens in several incidents in North Korea that closely resembled Japanese BW attacks from the 1940s. In early March 1952 China alleged BW attacks on its territory as well.[14] In an effort to characterize these attacks and produce evidence, Chinese scientists conducted extensive investigations.[15] The findings did not prove compelling to neutral observers. After mid-1952 the allegations relied upon confessions (later retracted) from US airmen. Early confessions described limited clandestine BW missions;[16] later ones portrayed a widespread BW campaign ordered by the US Joint Chiefs of Staff.[17]

It is clear that the extensive BW campaign alleged in the later confessions could not have occurred. The US BW program did not have sufficient production capacity to mount such a massive campaign, and the agents and munitions in its standardized arsenal did not correspond to those described in the allegations.[18] Such a large campaign would have required the direct participation of many hundreds of US military personnel, and the direct knowledge of thousands more. Such a campaign could not have been concealed for 50 years.

Furthermore, transcripts of Soviet documents indicate that some data originating in North Korea had been fabricated with Soviet aid, and that after May 1953 the USSR considered the allegations to have been fabricated.[19] However, these documents are fragmentary, and they do not justify the conclusion that all the evidence was fabricated, although they raise that possibility.

The Chinese Threat Analysis

The allegations were triggered by reports of suspicious events from North Korea, but it is clear that these events were considered significant largely because intelligence in Chinese hands was interpreted as showing that the US had for some time pursued progressively more threatening preparations to mount a BW attack.[20]

It was clear to the Chinese that the US had acquired Japanese BW data in the late 1940s, and in 1949 the US confirmed that its World War II BW program was continuing.[21] In the May 1951 allegations General Douglas MacArthur was charged with using the Japanese government as a front for BW preparations.[22] Later allegations maintained that the US brought Japanese BW war criminals to Korea as consultants on BW.[23]

The LCI1091 Navy epidemic control ship was alleged to have been used to kidnap North Korean and Chinese soldiers for BW tests. The ship was then said to have traveled to the POW camps at Koje Island to test BW.[24] A BW production facility in Japan was alleged—almost certainly the US Army's 406th General Medical Laboratory (GML), which because of the presence of insect breeding facilities and large numbers of rodents could easily have been mistaken for a Japanese-inspired BW production facility. Captured spies confessed to having been sent by the US to report on the progress of epidemics.[25]

The Chinese further claimed that the US had begun immunizing its troops against diseases not present in Korea, and had begun a seasonally inappropriate, aggressive campaign of insect and rodent control along the front lines. These were interpreted as measures to protect US troops from an impending Japanese-style BW attack.[26]

The Chinese recounting of these events is accurate, although the interpretation is not. The US had issued large medical supply contracts to Japanese companies when the Korean War broke out, including a contract to supply blood from a company (later the Green Cross company) headed by one of the major Japanese BW war criminals.[27] The 406th GML had renewed contact with the Japanese BW war criminals in November 1950 to gather information on hemorrhagic fever, a disease unrelated to contemporary BW.[28] In 1951 they were taken to Korea near the front lines to assist in controlling an outbreak of hemorrhagic fever among US troops.[29] The insect and rodent control measures were also taken for this purpose. Changes in vaccination policy in early 1952 did occur, but they did not correspond to US BW agents in development.[30]

The LCI1091 naval epidemic control vessel did participate in clandestine forays to monitor the North Korean smallpox epidemic, and did at least attempt kidnappings of enemy patients.[31] The US had placed agents in North Korea to report on the epidemics. However, this medical intelligence was developed for the conventional medical service rather than for the BW program.[32]

The LCI1091 did then go to the POW complex on Koje Island from March to November 1951, and supported the Commission on Enteric Diseases (CED), a group of experts in infectious disease sent from the US to perform clinical therapeutic trials on POWs while attempting to control a massive dysentery epidemic with an appalling death rate.[33] The Chinese were quick to see a parallel with the human experiments of the Japanese BW program. However, the activities at Koje Island were not associated with the US BW program. CED records were never classified, and the results of the investigations were published in the open literature in a timely fashion.[34]

In retrospect, the Chinese were in error in concluding that these activities indicated an immediate danger of a BW attack. However, in 1952 they would have appeared completely consistent with a progressive expansion of activities needed for BW attacks modeled after the Japanese BW program. This continuity would have been compelling from the Chinese point of view.

Public statements in early 1952 also appeared to indicate an imminent CW or BW attack. Threats of the use in Korea of "new" weapons had begun in late 1951.[35] In early 1952 high-ranking active-duty US military leaders made statements praising the value of BW. These statements appeared in the US press and were reprinted in official government publications, suggesting an official change of policy. On 8 February 1952 Major General E. F. Bullene, chief chemical officer, made a "hawkish" speech supporting the military values of CW and BW, and this was reprinted in the *Congressional Record* at the request of an influential congressman.[36] Brigadier General Creasy, chief of the Chemical Corps Research Command, gave a speech in Washington on 25 January praising "germs, gas and radio-active materials" as efficient weapons of war.[37] His comments were reprinted in the US Army newspaper *Stars and Stripes*. On 2 February 1952 a political columnist wrote that secret projects to increase the insecurity of China had powerful backing in the US government.[38]

Given this context, it is not surprising that the Chinese issued allegations at the first indication of a BW attack with reported laboratory confirmation. The risk appeared real, and restraint offered no benefit. Issuing allegations allowed a diplomatic offensive internationally and a defensive mobilization internally. The international attention might truncate or prevent a nascent BW offensive. Beginning immunizations and enforcing hygiene might blunt any BW attack in progress, and might deter a future one. Galvanizing the civilian population with the threat of a BW attack allowed the Chinese government to proceed with massive immunization campaigns and public health measures that had previously been unpopular and unsuccessful.[39]

The Question of Limited Use of BW

Because widespread BW use did not occur, the historical question contracts to that of limited, clandestine BW field trials.[40] The practical and political barriers that make the alleged full-scale BW campaign impossible do not necessarily apply to limited BW field tests. Sufficient munitions were available, and pilot plant production could have provided sufficient agent. Clandestine operations and plausible deniability could have allowed freedom from the policy restrictions that applied to the mainline US BW program.

If any such trials occurred, they would have been conducted by the Special Operations Division (SOD) of the US Army Biological Laboratory at Camp Detrick, which collaborated with the CIA to develop biological weapons for sabotage and clandestine use.[41] Most documents describing the projects of the SOD remain classified, and accounts of its activities have been redacted from official histories of the US BW program. Congressional investigations determined that clandestine BW use could not be excluded, because the CIA had destroyed the pertinent records.[42] Since the documentary evidence remains unavailable, evidence of US BW field trials must be sought in the available scientific data regarding the allegations.

The International Scientific Commission Report

The most important document detailing the scientific investigation of the allegations is the report of the International Scientific Commission (ISC).[43] This group consisted of sympathetic Western scientists, assembled by China after it rejected Western proposals that the World Health Organization or International Committee of the Red Cross (ICRC) investigate.[44] The ISC compiled the scientific data that had been gathered by the Chinese scientists in March and April 1952, but it did no field or laboratory work itself, nor did it witness any BW attacks. It collected testimony and scientific reports in China and North Korea from 22 June to 6 August 1952, completing a report on 31 August that was published on 1 October 1952. The report contained over 665 pages of text and 100 pages of photographs. The report proper ran to 61 pages, with the remainder consisting of 48 scientific appendices.

The ISC report detailed 37 incidents in China and 13 in North Korea that were alleged to be biological attacks. The North Korean and Chinese portions differed significantly. In North Korea, Chinese scientific teams relied heavily on North Korean officials and on Soviet-advised North Korean laboratories. In northeastern China, Chinese scientists worked with Chinese epidemic control teams and used Chinese civilian laboratories.

All the Korean reports are suspect. Soviet documents and inconsistencies in ISC sources reveal most to be fabrications. Because of their patent unreliability, the Korean-alleged incidents are unsuitable for scientific evaluation.

In marked contrast, the Chinese reports are documented with detailed analyses by identified scientists, who appeared to have been scrupulously accurate. For instance, Chinese scientists retracted reports that they had isolated pathogens from two early events in China. These were critical events, the retractions embarrassing, and supporting evidence could easily have been fabricated.

The striking feature of the reports is that most were not very convincing as BW attacks. A wide variety of insects were reported, often ones without obvious BW vector potential: springtails were reported from 6 events, spiders from 10, and crickets from 3. None of these naturally carry human disease, and the Chinese scientists were well aware of this. Mosquitoes were reported 15 times, but when species were identified, they were ones without disease vector potential. Much of this entomological diversity can be attributed to hypervigilant citizens reporting unusual but natural insect behavior, or fabricating or exaggerating events to appear politically conscientious. Internal documents of the Chinese public health officials charged with investigating these reports commented on this problem.[45]

Not surprisingly, these inauspicious entomological events produced little evidence of biological attack; only nine produced human pathogens on culture (two produced an animal pathogen), and in only six did the pathogen and the vector coincide in an expected way (two of *Salmonella* in flies, one of plague in rodents, three of anthrax in flies/beetles). Only two outbreaks of disease were alleged to have resulted from BW attacks in China: an unidentified form of encephalitis and a cluster of inhalational anthrax cases in northeastern China. The encephalitis outbreak showed anomalous seasonality and virology, but appears to have been a natural outbreak because similarly anomalous outbreaks occurred in South Korea and Pacific islands at the same time. The anthrax cases are the most plausible candidates for clandestine BW attack.

The Anthrax Cases

The ISC report contained case histories and autopsy findings from five cases of inhalation anthrax in northeastern China during March and April 1952, in five settlements separated by 35 to 400 kilometers.

Witnesses in four of the locations reported overflights of US aircraft that dropped objects, two of which were observed to burst in the air. Later, masses of insects or feathers were collected near the drop zones and reported to be contaminated with *Bacillus anthracis*. Four of the five fatal cases had helped to collect and destroy contaminated material. Gross autopsy results, including postmortem cultures, were presented for all five, and histopathology was presented from three. Photographs of preserved gross specimens were submitted for two. The clinical accounts and postmortem pathology were fully consistent with the current understanding of human inhalation anthrax, including sporadic cases caused by manual handling of contaminated material, such as occurred in Florida during the 2001 US postal outbreak.

The close correlation of the 1952 ISC inhalation anthrax cases to the current understanding of this disease makes it clear that the data were not fabricated. In 1952 descriptions of the pathology of inhalation anthrax were scant, and they differed substantially from the modern understanding. In about 30 autopsies between 1890 and 1924, which formed the textbook understanding of this disease until after the Korean War, involvement of the lung itself was considered the fundamental lesion, with bronchial ulceration reported in almost all cases. Mediastinal lymph-node involvement was a minor finding, inconsistently described, and meningitis was not described. The pathology texts cited in the ISC report stressed lung and bronchus involvement as the fundamental process.[46]

In marked contrast, in more than 60 autopsies since 1957 (including two BW-related events: the 1979 Sverdlovsk and 2001 US postal outbreaks), lung involvement was absent or minimal, bronchial ulceration was not reported, mediastinal enlargement was reported in almost all, and meningitis was common.[47] This is the "modern" pattern, first described in 1960.

The ISC cases clearly presented the "modern" pattern: lung involvement was minimal, none showed bronchial ulceration, four of five showed mediastinal enlargement, and all showed meningitis. These results must have been from real human inhalation anthrax cases, since the information did not exist in 1952 to have allowed fabrication using textbook or medical literature sources.

Further evidence against fabrication of the data lies in the apparent incubation periods. The intervals between US overflights and the onset of disease ranged from 3 to 35 days. In 1952 there were no reports of incubation periods for inhalational anthrax, and it is unlikely that the Chinese would have fabricated values with such an unusually wide range. It was not until the 1994 publication of data from the 1979 Sverdlovsk outbreak that a wide range of incubation periods became known as characteristic of this disease.[48]

There is nothing inconsistent between the Chinese contention that the ISC anthrax cases were US BW attacks, and known US capabilities. US aircraft did overfly the alleged target area. Anthrax had been developed as a BW agent at Camp Detrick during World War II,[49] and the pilot plant for anthrax production, concentration, and drying had been reactivated in late 1950,[50] almost certainly to supply the Special Operations Division. The use of feathers as a carrier for cereal rust spores was already a feature of a standardized US BW munition,[51] and it had been successfully tested with two animal pathogens as well.[52] Tests of flies as carriers for animal pathogens had also been successful. In 1952 externally contaminated insects were considered an effective dispersal method for BW agents.[53]

Furthermore, inhalation anthrax has been reported only after readily identifiable occupational environmental exposures or as the result of BW.[54] No suitable environmental sources were described in northeastern China in 1952, and the wide geographic separation of cases would have required multiple sources. Only one of the victims (a farmer) had an occupational risk factor, and this would have been to cutaneous, not inhalation, anthrax. In the entire Chinese medical literature there were no reports of inhalation anthrax, and only three case reports of isolated anthrax meningitis,[55] none with autopsy examination.

Despite the highly suspicious nature of this outbreak and the coincidence of US capabilities and the alleged attack method, it appears to us that this outbreak was natural after all, and not a result of BW use. More than a dozen isolates of *B. anthracis* obtained in northeastern China in

1952 have recently been subjected to genomic analysis, and all were clearly indigenous to China.[56] It seems almost certain to us that this collection includes isolates from victims and alleged BW fomites. Had this been a US BW attack, the *B. anthracis* strain would have been Vollum or Vollum 1B. That all isolates from the region and year of the alleged attacks were indigenous, and not the strain used in the US BW program, makes it certain that the outbreak was natural, despite its unusual features. We suspect that the intensive civilian searches for unusual concentrations of insects, feathers, and the like, which characterized Chinese civil defense activities, may have disturbed burial sites containing remains of veterinary anthrax cases. While such an etiology has never been reported for inhalational anthrax, it appears to be the most plausible explanation for this unusual outbreak.

Allegations of US BW Use against Cuba, 1962–1997

Shortly after the Cuban revolution that installed Fidel Castro in early 1959, the CIA began supporting expatriate Cuban forces and dissident elements within Cuba that for more than a decade mounted a campaign of covert and overt attacks on Cuban targets. Support of covert sabotage activities was approved in July 1960,[57] five months before the US broke diplomatic relations with Cuba on 3 January 1961 (over restrictions Cuba had placed on the size of the US diplomatic presence in Cuba). Within this context, Cuba has suggested periodically that diseases were deliberately introduced onto the island by the US. Formal presentation of a list of allegations to the international community has occurred three times;[58] the most recent, in 2000, alleged 19 specific instances of biological aggression against domestic animals, crop plants, and humans.[59]

No evidence other than purported unusual epidemiological features was available to support most of these allegations; however, natural outbreaks frequently display unusual epidemiology, so this feature by itself is rarely definitive, and may not be even suggestive.[60] In most cases the epidemiological features that were claimed to indicate unnatural etiology were not provided in detail, nor were they independently published. In a few cases other evidence was adduced (such as confession of captured perpetrators), but again there were no details and no opportunity for independent confirmation. This lack of evidence makes it difficult to take most of the allegations seriously.

Following a thorough review of eight of these allegations, Raymond Zilinskas has concluded that natural routes of disease introduction are plausible for all the outbreaks considered, and that all the allegations are probably incorrect.[61] However, the two earliest allegations, both alleging biological attack on domestic animals, are supported by some independent evidence, and in our view warrant more serious consideration.

Context of the 1962 and 1971 Allegations

Following the failure of the CIA-trained and supported Cuban Expeditionary Force landings at the Bay of Pigs in April 1961, the CIA proposed a program of covert action designed to weaken the Castro government.[62] The proposal included a number of elements, one of which was expanded sabotage against selected targets; sugar refineries were mentioned, but no other agricultural targets were listed. On 3 November 1961 President Kennedy established a new program targeting Cuba, named Operation Mongoose, overseen by the "Special Group (Augmented)" (SGA). The SGA consisted of an expansion of the existing National Security Council 5412 Special Group, which oversaw all covert operations.[63] Brigadier General Edward Lansdale was appointed chief of operations, with instructions to conduct activities through existing agencies of the government—the CIA, DOD, US Information Agency (USIA), and the State Department—each of which appointed a full-time operational representative to assist him.[64] By the middle of 1962 Operation Mongoose had a budget of more than $10 million, employed nearly 500 CIA staff full-time (and many more part-time), and had about 90 intelligence agents and more than 10 operational teams inside Cuba.[65]

If a US biological attack on Cuban agriculture did take place in this period, it would almost certainly have been part of Operation Mongoose. Fortunately, thousands of documents relating to this operation were declassified in 1998 by the Assassination Records Review Board (ARRB).

By 1971, the year of the second alleged attack, Operation Mongoose had long since ceased, and any covert actions involving biological agents would have been conducted directly by the CIA. Unfortunately, the documentary record of covert actions from this time is very sparse, as these records did not fall into the scope of the ARRB. Sabotage in Cuba had been largely discontinued for many years, although it reportedly had a brief revival during the Nixon administration.[66] The nature of sabotage targets is not known.

Of course, 1971 was after President Nixon's executive orders renouncing offensive biological and toxin warfare and ordering all weapons stockpiles destroyed (see Chapters 2 and 15). However, in 1975 the CIA was discovered to have ignored the executive orders and to have retained stocks of saxitoxin and cobra venom.[67] Given the agency's willingness to disregard one element of the presidential orders, it is not unreasonable to consider that it might also have ignored another.

Furthermore, there was long-standing sentiment, supported by a 1945 legal opinion by the judge advocate general, that attack on enemy agricultural targets with chemical or biological agents did not constitute chemical or biological warfare, but was rather a legitimate extension of traditional methods of economic blockade.[68] Such a belief would have found ready application to a covert biological attack on Cuban agriculture.

However, after the 1975 embarrassment of being found to have violated a presidential executive order, and the coming into force of the BWC in the same year, it is highly unlikely that the CIA would have felt free to conduct any covert biological attacks. Given their political implausibility, the lack of any verified supporting evidence, and the existence of credible natural means of introduction of the disease agents, we believe that all the post-1975 allegations are almost certainly mistaken.

The 1969 and 1971 outbreaks involved animal pathogens in which the US BW program had long been interested. Newcastle disease (ND) had been of interest in the BW program since World War II,[69] and African swine fever since the early 1950s (see Chapter 2). Work on these agents had originally been conducted at Fort Detrick, but had been moved to the US Department of Agriculture site at Plum Island, New York, by the 1960s. Neither introduction would have required more than a small amount of agent (a few milliliters of liquid suspension would have sufficed), and no specialized devices would have been needed for dissemination.

The 1962 Newcastle Disease (ND) Outbreak

Cuba alleged that a 1962 outbreak of ND among fowl in four provinces was the result of sabotage at a vaccine facility, leading to the death by disease or culling of more than a million birds.[70] Zilinskas speculated that this outbreak was due to inadequate attenuation of the vaccine; however, we consider this unlikely. It is not consistent with the Cuban allegation in 2000, in which the ND virus was alleged to have been communicated via a vaccine against avian influenza, not against ND.[71] Furthermore, most ND vaccines, including the one common in Cuba, used live viruses of intrinsically low virulence, whose production does not involve attenuation.[72]

No details have been published about this outbreak. However, Cuba described ND to the Food and Agriculture Organization (FAO) as "widespread throughout the country" for each year from 1957 through 1963; only in 1964 was ND described as "disease much reduced but still exists."[73] Given the extensive background of ND in the country, it is unlikely that any explanation other than natural spread would be necessary, especially since ND is highly contagious.[74] Even a pattern of ND that correlated closely with the use of avian influenza vaccine could result from the known mechanism of unwitting transmission by vaccine workers traveling among poultry farms.

Nevertheless, US interest in transmitting ND to Cuba is independently suggested by the testimony of an unnamed Canadian poultry expert who traveled frequently to Cuba. He claimed that in 1962 an agent from the DOD's Defense Intelligence Agency (DIA) had paid him $5,000 to infect Cuban turkeys with ND virus.[75] He said he had accepted the money and the virus in May, but destroyed the viral cultures before traveling to Cuba in June.

Available files from Operation Mongoose suggest that a covert biological attack on an agricultural target, had it been proposed, might have been seriously considered by Lansdale and his operational assistants; whether the SGA would have approved it is unclear. In January 1962 Lansdale tasked the DOD to develop plans to use insect-borne biological agents to incapacitate Cuban sugarcane workers.[76] The DOD completed the planning process in February, concluding that the proposal was not technically feasible; there was no mention of legal, political, or ethical obstacles.[77]

However, the only agricultural commodity identified as a target in any covert sabotage document we have seen is sugarcane,[78] a major export crop whose disruption would contribute to US economic destabilization goals. As late as October 1962 the CIA indicated that it did not have the capability to attack this crop.[79] In October 1962, in response to SGA demands for a more aggressive program of sabotage, specific target lists were developed; there were no crops or animals listed among the 30–40 identified targets.[80]

The absence of documentary evidence for this operation suggests that it did not happen. It is very unlikely that DOD would have attempted sabotage in Cuba outside of Operation Mongoose, or to have proceeded without SGA approval. Furthermore, we have found no instances in the record of the involvement of any agency other than the CIA in covert sabotage operations inside Cuba. A Freedom of Information Act (FOIA) search of CIA records produced no documents on ND in Cuba in 1961 and 1962.[81] A FOIA request for DIA documents (submitted in December 2003) on this topic has yet to be answered.

In a telegram two days after the story appeared, apparently from CIA headquarters to Latin American CIA stations, the news story was summarized and the claim made, "there is no record in HQS of any CIA connection with the matter described in the article. We also checked with the DIA. Officials there have no record of the matter described in the article and to date have been unable to identify the 'DIA source' mentioned in the article."[82] Shortly thereafter Stansfield Turner, then director of the CIA, wrote in answer to a query from Senator Inouye, chair of the Select Committee on Intelligence, that "DIA has informed my staff that they can find no basis in fact for the allegations made in the article. In addition, CIA has no information of any kind connecting with the alleged incident."[83]

All of this suggests that no formal CIA-directed sabotage was involved in the 1962 ND outbreaks. However, during this period there were many sabotage actions initiated by militant expatriate Cuban organizations, with CIA materiel support, but not under CIA control.[84] There is little mention in declassified documents of these independent sabotage activities, and it is not clear that it would have been recorded if ND had been introduced by such an action. There is documentation that on occasion this independent sabotage was aimed at agricultural targets; sabotage of both chicken and pig farms in March 1962 by unspecified (probably in-cendiary) means is recorded.[85]

In addition, a declassified telephone memo suggests that the CIA may have known something about the incident after all. Donald Massey, of the CIA Office of Legislative Counsel, left a telephone message for John Waller, inspector general of the CIA, indicating that he had the signed letter from Turner to Inouye (mentioned above), but wished to talk to Waller before sending it. In a handwritten note on the memo, Waller wrote: "Called Massey—1620, 15 June '77—and said we had records indicating CIA knowledge or connection."[86] While this may be the result of a slip of the pen, leaving out a "no," it does cast some doubt on the denials.

The 1971 African Swine Fever Outbreak

The second allegation is that African swine fever (ASF) was deliberately introduced into Cuba in 1971.[87] This was the first time the disease had been seen in the Western Hemisphere (although it has since emerged in Central and South America several times), so it was a noteworthy event in world animal health circles. The details of the outbreak have been de-scribed in the scientific literature.[88]

The disease was first identified on 6 May in a fattening piggery, and it then spread throughout Havana Province. The outbreak affected both state-run piggeries, containing thousands of animals, and small-scale ru-ral, suburban, and urban backyard husbandry. Over 12,000 animals died of disease, and more than 20,000 exposed animals were culled. Preventa-tive slaughter of all other pigs in Havana Province added another 380,000 privately owned animals and 30,000 state-owned animals. The cumula-tive pig mortality due to disease, culling, and forced slaughter was in ex-cess of 445,000 animals, over a period of about a month.

ASF is a highly contagious disease, caused by a virus that is stable for long periods in infected tissue and fluids. It is readily transmitted among animals by direct contact and by fomites. Outbreaks in new areas are caused by the practice of including food scraps from restaurants, ships, and airplanes in pig swill; disease transmission results if garbage includes infected material and if it is inadequately sterilized. In 1971 ASF was found sporadically in domestic swine in southern and eastern Africa, and at moderate to high incidence in Spain and Portugal. A natural introduction, most likely from Spain, is thus plausible.[89]

However, six years later US investigative journalists published an account based on the testimony of several people who claimed to have been involved in an operation in which the ASF virus was transferred to Cuba from a CIA training camp at Fort Gulick, in the Panama Canal Zone.[90] The account was based on testimony of two alleged participants (CIA-trained Cuban expatriates), with confirmation from a third CIA source in Miami. None of the sources was named.

FOIA requests for 1971 CIA documents regarding ASF yielded only translations of Cuban media accounts of the epidemic, intelligence source reports on the epidemic, and internal intelligence summaries about the epidemic. The last mentioned how dangerous this outbreak was to North America. The apparent lack of any policy or planning documents, even heavily redacted, suggests that there was no direct CIA involvement.

Genesis of the Allegations

Several scholars have interpreted the repeated Cuban allegations of US biological attack as fraudulent.[91] They recognize that such allegations may serve a political agenda to keep domestic loyalties strong by identifying an external enemy.

However, we suggest that, as for the Chinese, the Cubans' belief that they have been targets of US covert biological attack may be understandable. The broader context is, of course, the long history of hostile US actions against Cuba, including training and financing an invasion force, providing training and materiel for violent sabotage in Cuba, mounting efforts to assassinate Cuban leaders, working to isolate Cuba diplomatically and economically, and imposing a total embargo on US trade with Cuba. Thus there is a history of hostile activity, some of which (elements of the trade embargo) continues today, much of which was covert, and some of which was illegal.

In 1971, during the ASF outbreak, Havana's newspaper *Granma* speculated that the outbreak might have been deliberately introduced from the US.[92] However, the idea appears not to have resonated within the government, and the allegation was not repeated. In a speech in 1980 Castro referred to several disease outbreaks over the previous two years but attributed them to natural causes.[93] A year later, however, he alleged that the same outbreaks were instances of biological warfare by the US. In support, he detailed the US BW program as described in books by US journalists, mentioned the Korean allegations, and quoted several congressional hearings.[94] Later in 1981 he added the 1971 outbreak of ASF as a US attack, again quoting US sources.[95] It thus appears that the Cubans' belief that they were targets of biological warfare was largely suggested by, and based on, US publications, especially congressional reports. If so, the idea is likely to have had great credibility.

Although we believe that the US is highly unlikely to have countenanced covert biological attack after the mid-1970s, this conclusion would not have been obvious to the Cuban government. The US still maintains very tight secrecy over the activities of the CIA, and rogue elements within the US government are known to have conducted illegal activities. The US has on several occasions publicly named Cuba as having an illegal BW program, and such claims may look to the Cuban government like an indirect justification for continuing biological attack.

Certainly, these allegations may also serve domestic and international political purposes. But the considerations above suggest to us that the suspicions themselves are understandable, and may still be believed within the Cuban government.

Yellow Rain

Beginning in September 1978, the US embassy in Bangkok, Thailand, sent a series of telegrams to Washington relating accounts from Hmong refugees that suggested possible use of chemical weapons. The Hmong, of the highlands of northern Laos, had contributed heavily to a CIA-organized and -financed army countering the Communist Pathet Lao and their allies the North Vietnamese Army, and had been resisting pacification since the Pathet Lao had taken power. Although the reports suggested CW use, the alleged agents came to include toxins that are BW as well.[96]

At nearly the same time (beginning October 1978), representatives of the Democratic Kampuchea (DK) insurgents (supporters of Pol Pot's deposed Khmer Rouge regime) began to charge that Vietnamese troops, fighting DK forces in support of the Hanoi-backed People's Republic of

Kampuchea government, were using CW. These charges were later supported by refugee accounts.

Finally, *mujahideen* guerrillas in Afghanistan reported being attacked by Afghan forces with CW in the summer of 1979, six months before the Soviet invasion in December. After the invasion, the frequency of reported incidents rose sharply.[97]

These reports led the US to charge that the Soviets were supplying CW and expertise to the Vietnamese and Lao governments, and using the weapons themselves in Afghanistan. The charges were pursued through a series of démarches and in the UN, and were made publicly by high-level US officials. One result was the formation by the UN General Assembly of an Experts Group to investigate the alleged use; however, it was unable to gain access to the sites of the alleged attacks, and its report was inconclusive.[98]

The Allegations

The charges were detailed and voluminous. In essence, the US charged that the Soviet Union was systematically using a range of harassing, incapacitating, and lethal CW and BW against *mujahideen* in Afghanistan, and that it was supervising the use of the same (Soviet-provided) agents by the Vietnamese in Kampuchea, and by Vietnamese-advised Laotians in Laos.

The reported symptoms were numerous and occurred in a bewildering variety of combinations: itching, blistering, eye irritation, nausea and vomiting, diarrhea, gastrointestinal hemorrhage, respiratory difficulty, headache, vertigo, rapid unconsciousness, fatigue, paralysis, and others. Many deaths were alleged. Because no one chemical agent could explain all, or even most, of these symptoms, the US postulated that multiple chemical agents were being used, sometimes singly, sometimes in combination. The alleged agents included tear gas, phosgene oxime, a rapid-acting incapacitant of unknown identity, nerve agents, and trichothecene mycotoxins. The last is our concern here, as mycotoxins, when used as weapons, are considered both biological and chemical weapons.[99]

The trichothecenes are a family of toxins produced by fungi, principally *Fusarium* (figure 13.1). Ingested doses in the milligram range can cause

immediate onset of nausea, vomiting, diarrhea, mental confusion, hypotension, and other symptoms. They cause skin blistering and necrosis (reportedly more effectively on a weight basis than mustard). Gastrointestinal hemorrhage is common in severe cases, causing bloody vomit and
diarrhea, and other hemorrhages (lungs, skin) are also seen. LD_{50} values
(the dose that causes death in 50 percent of exposed individuals) for in-
gestion in animals are in the range of 1–10 milligrams per kilogram of
body weight. The adult human lethal ingested dose has been estimated,
very roughly, from natural outbreaks caused by consumption of spoiled

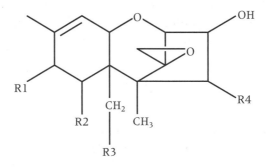

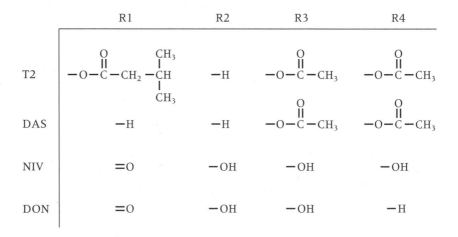

	R1	R2	R3	R4
T2	$-O-\overset{O}{\overset{\|}{C}}-CH_2-\overset{CH_3}{\underset{CH_3}{\overset{\|}{CH}}}$	$-H$	$-O-\overset{O}{\overset{\|}{C}}-CH_3$	$-O-\overset{O}{\overset{\|}{C}}-CH_3$
DAS	$-H$	$-H$	$-O-\overset{O}{\overset{\|}{C}}-CH_3$	$-O-\overset{O}{\overset{\|}{C}}-CH_3$
NIV	$=O$	$-OH$	$-OH$	$-OH$
DON	$=O$	$-OH$	$-OH$	$-H$

Figure 13.1 The major trichothecene mycotoxins: diacetoxyscirpenol
(DAS), nivalenol (NIV), and deoxynivalenol (DON). (Source: Committee on
Protection against Mycotoxins, *Protection against Trichothecene Mycotoxins*
[Washington, D.C.: National Academy Press, 1983], pp. 18–19.)

grain, to be 20 milligrams or more.[100] Trichothecenes can be absorbed through the skin, inhaled, or ingested.

Alleged delivery devices were equally varied: aerial spraying, bombing, and air-to-surface rockets in all three countries; mines and shells additionally in Kampuchea and Afghanistan; and mortars and rocket-propelled grenades as well in Kampuchea.

The agents were also described variously: green, yellow, red, black, or white smoke; powder or gas; or, most commonly, a rain of sticky yellow droplets. This last description gave rise to the term "yellow rain," by which the allegations came to be known.

In total, in Laos, Kampuchea, and Afghanistan combined, nearly 500 separate attacks were alleged through 1982, with about 11,000 people said to have been killed.[101] Continuing attacks, at much reduced frequency, were alleged into the 1990s in Laos.

The Evidence

The evidence was voluminous and varied. It included testimony from dozens of alleged victims and other eyewitnesses, testimony from defectors who claimed to have participated in chemical attacks, and reports of trichothecenes in environmental samples from attack sites in all three countries, and in biomedical samples taken from alleged victims.

Independent experts, however, expressed serious reservations, and their critique of the evidence has established substantial grounds to doubt its reliability. Nevertheless, the US government has never retracted the allegations, and in US governmental and military circles they are still widely believed.[102]

VICTIM ACCOUNTS A large number of firsthand accounts from alleged victims and from other eyewitnesses were accumulated between 1976 and 1983 by interviewers from the US embassy in Bangkok, a team from the State Department, and a team from the US Army. The sheer volume of this testimony has convinced some that chemical attacks of some kind took place in Southeast Asia.

However, this testimony is now considered highly unreliable. It was gathered by interviewers untrained in social science methods, through interpreters, interviewing subjects preselected for having claimed to have witnessed attacks, with no attempt at cross-checking, and no reinterviewing to check consistency of accounts, with the interest of the interviewers in chemical attacks known in advance to the interviewees, who were refugees strongly motivated to please their interviewers. Many accounts were of events months or years past, and their dating was unreliable.

Later, from November 1983 to October 1985, a joint State and Defense Departments team, with medical and CW expertise, was sent to Thailand. It conducted more-systematic attempts to confirm some witness accounts. When reinterviewed, refugees who had previously reported chemical attacks denied having witnessed any or admitted that they had only heard such accounts secondhand. Only 5 of the 217 people interviewed claimed to have suffered the constellation of symptoms that the CIA claimed to be indicative of trichothecene poisoning. On-site investigation of alleged attack sites on the Thai-Kampuchean border, complete with yellow spots, failed to confirm them as chemical attacks. Prophetically, the team wrote of one of these allegations: "the incident appears to have been caused by insects or some other natural phenomenon."[103] Medical examination of alleged victims did not confirm exposure to chemical or toxin agents.

PARTICIPANT ACCOUNTS   Some defectors from the Soviet, Vietnamese, and Lao armed forces claimed to have participated in chemical attacks, or to have been involved in logistical operations with special munitions thought to contain chemicals or toxins. However, these accounts suffer from many of the same methodological deficiencies as the victim accounts, and some of them have their credibility reduced by obvious errors. For instance, a former Soviet soldier reported that chemical attacks took place in two stages—in the first, a container was dropped, and then in a second pass a bomb was dropped at the same spot to cause two chemicals to mix, forming a lethal chemical agent.[104]

The star testimony came from a Lao pilot who claimed to have flown multiple missions against the Hmong, carrying standard high-explosive rockets and "smoke" rockets. The CIA concluded from circumstantial evidence that the latter were chemical munitions: the pilot was instructed to fly higher than usual, to keep the missions secret, not to fire "smoke" rockets near Lao forces; flights were accompanied by Vietnamese advisors; and the pilot was examined medically after each mission and received extra pay. This testimony was obtained by interviewers from the US embassy who were not themselves pilots, and they requested that Washington send a pilot to conduct further interviews. This was done, and the US Air Force pilot who conducted the interviews concluded, on the basis of technical information about arming the rockets, targeting practices, and the like, that "it seems very unlikely that the rockets fired contained toxic [illegible word] CW agents."[105]

Reports of Trichothecene Detection

The centerpiece of the US allegations was the claim to have detected trichothecene mycotoxins[106] in environmental samples taken from areas of alleged attack, and in biomedical samples from alleged victims. However, this evidence, too, is badly compromised by procedural flaws. Provenance and handling of the samples was undocumented, as they were collected by refugees or guerrillas, and only much later handed over to US personnel.

More seriously, the presence of trichothecenes could not be replicated. Most of the positive reports for environmental samples (four, with trichothecene concentrations ranging from a few parts per billion, to 150 parts per million) were obtained by a University of Minnesota laboratory commissioned to do the tests, as the Army's Chemical Systems Laboratory lacked the capability until October 1982.[107] When the Army's analytical capability was developed, several of the environmental samples that had tested positive were retested, with negative results. Using gas chromatography and high-resolution mass spectrometry, (state of the art for detecting trichothecenes), coupled with scrupulous controls against contamination, the Chemical Systems Laboratory tested over 250 samples from alleged attack sites in Southeast Asia without finding toxin in any of them.[108] Similar failures were later reported by British, Swedish, Australian, and French laboratories. No positive test for trichothecenes from Southeast Asia has been independently validated, and several have been negated. Almost certainly the early results were false positives, disclosed prematurely before confirmation.

Not only were the positive results unable to be replicated, but the US never published the results of control samples and blanks, although the US asserted that control samples were negative. It is thus not clear whether testing in the Minnesota laboratory included suitable blanks (samples known not to contain trichothecenes, to control for false-positive results) or matched control samples (environmental and biomedical samples taken from locations and people not exposed to alleged attacks, but otherwise similar or identical). Without this information, it is not possible to evaluate the reported positive results.

Even more embarrassing, samples of yellow rain—dried yellow spots 2–6 millimeters in diameter on samples of vegetation or rocks—were shown to be honeybee feces, consisting largely of digested pollen residue. Many honeybees, including ones common in Southeast Asia, conduct "purging flights" in which they periodically leave the hive *en masse* and defecate collectively. The purging flights are conducted up to several hundred feet, where the swarm is usually invisible and inaudible, and the result is a shower of sticky yellow or brownish droplets that can cover an area of thousands of square meters. with yellow spots reaching densities of several hundred per square meter.[109]

Biomedical samples reported by the Minnesota laboratory as containing trichothecenes (29 out of 80 samples from alleged victims) may also have been false positives. It is impossible to know, in the absence of complete information about controls. However, even if the biomedical results were credible, their significance would be unclear. Most positive samples had been taken many weeks or even months after the alleged attack. Animal studies show that trichothecenes are cleared from the body rapidly, and there is no reason to believe humans to be significantly different. That toxins were claimed after such an interval should have suggested recent dietary exposure as a more likely explanation. In the most thoroughly studied case, tissue samples from an autopsy of a Democratic Kampuchea soldier, who died a month after an alleged attack, were reported positive for trichothecenes, and this finding was offered as evidence of weapons use. However, very high concentrations of another group of mycotoxins—aflatoxins—were also reported, indicative of dietary mycotoxin sources. Of the five tissues analyzed, the greatest trichothecene concentration was reported in stomach tissue, again suggesting a dietary source. No effort was made to test the victim's food or messmates.[110]

A later serological study by a Canadian team took 270 biomedical samples in Thailand and analyzed them for trichothecenes, reporting toxin in 5 people with no exposure to alleged chemical attacks. British government scientists reported trichothecenes in the environment and in stored foodstuff in Thailand. Thus there is no reason to doubt that trichothecenes can be found naturally in Southeast Asia, and that trichothecene poisoning may be a seasonally significant medical problem in refugee camps, combat zones, and other disrupted areas.[111]

There also seems to have been insufficient effort to consider infectious disease or deficiency disorders as alternative explanations for symptoms attributed to trichothecenes. For instance, gastrointestinal hemorrhage can be caused by bacterial necrotic enteritis, a disease mainly of children but occasionally of adults, which has been reported in Southeast Asia since 1967. An outbreak in a Thai refugee camp in 1982 sickened 62 Khmer (Cambodian) children, of whom 36 died. Bloody diarrhea, one of the symptoms that the US attributed to trichothecenes, was nearly universal among victims; vomiting was common, and bloody vomit (another alleged symptom of trichothecene poisoning) was present in about 10 percent.[112]

Conclusion

The evidence for toxin weapon use by the USSR and client states was voluminous but unreliable—and masses of poor evidence do not add up to good evidence. Much eyewitness testimony appears to be badly contaminated by interviewer biases and expectations, and by cultural and linguistic ambiguities. No physical evidence from attack sites of chemical agents other than trichothecenes was reported, and the trichothecene results could not be replicated, an outcome that negates their significance. No delivery devices were recovered. No medical evidence unambiguously supports the claims. The trichothecenes detected in biomedical samples may have also been false positives; if not, they could readily have been the result of dietary intoxication, not toxin warfare. Infections may have caused symptoms attributed to trichothecene intoxication.

Clearly, the evidence for toxin warfare in Southeast Asia or Afghanistan, or even for chemical attacks more generally, is insufficiently robust to sustain such serious charges. Nevertheless, the absence of definitive evidence is not proof that attacks did not take place, and many analysts have been persuaded by the great volumes of firsthand testimony from Hmong refugees that at least in Laos there was systematic use of CW, and some remain persuaded that mycotoxins were among the agents used. A large volume of material has recently been declassified by the US government, and there will be reevaluations of the yellow rain allegations in the near future,[113] but it is unlikely that any of this will be definitive. It is also unlikely that further scientific work could retrospectively address remaining suspicions of chemical or toxin warfare in Southeast Asia, although better understanding of the ecology of trichothecene-producing fungi in the region would be helpful in evaluating alternatives. Final resolution will come only with the opening of Russian, Vietnamese, and Laotian archives. Until then, the lack of credible evidence forces us to reject the allegations.

Sverdlovsk

In April 1979 there was a large outbreak of anthrax in the Soviet city of Sverdlovsk (formerly, and now again, Ekaterinburg). Initial speculations were that perhaps as many as 1,000 people died, but later investigations showed that the number was probably under 100. The outbreak was first mentioned in the Russian émigré press in Germany, but shortly thereafter the US charged publicly that the outbreak was the result of an accidental release of an anthrax aerosol from a military microbiology facility (Compound 19) in southern Sverdlovsk. The evidence that convinced US analysts was a combination of long-standing suspicions that the Sverdlovsk facility was a component of a covert BW program (see Chapter 6), combined with human intelligence from Soviet citizens, the most important of whom was a physician in Sverdlovsk.[114] Although this is not technically an allegation of BW use, it is an allegation that the outbreak was the result of illegal BW activity, and hence we consider it here.

The Soviet Union responded that the outbreak was of the intestinal form of the disease, with some cutaneous cases, caused by eating contaminated meat from animals slaughtered in the late stages of anthrax. It was said to have followed a veterinary outbreak caused by contaminated bonemeal used in feed. The outbreak was described in the scientific literature as gastrointestinal disease.[115] Neither country would budge from its position, and the accusations became quite divisive, especially since they came on the eve of the First Review Conference of the BWC, and were still clouding perceptions at the Second.[116]

Major political changes a few years thereafter led to the collapse of the Soviet Union, and a window of opportunity opened for independent investigation of the charges. A team of Western scientists and social scientists, led by Matthew Meselson of Harvard University, was allowed to visit Sverdlovsk and to interview surviving family members, physicians, and public health authorities. The team made two trips to Russia, and was in the end able to provide convincing evidence that the allegations were in fact correct, and that the outbreak was indeed inhalational anthrax, due to an accidental release from Compound 19.[117]

There were two principal lines of evidence supporting this conclusion. The first was a careful reexamination of preserved pathology specimens retained from the outbreak, done in collaboration with the Russian pathologists who had performed the original autopsies of the victims, and who had concealed from KGB confiscation a large number of tissue samples and detailed notes from a total of 42 autopsies. These clearly indicated inhalational anthrax, not the gastrointestinal or cutaneous forms.[118]

The second line of evidence was epidemiological. Sixty-six fatal cases of inhalational anthrax were documented, with 9 survivors, and 2 cases of cutaneous anthrax. Through a laborious process of interviewing surviving family members and reviewing documentary sources, the team was able to identify the probable daytime and nighttime locations of 66 of the victims. Of these, 61 were in a narrow strip extending southeast of Compound 19 on the afternoon of 2 April (four days before the first case was recorded). Fifty-two victims were routinely there during the day; another 5, Army reservists, had that day started a training course at a military facility just south of Compound 19; and one was working temporarily in the area. Of the remaining 8 victims, 3 had occupations (such as truck driver) that could well have taken them through the same zone during the day; and 3 lived there, although they worked elsewhere. Thus 64 of 66 victims were either certainly or plausibly in a narrow corridor southeast of Compound 19 on 2 April. This is a distribution that clearly cannot occur by chance, and convincingly argues for aerosol exposure. Meteorological data confirm that the prevailing winds were from the northwest for all of 2 April, confirming an aerosol exposure on the afternoon of 2 April.[119]

This conclusion is further confirmed by the location of animal anthrax cases. The animal cases had originally been used as evidence in support of the intestinal anthrax scenario, but careful investigation showed that they occurred at the same time as, not before, the human outbreak. In striking confirmation of aerosol exposure, all lay on an extension of the same axis along which the human cases were found (figure 13.2).

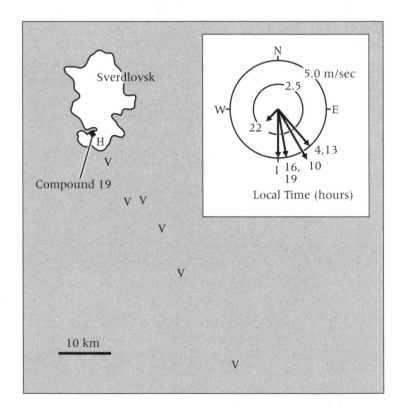

Figure 13.2 Location of human (H) and veterinary (V) cases of anthrax in the Sverdlovsk area in April–May 1979. Inset: direction and strength of winds at Sverdlovsk airport on 2 April 1979. (Source: Matthew Meselson et al., "The Sverdlovsk Anthrax Outbreak of 1979," *Science*, 266 [1994], 1202–08.)

The event that caused the aerosol release remains obscure. A high-level defector from the Soviet BW program has stated that Compound 19 was a major production facility for the USSR's BW program, producing dried anthrax spores in large amounts. He was told that the accident was the result of an inadvertent removal of an exhaust filter during a production run. However, he ascribed this event to a different date, so the reliability of his sources is questionable.[120]

It is also unclear how much material was released. Estimates by the CIA and DIA assumed that "large quantities" must have been released in order to cause such an outbreak, but this inference seems to have been based more on intuition than on science. Estimates based on different aerobiological models and different assumptions about infectious dose gave very different results, ranging from 4 to 600 milligrams.[121]

The responsibility of the Soviet government for the outbreak was admitted by Boris Yeltsin, who said that the military had caused the outbreak. However, many Russian military personnel still deny that the event was an accidental release, and instead cling to the infected-meat scenario, or allege US sabotage.[122]

Conclusion

Allegations of BW use are very serious matters, asserting as they do that the accused has violated international law in using weapons banned by international treaty and by customary law. Allegations of violation of the BWC short of use are also serious matters. Both call into question the effectiveness of international norms and legal prohibitions, and they reduce the trust upon which compliance with these norms and laws partially depends. And so long as allegations remain neither verified nor disproven, they suggest that covert biowarfare may be waged without significant political cost.

Given this context, nations have a responsibility not to make such allegations without a high level of certainty that they are correct, and without adequate evidence to present to the international community. We believe that all the accusing states we have considered here fall short, to varying degrees, of that standard of responsible behavior.

The historical evidence suggests to us that all the accusers believed that their accusations were correct. Even the Korean War allegations, long dismissed as mere propaganda, seem understandable: China knew that the US had debriefed Japanese BW war criminals but denied this to the international community, Chinese intelligence had identified all the elements of a Japanese-style BW program at forward locations, the US was publicly hinting that it was going to use BW in Korea, and evidence suggestive of such attacks started to accumulate. In hindsight the allegations were incorrect, but in context they were understandable.

Cuba as well, whose allegations have little evidence to support them, may well be convinced, from a long history of US-sponsored illegal attacks, that BW use has been among the dirty tricks employed. This belief is especially plausible for the earlier allegations, stimulated by US congressional hearings that revealed US efforts to assassinate Castro, and CIA retention of BW agents in violation of executive orders. However, it is possible that more recent allegations reflect more habit than conviction.

While the accusing states appear to have been convinced that their accusations were correct, none of them did an adequate job of assembling evidence. There are two natural groups of accusers in this regard: Cuba, and the US in the Sverdlovsk case, presented little or no evidence; the Chinese, and the US in the yellow rain case, presented flawed evidence. Cuba has claimed to have evidence of US BW attacks, but has published none of it except for evidence in support of accusations of the introduction of the insect pest *Thrips palmi*, which was highly circumstantial and unconvincing. For the most plausible accusations—those of 1962 and 1971 attacks on domestic animals—no effort has been made to produce solid evidence. The US, in the Sverdlovsk case, claimed to have classified evidence to support its case, but did not publish it. With the later declassification of much of it, the evidence appears to have been startlingly scant and ambiguous. In failing to produce scrupulously collected, rigorously analyzed evidence to support their serious charges, these two countries acted irresponsibly.

The Chinese, and the US in the yellow rain case, presented voluminous eyewitness testimony, confessions from participants, and isolation of agent from environmental and biomedical samples. Both accused countries had serious, large-scale BW and CW programs. Both accusing countries assembled hundreds of pages of detailed scientific documentation of their cases. However, in both cases the evidence appears to have been misleading, with natural phenomena misattributed to deliberate human action. In these two instances, the accusing countries made a serious attempt to meet their responsibilities, but failed because of serious problems with the provenance of evidence, with investigator bias, and with ambiguities about what might be natural. These were foreseeable problems, and could have been anticipated and coped with.

Clearly, extraordinary care should be taken when alleging BW use, as evidence sufficient to convince the predisposed may be routinely available to the (consciously or unconsciously) selective data gatherer. Scrupulous attention to best scientific practice and to forensic standards in the gathering and testing of evidence is necessary to sustain serious charges. It is one thing to collect evidence that confirms prior convictions, but quite another to collect evidence sufficient to convince skeptics.

Mechanisms of investigation of disease outbreaks that will allow a robust determination of the origin of suspicious outbreaks should be an important priority for the international community, as the primary stakeholder in resolving allegations. Whether such mechanisms are established via binding international agreements or through less formal arrangements matters less than their effectiveness. The technical capacity to investigate outbreaks so as to clarify their origin has existed for a number of years now,[123] so serious attention to a workable mechanism for applying this capability should be productive.

One possibility would be to use the existing authority of the secretary general of the UN to assemble lists of experts and competent laboratories and to mount investigations of allegations of use of CW or BW.[124] Unfortunately, the few attempts to invoke this authorization have been failures, as UN teams have been denied necessary access.[125] To be effective, this mechanism would need serious political commitment by the UN community to insist on rights of access of investigation teams sponsored by the secretary general. Alternatively, the Security Council could establish a permanent body with responsibility for preventing BW development complementary to the International Atomic Energy Agency and the Organization for the Prevention of Chemical Weapons.

For concerned states that are parties to the BWC, it would be possible to use Article V of the Convention (see the appendix) as a vehicle for cooperative efforts to resolve allegations. Not only would Article V provide a context for international consultations, but using it would reinvigorate this consultative provision of the BWC which has been so rarely invoked,[126] yet which is centrally important to a treaty that lacks an effective compliance regime.

Article V could also be used for retrospective resolution of allegations. In the case of the Korean War allegations, there are only a few specific outbreaks in which the allegations matched US capabilities for small-scale, covert attack,[127] and scientific study of retained autopsy and other material could definitively prove or disprove the allegations. Indeed, it appears that this has already been done for one specific instance—a cluster of inhalational anthrax cases—in which the agent appears to have been indigenous, not the one used in the US BW program. Collaboration between China and the US, under the aegis of Article V, could finally lay to rest any remaining suspicions. Similar collaboration between Cuba and the US has the potential to yield decisive results. However, given the long and bitter relations between these two countries, it may be unreasonable to hope for such collaboration in the near future.

Despite the lack of credible evidence, there are lingering suspicions over the yellow rain allegations, and it is unlikely that any retrospective scientific study could resolve them. It would be of great benefit to the international community if Russia, Vietnam, and Laos (all parties to the BWC) would provide independent scholars with archival access adequate to clarify the issue.

Even in the absence of formal international arrangements, independent collaboration among scientists can be quite productive. This fact was demonstrated dramatically in the Sverdlovsk case, in which definitive retrospective resolution of the allegations was achieved by a collaboration of independent Western and Russian experts. Of course, this required governmental consent and access, but did not require the more substantial political and diplomatic commitment of a formal international or bilateral effort. Such efforts could be productively encouraged by non-governmental organizations, scientific societies, and individual scientists.

It is clear that the allegations we have considered here were stimulated in part by the secrecy that surrounds military biology programs. When accused countries had offensive BW programs, secrecy was understandable, but since 1975 offensive BW programs have been illegal for States Parties to the BWC. For any country with a defensive program only, secrecy is less obviously necessary. We believe that national and international security would be enhanced by a reduction of secrecy to the minimal level necessary to prevent vulnerabilities from being advertised, or research results from being misapplied. This would be especially important in states, like the US and Russia, with advanced military biology capabilities, a history of offensive BW programs, and a legacy of allegations of covert BW use. Such a policy could, in the long run, protect states better than continued secrecy, as it would discourage and discredit false allegations, as well as reduce the suspicions that risk stimulating BW proliferation. The confidence-building measures adopted for the BWC are available as a vehicle for greater transparency.

Overt lies have also contributed to the climate of suspicion in which such allegations are formed. The US denials that it shielded Japanese BW war criminals were known by China to be false; it is not surprising that the Chinese would disbelieve other US denials, and to believe that these lies concealed malign intentions. A similar context obtained in Cuba, when the US publicly denied its program of covert actions in Cuba, despite Cuban certainty based on capture and interrogation of agents. It is not surprising that US denials of BW use are viewed skeptically. And the Soviet duplicity in maintaining a very large, secret offensive BW program long past the entry into force of the BWC was a substantial contributor to the yellow rain and Sverdlovsk allegations.

Secrets and lies are integral parts of the ways in which all nations protect their national security. Yet they have the consequence of stimulating suspicions that can provoke reactions that reduce international security. These consequences are particularly true of BW, which can so readily be hidden from technical means of surveillance. Countries concerned about the possibility of proliferation of these weapons, and the dangers of their covert use, should consider carefully whether their long-term security would not be best served by demonstrated transparency and candor.

CHAPTER 14

Terrorist Use of
Biological Weapons

MARK WHEELIS

MASAAKI SUGISHIMA

Biological weapons (BW) might be attractive not only to states, but to criminals and to terrorists as well. The history of terrorist use of BW is, like the history of state use, scant, presumably because of the high degree of expertise required and the difficulty of obtaining necessary materials. However, these barriers may be diminishing as widespread discussion of the threat of bioterrorism makes the requisites explicit. Thus it would be a mistake to assume that the sparse record of bioterrorism to date accurately predicts the future.

There is no generally agreed definition of bioterrorism, or even of terrorism. Here we consider bioterrorism to have four key elements: (1) the deliberate use, or the threat of use, of biological agents or toxins (2) by individuals or groups (but not states) (3) against nonmilitary targets (such as civilians or agricultural targets) (4) to achieve a political, ideological, or religious goal. The intent may be to cause disease, fear of disease, or both.

The use of BW for entirely nonpolitical goals (such as revenge or profit) is not considered bioterrorism, but rather biocriminality. The extensive history of the use of toxins for criminal purposes is beyond the scope of this chapter.[1]

There have been only two confirmed attempts to use BW for terrorist purposes targeting humans: the 1984 use of *Salmonella* by the Rajneesh cult in Oregon, and the 1990–1995 attempted use of anthrax and botulinum toxin by the Aum Shinrikyo cult in Tokyo. A third incident, the alleged use of a variety of infectious diseases against Native Americans in the Amazon basin during the 1950s and 1960s, is probable but not yet firmly established. A fourth incident, the 2001 anthrax letter attacks in the US, may be bioterrorism or biocriminality, determination of which must await the identification of the perpetrator(s) and motive.

In addition to these several known incidents of BW use by terrorists, there have been thousands of hoaxes—false claims that a biological attack has been perpetrated. Some of these have been ideologically motivated (most clearly when abortion clinics were the targets) and were intended to cause fear and to intimidate; they are thus incidents of bioterrorism.

Finally, there has been a long history of interest by terrorist groups in chemical and biological weapons (CBW), but few have had the expertise necessary to gain actual possession of a bioweapon. Most of these have been individuals or small groups on the far right or the far left. More recently there has been evidence of interest in such weapons by al Qaeda. However, so far lack of expertise appears to have prevented even large, well-funded groups from implementing their interest. How long this state of affairs will continue is unclear.

There are also a handful of allegations of terrorist use of BW against plants or animals; these are mentioned in Chapter 11.

1957–1963 Attacks on Indigenous Brazilians

In 1964 the Brazilian military seized power, claiming among other things a goal of ridding the country of corruption. As part of that agenda, in 1967 Attorney General Jader Figueiredo was commissioned by General Albuquerque Lima, minister of the interior, to investigate charges of rampant corruption in the Indian Protective Service (SPI). His 5,000-page report, issued in 1968, alleged that among other things, SPI officials had collaborated with landowners to displace indigenous Native Americans from their land. The means were largely conventional violence, including aerial attack, mass murder with firearms, and other forms of terror, but they included alleged attempts to use chemical agents (arsenic-laced sugar) and to introduce a variety of diseases into indigenous communities. Two hundred SPI officials were dismissed and another 134 were indicted for the alleged crimes, but whether they went to trial, and if so what the outcomes were, is not clear.[2] Apparently some received administrative discipline, but none appear to have been convicted of crimes. Ultimately the SPI was disbanded and a new organization (the National Indian Foundation, or FUNAI) formed, integrating the functions of the SPI and several other departments.

By exterminating the indigenes, landowners hoped to free the land for sale. Most of the attacks were on tribes living in the Matto Grosso, a vast savanna and scrubland south of the Amazon River, which was prized grazing land. There were also attacks alleged in the more heavily forested northern slope of the Amazon basin. The alleged attacks involved the introduction of smallpox, measles, influenza, and tuberculosis into tribes of the Matto Grosso from 1957 to 1963, and tuberculosis into tribes of the northern part of the Amazonian basin in 1964–65. Smallpox was alleged to have been introduced by variolation (the deliberate inoculation with smallpox virus): "a doctor—now alleged to have been sent by the Indian Protection Service of those days—instead of vaccinating them, inoculated them with the virus of smallpox. This operation was totally successful in its aim, and the vacant land was immediately absorbed into the neighbouring white estates."[3] Tuberculosis, and probably the other diseases as well, were said to have been introduced by deliberate exposure of Native Americans to outsiders with active, symptomatic disease. The results of these attempts are not documented, but are claimed to have been devastating.

The motivation for these alleged attacks was clearly greed, and they are thus examples of biocriminality. We include them here as bioterrorism because of their overtly and explicitly genocidal intent. Of course disease has for centuries been responsible for enormous mortality in the indigenes of the New World and the islands of the Pacific and Indian Oceans, with massive genocidal effect. However, most of this mortality was the result of inadvertent disease introductions; only rarely were there deliberate attempts to use disease as a weapon.[4] The Brazilian case is the only instance recorded in the 20th century, and is thus of considerable importance. Unfortunately, it has yet to receive the scholarly study that it deserves.

The 1984 Rajneesh Attacks

The only bioterrorist attack known to have caused an actual outbreak of disease was perpetrated by a religious commune in the US in 1984. In contrast to the current focus on bioterrorism, involving aerosols of lethal agents causing mass fatalities, these attacks used food contamination at restaurant salad bars, with an agent of low lethality but capable of causing mass morbidity. The success of this technologically undemanding method of attack is worth noting, especially as some lethal agents could be disseminated in the same fashion.

Interestingly, the outbreak was not recognized as a deliberately instigated event, although there was ample evidence to support such a conclusion. Only later was it discovered to have been an attack.

The Outbreak

In the fall of 1984 there was an outbreak of restaurant-acquired salmonellosis in The Dalles, Oregon, the county seat of rural Wasco County. With 751 identified cases, it was the largest outbreak of food-borne disease in Oregon's history, and the largest in the US for 1984.[5] However, it undoubtedly afflicted many more than the Centers for Disease Control and Prevention (CDC) reported. The Dalles is on a major highway out of Portland, a heavily used scenic tourist route along the Columbia River. The town, with a population of only about 10,000, had almost 40 restaurants in 1984; clearly, most meals were served to travelers, and thus most cases must have occurred in travelers. The geographic dispersal of restaurant patrons, coupled with the fact that most cases of salmonellosis are never detected—as few as 1 percent of cases may be reported[6]—makes it certain that a large number, perhaps thousands, of cases among travelers were missed.

Salmonellosis is normally a brief illness, lasting a few days. It can present a variety of symptoms, but is usually associated with diarrhea and fever, often preceded by vomiting. Symptoms can range from barely detectable to fatal. Transient, asymptomatic infection occurs. In the outbreak in The Dalles, 45 of the 751 known victims were sick enough to be hospitalized, but none died.

The Wasco-Sherman Public Health Department began receiving physician reports of gastrointestinal illness on 17 September[7] and rapidly associated them with consumption of meals in one of two restaurants. By 25 September the Oregon Health Division and local authorities had identified hundreds of cases, the number of restaurants involved had increased substantially, and salad bars had been implicated as a major source of infection. On that day, state health officials ordered all salad bars in the city closed and requested assistance from the CDC. The epidemic ended with the last case on 10 October.

The outbreak was caused by a strain of *Salmonella enteritica* serotype typhimurium that has rarely been implicated in outbreaks. However, it caused two (and possibly three) other outbreaks in Oregon in late 1984, so its occurrence in The Dalles was relatively unremarkable.

The outbreak occurred in two waves (figure 14.1). Thirteen percent of the cases were in the first wave, from 9 to 18 September (peaking on 15 September), and were linked to only 2 restaurants. The second wave, from 22 September to 10 October (peaking on 24 September), produced 87 percent of the cases and was associated with 10 to 22 restaurants (in-

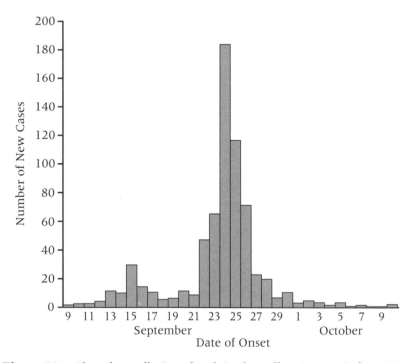

Figure 14.1 The salmonellosis outbreak in The Dalles, Oregon, in late 1984. (Source: CDC, "Restaurant-Associated *Salmonella typhimurium* Gastroenteritis, The Dalles, Oregon, 1984," limited-distribution report, EPI-84-93-2, 1 November 1988.)

cluding both first-wave restaurants).[8] Only 11 cases resulting from secondary transmission were confirmed.

The two waves of infection appear to have been the result of at least two (probably more) separate series of deliberate contamination of foods at salad bars: the first at 2 restaurants around 8 September, and the other at 10 or more restaurants around 21 September. New cases continued to appear for a week or two afterward, probably the result of multiple factors: variable incubation periods (normally 6–48 hours, but sometimes a bit longer); a common practice of reusing foods from salad bars over several days rather than discarding them at closing time; insufficiently low temperatures to prevent bacterial proliferation, allowing fresh food mixed with a small amount of old to develop high levels of contamination; and some secondary transmission. However, continuing incidents of deliberate contamination may have been responsible for some of the continuing illness, as suggested by detailed analysis of cases associated with individual restaurants.

Although restaurant-acquired salmonellosis is common, it is rarely associated with salad bars. Furthermore, this outbreak had a number of anomalous features that should have suggested deliberate contamination as the likely source. These included:

- In only one of the 10 restaurants definitely identified as sources of the *Salmonella* did employees become ill before customers (in contrast to the usual pattern, in which employee infection is the source of customer infection).
- Two of the involved restaurants served a total of 20 private banquets during the outbreak, to nearly 900 participants, none of whom became ill. The banquet food was prepared in the same kitchens by the same people as material for the public rooms, but banquet rooms contained dedicated salad bars.
- Illness, though associated in most instances with salad bars, was associated with a range of food items at each restaurant.
- No source of any food item was common to all affected restaurants, and no food distributor serviced more than 4 affected restaurants.
- Affected restaurant employees had little social contact with one another, but most had eaten at their own restaurant's salad bar or at that of another affected restaurant.

- Customer cases declined dramatically when salad bars were closed, despite the fact that ill employees continued to work.

These findings indicate clearly that the outbreak was due to consumption of food from salad bars, which had been contaminated after it left the kitchen. All but two of the definitely implicated restaurants had salad bars. One of the two remaining restaurants clearly had cases as a result of transmission of disease from an ill employee (infected at another restaurant), and secondary transmission is also probable for the other one.

Yet state and national health officials did not draw the obvious conclusion, and it is not clear why. Apparently the preliminary data analysis did not very clearly identify some critical findings. A preliminary report filed by the Oregon deputy state epidemiologist claimed, in contrast to later analysis, that some cases were clearly caused by contamination in the kitchen, and that in many cases employees were ill before customers.[9] The report went on to state that intentional contamination had been considered but rejected because there was no evidence for it, and concluded that infected food handlers had been the cause of customer cases. The source of their infection was speculated to be unidentified common contacts in the community.

The later CDC accounts of the outbreak, a limited-distribution internal report and the published version (both written after it became clear that the causation was deliberate), claimed that deliberate contamination was rejected largely because it didn't fit expectations of motivation and behavior: no motive was apparent, no one had claimed responsibility or issued any demands, a single attack could not explain the pattern of infections, and no disgruntled employees were identified.

It should be remembered that this event occurred before there was widespread concern over bioterrorism, and it is perhaps understandable that public health personnel were reluctant to draw such a controversial conclusion on the basis of epidemiological evidence alone. It is unlikely that there would be any such hesitancy today.

The Rajneesh

The salmonellosis outbreak occurred during an increasingly contentious stay in Wasco County of a commune organized around the East Indian guru Bagwan Shree Rajneesh. The Bagwan ("enlightened one") had built a large following in India and had recruited many followers in Europe and the US as well. In 1981, because of legal difficulties in India, he moved his group to Oregon, to the Big Muddy Ranch, straddling the Wasco/Jefferson County line.[10] They set about building a town on the site, incorporated as Rajneeshpuram, and within three years had a population of about 2,000. In so doing, they ran into restrictive county zoning laws, and relations between the commune and the local authorities rapidly became difficult. The cult was also under scrutiny by the federal government for violations of immigration laws (many of its members were foreign nationals).

The troubled relations between the Rajneesh and the county led the commune to attempt to take control of the county government by electing their own candidates to the county commission; they had already done this in the town of Antelope, adjacent to their commune. However, the 15,000 registered voters in the county greatly outnumbered the commune members. The Rajneesh responded to this obstacle with an initiative to bring thousands of homeless people to the ranch from cities all over the US and register them to vote.

In anticipation of a very close vote, the Rajneesh also had the idea of making potentially opposed voters so ill that they would stay away from the polls on election day. The biological attacks on The Dalles were apparently trial runs of this scheme. Despite the success of the trials, the scheme was not implemented during the November elections. The homeless recruitment fell far behind target, and it became clear that favorable election results were out of reach; the Rajneesh abandoned their campaign and boycotted the elections.

The plan to use BW originated with the sect's de facto administrator, Sheela, personal secretary to the Bagwan.[11] Sheela worked closely with Puja, a registered nurse who was head of the health center at Rajneeshpuram.[12] Fewer than a dozen people, all senior members of the Rajneesh insiders, participated in planning the scheme. The Bagwan appears not to have been involved.

A secret laboratory was established in an out-of-the-way spot on the commune. Puja obtained the *Salmonella* by using her credentials as head of the Rajneesh Medical Corporation. Three people cultivated the bacteria, and about a half-dozen disseminated the cultures. Only small-scale culturing was required, using ordinary laboratory glassware and media.

The cultures were first used against individuals. On a routine visit to Rajneeshpuram by the three county commissioners on 29 August 1984, two of them were given glasses of water laced with *Salmonella* (the third was given pure water, perhaps because he was less hostile to the Rajneesh, or perhaps to deflect suspicion). Both men became ill, one of them requiring hospitalization.

There were several attempts to infect large numbers of community members. In addition to the spectacularly successful September attacks on restaurants in The Dalles, *Salmonella* cultures were used on several occasions without apparent effect. They were apparently dumped into one of the water systems in The Dalles, spread on produce in a local supermarket, and spread on doorknobs and urinal handles in the county courthouse. Only the restaurant contamination proved effective.[13]

A few months after the election, the commune self-destructed. Sheela and Puja had attempted to poison the Bagwan's personal physician with arsenic. A large number of lawsuits were about to be lost, and Rajneeshpuram's incorporation was being challenged. It looked as though a number of civil and criminal warrants would be issued imminently. Sheela, Puja, and several others among the top leadership fled to Europe in September 1985. These events brought teams of state and federal agents to investigate the commune, during which the *Salmonella* culture was found and matched to the outbreak. Several participants in the food poisoning program described their activities, confirming the deliberate nature of the outbreak.

Puja and Sheela returned to the US and agreed to a plea bargain in which Sheela got concurrent sentences of 20 years for attempted murder (the arsenic poisoning), 20 years for first degree assault (the *Salmonella* poisoning of one of the county commissioners), 10 years for second-degree assault (*Salmonella* poisoning of the other county commissioner), 4.5 years for product tampering (the *Salmonella* poisonings in The Dalles), and several other sentences. Puja received slightly lighter sentences for the same charges. In the end, both women served only 2.5 years in a minimum-security federal prison and were released early for good behavior.[14] Such light sentences for such a major biological attack and such early releases would almost certainly not occur now, given the widespread concern about bioterrorism, and given the expanded federal criminal laws addressing possession and use of biological agents.[15]

The 1990–1995 Aum Shinrikyo Attacks

The only other known case of attempted mass-casualty bioterrorism occurred in Japan over several years. Like the Rajneesh attacks, it was perpetrated by a religious cult. However, unlike the Rajneesh, the Aum elected to try aerosol dissemination of lethal agents. Despite repeated attempts, they were unsuccessful in causing any disease, and in retrospect it is clear that they did not even make the first substantive step toward an effective bioweapon.

Rise of the Aum

Chizuo Matsumoto was born in 1955 in Kongo, a village in Kumamoto Prefecture. He suffered from congenital glaucoma and went to a school for the blind, eventually obtaining his license for acupuncture. In 1984 he established a small yoga school, Aum No Kai ("Circle of Aum"; *Aum* is a Sanskrit syllable taken to symbolize the creation, sustenance, and destruction of the world), and began to call himself Shoko Asahara (no specific meaning).[16] Two years later he declared himself the only person in Japan to have achieved the highest level of enlightenment, and he reorganized his yoga circle into a religious group. The name of the group was changed to Aum Shinrikyo ("Religion of the Supreme Truth of Aum") in 1987, and it obtained official religious corporation status from the Tokyo metropolitan government in 1989.

Asahara became famous among young Japanese interested in the supernatural when a magazine featured a photo of him levitating. The Aum's membership increased rapidly, to more than 15,000 in Japan (plus many thousands of foreign members), at the time of the sarin incident in Tokyo. Asahara designed a step-by-step curriculum for developing supernatural power and told his followers that they would invariably succeed if they sincerely followed the training course.

Asahara's doctrine was based mainly on Tibetan Buddhism and Yoga, but other influences, such as the book of Revelation and the prophecies of Nostradamus, also helped shape his apocalyptic view. Shiva, the Hindu god of destruction, was worshipped. Asahara told followers that in order to attain emancipation, they should be empty vessels to be filled with the will of the guru.

The Aum emphasized salvation, a notion found in several Japanese Buddhist sects, in order to reduce the number of victims of an imminent Armageddon, the divine purification of bad karma accumulated by humans. Asahara conceived of two strategies for facilitating widespread salvation. One was to form a majority in the Diet (parliament) and take control of the Japanese political system. The Aum's practices could then be implemented nationally. The second strategy was to destroy Japan by forceful means for the purpose of purification. Asahara had argued for the morality of salvation-promoting destructive acts since at least 1989: "If there is a man accumulating bad karma, transforming his life to help his soul reincarnate in a higher level is a righteous act."[17]

Asahara attempted the legitimate route first, and established his political party (Shinrito) to run for the lower Diet election campaign in February 1990. Confident of victory, he was shocked when all 25 Shinrito candidates, including himself, were defeated. He blamed a conspiracy of the Liberal Democratic Party and the Japanese government for the defeat. Soon after the election campaign, Asahara secretly told senior followers that in the light of the desperate results of the election campaign, it would be no longer be possible to save Japan through legitimate means, and announced the campaign for salvation through force.

Bioterrorism Plots of the Aum

The Aum was interested in a variety of weapons to implement its apocalyptic plans, but the only progress it made involved CBW.[18] The Aum's use of CW in Matsumoto and Tokyo is well known; its extensive efforts to develop BW is less familiar, because all the attempts failed. Efforts were made to develop two agents as BW (botulinum toxin and *Bacillus anthracis*), and there were numerous attempts to disseminate them (table 14.1).

BOTULINUM TOXIN   Asahara first became interested in botulinum toxin as a weapon, having learned about its extreme toxicity from Tomomasa Nakagawa, a physician and senior follower of the Aum. Asahara named Sei-ichi Endo leader of the Aum's biological and toxin weapons program. Endo was a microbiologist who had dropped his doctoral work at the Institute of Virus Research of Kyoto University to follow

Table 14.1 Bioterroism plots of the Aum

Time	Place	Modality
April 1990	Throughout Japan	Plot to disseminate botulinum toxin from balloons and ships; not implemented because of inability to culture the organism in time
May 1990 (several attempts)	US naval base at Yokosuka, the Imperial Palace, headquaters of Soka Gakkai Institute, and governmental buildings in Kasumigaseki	Disseminated putative botulinum toxin
July 1990	Unidentified water purification plant	Tried to contaminate water supply with putative botulinum toxin; discovered by police guarding the plant, and substance confiscated
May 1993	Wedding parade of Prince Hironomiya	Plot to disseminate *B. anthracis;* not implemented because of inability to culture the organism in time
End of June–early July 1993	Aum's new headquarters building in Kameido, Tokyo	Disseminated large volume of *B. anthracis* culture from rooftop; strain was avirulent, spore concentration low, and droplet size large
Summer of 1993	Around the Kanagawa prefectural office, Imperial Palace, etc.	Disseminated *B. anthracis* from customized vehicles; strain was avirulent, and sprayer nozzle clogged
November 1993	Headquarters of Soka Gakkai Institute, Tokyo	Disseminated putative botulinum toxin from customized vehicle; sprayer nozzle clogged
June 1994		Plan to equip a helicopter purchased from Russia with sprayers for biological agents; not implemented
March 1995	Kasumigaseki Station, Tokyo	Disseminated putative botulinum toxin from three sprayers hidden in customized attaché cases

Source: Masaaki Sugishima, "Biocrimes in Japan," in *A Comprehensive Study on Bioterrorism,* ed. Sugishima (Legal Research Institute of Asahi University, 2003), p. 98.

Asahara. Around March 1990 the Aum built a laboratory of about 200 square meters in their Kamiku-isshiki compound at the base of Mount Fuji. After the Tokyo sarin incident, police raided the laboratory and confiscated the equipment. It included a glove box, incubator, centrifuge, drier, DNA/RNA synthesizer, electron microscope, two fermenters each having about a 2,000-liter capacity, and an extensive scientific library.[19]

Endo faced a daunting task. He first had to obtain a strain of *Clostridium botulinum* that produced the form of the toxin that is deadly to humans,[20] and second, he had to develop methods of producing and weaponizing toxin from this strain. This involves growing the organism under anaerobic conditions (*Clostridium* will not grow in the presence of air), extracting and concentrating the toxin from culture supernatants, preparing it in a form suitable for dissemination, developing a mechanism of dispersal, and field-testing it to verify its effectiveness.

Endo attempted to isolate *C. botulinum* from soil samples collected in Hokkaido, but there is no evidence that he was successful. Where the strain that was ultimately used came from is unknown, as are its identity and toxigenicity. The fact that multiple dissemination events led to no casualties suggests that the Aum was using an avirulent strain or was unable to cultivate *C. botulinum* properly.

The second obstacle was even greater. Several nations had developed botulinum toxin as a weapon, but their techniques were not publicly available, and the Aum had to produce bioweapons from scratch. It seems doubtful that the group ever really seriously grappled with the problems. Given the authoritarian structure of the Aum, it is likely that members found it easier to please Asahara in the present, with agreement and optimism, than to raise substantive problems. Nakagawa told another senior follower that the substance his team produced had proved ineffective in mice.

An ex-follower who participated in the production of botulinum toxin in the spring of 1990 recalled that after the agent was cultured, the culture liquid was dried and the residue powdered. Six followers with protective suits worked in the laboratory.[21]

In March 1990 Fumihiro Joyu, who steered the external affairs of the Aum, suggested disseminating botulinum toxin from balloons, an idea probably derived from the World War II Japanese balloon-bomb program.[22] Asahara approved the plan and instructed followers to commence the operation in mid-April. However, the plan proved too ambitious and was soon abandoned.

After this failure the Aum attempted to disseminate botulinum toxin from vehicles equipped with a spray device, targeting the US naval base at Yokosuka, the Imperial Palace, and government buildings. Since the agent was harmless, there is no way of knowing if the dissemination device was effective.

In July 1990 the Aum tried to contaminate a water purification plant with botulinum toxin. The plan was interrupted by police guarding the plant, who confiscated the substance but did not prosecute the perpetrators.

Finally, about a week before the Tokyo sarin incident in 1995, the Aum tried to disseminate botulinum toxin from attaché cases with built-in spray devices. Three such cases were found abandoned at the Kasumigaseki subway station in Tokyo, but nothing harmful was found in them. Endo testified at Asahara's trial that he had been unable to produce botulinum toxin and had filled the devices with harmless liquid.[23]

ANTHRAX   In the summer of 1992 Asahara asked Endo to recommend another agent for the Aum's BW program, and *Bacillus anthracis* was chosen. Once again Endo and his colleagues were faced with obtaining a virulent pathogen, weaponizing it, and disseminating it. Evidence suggests that they failed at all three tasks.

Retrospective analysis showed that the Aum had successfully obtained *B. anthracis*, but that it was the avirulent veterinary vaccine ("Sterne") strain.[24] The source is not known, but as Endo had conducted research at the graduate level at Obihiro University of Agriculture and Veterinary Medicine, he might have known where to look.[25] It is also not clear whether Endo knew that his strain was avirulent; if he did, he presumably withheld this fact from Asahara.

The first intended target was the wedding parade of Prince Hironomiya; however, the production program could not provide cultures in time, and the plan was cancelled. Then, from 29 June to 2 July 1993 the Aum conducted its most ambitious BW operation, aimed at causing massive numbers of deaths in Tokyo. Unknown, but probably very large, quantities of a liquid aerosol (most likely crude culture, unprocessed in any way) of *B. anthracis* was sprayed from the roof of the Aum's headquarters building in Koto Ward, Tokyo. Joyu, who led the headquarters at the time, supervised the operation. There were scores of complaints from local residents about foul odors, and Asahara convened a press conference near the building to explain that the odors were generated during the group's religious rituals. However, the complaints became so numerous, and the police attention so troubling, that the Aum ceased the spraying, removed all equipment from the building, and vacated the site.

For dissemination in this operation, the Aum set up two sprayers on the roof of the eight-story building, each within a large round cooling tower. Pipes were extended from the cooling tower to tanks below, filled with a liquid suspension of *B. anthracis*. The device worked poorly, producing large droplets rather than the very fine aerosol needed for effective transmission of anthrax. It also appears that the spore concentration was very low; much later analysis of a sample of liquid, retained by local authorities after their investigation of the odor complaints, detected fewer than 10^4 viable *B. anthracis* per milliliter,[26] a concentration at least five orders of magnitude below that necessary for a highly infectious wet aerosol. The apparatus was also reported to have leaked badly. Clearly, however, this was a major operation, which required a large volume of agent (probably thousands of liters) and was presumably intended to cause huge numbers of casualties. Even with the low quality of preparation and the crude aerosol, it is likely that many cases of cutaneous anthrax would have resulted had the strain been virulent.

After the failed operation in Kameido, the Aum continued production of *B. anthracis* at the Kamiku-isshiki compound. In the summer of 1993 the group tried to disseminate *B. anthracis* around the Kanagawa prefectural office and the Imperial Palace from vehicles equipped with spraying devices. According to prosecutors' statements, the nozzle of the sprayer clogged, and the operation failed.[27] This failure, and the failure of the spray device at the Tokyo headquarters, indicate that the Aum failed to develop workable agent formulations and dissemination devices, in addition to failing to obtain virulent agents.

Asahara and his followers had little expertise, and all their BW efforts failed. With CW it was different; the Aum had limited success in disseminating sarin in Matsumoto and Tokyo.[28] It was the Tokyo subway attacks with sarin that led to the group's demise. Several days after the Tokyo sarin attacks, 2,500 police raided Aum facilities and arrested more than 400 members. Ultimately close to 200 were charged with serious crimes, including more than 40 indicted for murder. To date, 12 senior Aum members, including Asahara, have been convicted and sentenced to death. The Aum was stripped of its religious status and declared bankruptcy. Nevertheless, its members have reorganized as a new group, Aleph, and may continue to pose a threat.

The 2001 US Anthrax Letter Attacks

In the fall of 2001, a series of letters containing *Bacillus anthracis* spores was mailed to media and congressional offices on the East Coast of the US. They resulted in at least 22 cases of anthrax—11 cutaneous and 11 inhalational. Five of the victims with inhalational disease died, and those who recovered suffered serious long-term disability. None of the victims with cutaneous disease died, but some suffered persistent disability. Thousands of people were put on prophylactic antibiotic therapy (none of whom became symptomatic), additional thousands had their lives disrupted, the US Congress shut down for several days, several federal buildings were closed for months to years because of persistent contamination, and cleanup costs, still accumulating, have been immense.

Despite the largest and most complex criminal investigation ever, the FBI has to date been unable to identify the perpetrator(s), and it is looking increasingly as though the case will never be solved. Without knowing the identity and motivation of the perpetrator(s), it is not possible to determine if these attacks were terrorist or criminal in nature, despite the nearly universal application of the bioterrorist label to them. We do not treat them in detail here, although in our conclusions we will draw some lessons from them for the evaluation of the bioterrorist threat.

The anthrax spore preparations in the last two of the mailings, out of probably five letters total, were of extraordinarily high quality[29] (earlier spore preparations appear to have been less sophisticated). They contained about 10^{12} spores per gram, near the theoretical maximum, given the approximately picogram weight of an individual spore. The particle size was quite uniform, about 3 micrometers (each particle consisting of a cluster of about two to six individual spores), and the preparation may have been treated with silica to prevent electrostatic clumping. This is a highly sophisticated method of spore preparation, widely thought to be beyond the capabilities of anyone not specifically trained to prepare it. Very few people in the world have such training; most of them are current or past Ph.D.-level employees of the US or Soviet BW programs or biodefense programs.[30]

Messages in the early letters were discarded; later ones contained crudely lettered warnings that identified the contents as anthrax, and wording apparently linking them to militant Islam. The latter feature is now widely thought to be an attempt to deflect suspicion from a domestic perpetrator. However, they probably were also intended to call attention to the contents of the letters, and thus to avoid casualties. Other features of the letters—they were tightly taped shut and contained a conspicuous amount of powder—also suggest an intent to alarm but not to kill.

Hoaxes

Although actual bioterrorist attacks have been extraordinarily rare, threats and hoaxes have not. From the first prominent bioterrorist hoax —the 1997 sending of a petri dish claiming to contain a *B. anthracis* culture to the B'nai B'rith center in Washington, D.C.—to the anthrax letter attacks in 2001, police and the FBI had probably responded to more than 1,000 hoaxes. There were occasional hoaxes overseas as well. The actual anthrax attacks in 2001 stimulated a resurgence of hoaxing and threatening, such that in the few years since there have been thousands of additional threats in the US and additional thousands overseas. The total is probably in the tens of thousands. The toll on law enforcement agencies and potential victims has clearly been substantial; domestic US costs have probably topped $100 million. Very few perpetrators of hoaxes have been identified and prosecuted.[31]

In addition to hoaxes, a high level of sensitivity to the possibility of bioterrorist attack has led to a large number of incidents in which powders of various kinds have provoked suspicion and stimulated formal responses to entirely innocent mail. For the Washington, D.C., area it was recently estimated that there are 3–10 alerts on Capitol Hill alone every day, and the FBI is called to investigate about 10 per week in the District.[32] Clearly, a huge investment of resources is required to deal with such suspicions.

Initial reactions to hoaxes were draconian;[33] potential victims were required to strip and undergo aggressive decontamination processes. The psychological toll was substantial, and confirmation that the incident was a hoax and not a real bioterrorist attack took days. More-reasoned approaches have since been instituted, based on speedier diagnostics and less aggressive treatment. These more routine responses have significantly reduced the intimidation potential of hoaxes. Nevertheless, for victims the period of ambiguity before a threat is determined to be a hoax can be highly stressful, and efforts to further reduce the time to a definitive diagnosis are important. It is probable that such hoaxes will continue to occur for the foreseeable future.

Interest in BW by Other Terrorist Groups

Although there has been little terrorist success in using biological agents as weapons, several groups besides the Rajneesh and the Aum have been interested in doing so.[34] Both bacterial and toxin agents have been involved in such interest, but in few cases have groups actually been able to obtain agents, much less develop a formulation that could be used. Several exceptions are known, however, suggesting that ricin in particular could be a serious threat in terrorist hands. For instance, the Minnesota Patriots Council (an ultraright, violent antigovernment group) successfully prepared a small quantity of low-purity ricin, intended for assassination of local law enforcement personnel.[35] And two letters containing ricin have been received by the US government. One, in October 2003, was addressed to the Department of Transportation; the other, in November, was to the White House. Both letters announced the presence of ricin and appeared to be intended to influence pending decisions on regulations affecting the trucking industry.[36]

Particular attention has been paid to hints that al Qaeda has been interested in developing anthrax as a biological weapon. However, there is no evidence that this interest has passed the exploratory stage, nor is there any indication that al Qaeda is even close to assembling the expertise and materials necessary to move beyond the exploratory stage.[37] On the other hand, a number of accounts suggest that several people associated with al Qaeda have been attempting to develop stockpiles of ricin.[38] However, these accounts are based largely on uncorroborated leaks from law enforcement personnel, and their reliability is unclear. Although it is alleged that ricin stockpiles were accumulated, none have been found. At present it seems prudent to take seriously the possibility that al Qaeda may be pursuing a limited BW capability, but it should be recognized that the evidence is weak and ambiguous, and there is no evidence that members of the group are close to making a usable weapon.

Conclusion

The history of terrorist development, stockpiling, and use of BW is very sparse. Only one group is known to have caused large numbers of casualties (the Rajneesh), one other (Brazilian land speculators) may have, and one (the Aum) attempted unsuccessfully to do so. Only a few other serious attempts to prepare BW have been recorded. If the anthrax letters are determined to constitute terrorism, and if further study of the Brazilian incidents shows that they caused outbreaks, the number of successful attacks will increase to three. This sparse record does not support the high levels of concern since the early 1990s, which increased greatly after the 2001 terrorist attacks and the anthrax letter attacks. These concerns are driven principally by consideration of worst-case scenarios, based on capabilities restricted (to date, at least) to nations with large, lavishly funded military BW development programs. There is no doubt that a military-style BW attack could have apocalyptic consequences, but there is no evidence that more than a few nations in the world could mount such an attack, much less any substate group. This state of affairs may change, but at least in the near term it appears that terrorist groups are unlikely to be able to develop a mass-casualty aerosol capability, short of recruiting personnel with the necessary military training and with current or past access to classified BW material.

Nevertheless, the concerns are not entirely absurd. Indications of increasing interest in BW by terrorist groups, and a demonstrated interest in causing mass indiscriminate casualties, legitimize some level of concern. The anthrax letter attacks hinted at what could be accomplished by a group should it ever get access to the expertise necessary to develop a "weaponized" preparation of anthrax spores; the perpetrator of these attacks probably could have killed hundreds, if not thousands, of people had he intended to do so (if there had been hundreds of letters, and if they had not contained warnings, and if the amount of powder in each had been too small to be obvious).

Fortunately, there is no evidence that any terrorist group currently has the requisite expertise to obtain a lethal BW agent, prepare it in a suitable formulation in quantity, and disseminate it via aerosol to cause mass casualties. The Aum's failure on all three counts, despite its unusually highly educated membership, and despite its expenditure of large sums (estimated to be around $20 million), is consistent with this evaluation. Nevertheless, the anthrax letter attacks remind us of the potential devastation that could result from the combination of the expertise of the anthrax letter perpetrator with the motivation to cause mass casualties of international terrorist groups.

It seems to us prudent to restrict the practical expertise of weaponization to the greatest extent possible. Doing so would entail ensuring that the expertise developed in the offensive BW programs of the countries described in earlier chapters is not available to terrorists. And it would suggest that biodefense programs be sharply limited, especially projects that entail activities which develop expertise that is useful for offense.

Although we think that the threat of BW aerosol attack by terrorists is exaggerated, we believe that the threat of attack via food contamination may be insufficiently appreciated. The success of the Rajneesh in causing (probably) thousands of casualties by contaminating restaurant salad bars is grounds for concern. There is no reason why the same method could not be used for some lethal BW agents, and it would avoid the necessity for special weaponized formulations and for sophisticated aerosol-generating technology. Unfortunately, there is no way to address this risk effectively short of prohibiting self-service restaurant facilities.

Regardless of the nature of potential attacks, no set of measures can ensure that a country is protected from bioterrorism. Every country will need to learn to live with some measure of vulnerability, and will have to balance the costs—financial, loss of openness, erosion of civil rights—of biodefense measures against their modest contributions to security.

CHAPTER 15

The Politics of
Biological Disarmament

MARIE ISABELLE CHEVRIER

The politics of biological weapons (BW) disarmament, culminating in the 1972 Biological Weapons Convention (BWC), are firmly embedded in the predominant political paradigm prevailing in the post–World War II world: the intense engrossment among major states with the Cold War. The BWC is often described and praised as the first international treaty banning possession of an entire class of weapons. The extraordinary nature of that achievement is rarely analyzed in detail. Moreover, in banning an entire class of weapons the states negotiating the BWC agreed upon an instrument that history has judged as remarkably weak and in need of strengthening.

By examining the principal political elements of the development of the BWC, this chapter seeks to answer two central questions: What was the thinking behind the UK Draft Convention, and why did the West agree to the weaker Soviet Draft Convention rather than insisting on elements in the stronger UK draft? Specifically, the UK Draft Convention included a ban on research and on the use of BW, as well as more stringent procedures to investigate alleged use, development, and production. Why were these elements weakened or eliminated?

Earlier studies of the history of the BWC have focused on other questions. As part of its six-volume seminal study *The Problem of Chemical and Biological Warfare,* SIPRI devoted an entire volume to CBW disarmament negotiations. That volume contains a detailed history of the negotiations in the United Nations General Assembly and the Geneva Disarmament forum.[1] The study ends before the Soviet Draft Convention was presented in March 1971. The Soviet draft broke the impasse at the Confer-

ence of the Committee on Disarmament (CCD) between those states fa-
voring the continued consideration of chemical and biological weapons
in a single convention and those, like the UK, that argued for the separation of the two. Ultimately the Soviet draft led to the BWC. Forrest Russell Frank, in his 1974 doctoral dissertation, focused on US policy regarding the BWC. He conducted interviews with more than 300 participants
in the negotiations within the US government and between the US and
other nations, but he did not have access to information that has since
been declassified.[2] Elisa Harris, more than a decade later, also focused on
the US with a retrospective analysis of how the BWC lived up to its expectations.[3]

This chapter has the advantage of being able to rely on internal government documents in the UK and US archives, many of which have been
recently declassified, as well as the official diplomatic record, which has
long been available. The declassified documents from the US and UK governments have given researchers access to internal government discussions of the disarmament processes in these countries and the discussions
of national priorities that shaped the language of the BWC. A thorough
analysis of the positions of all the countries contributing to the way in
which the Convention took form is beyond the scope of this chapter. Two
reasons for focusing on UK and US government documents are that the
UK Draft Convention of July 1969 formed the basis for all subsequent
drafts, and that the US association with the Soviet Draft Convention of
March 1971 provided the opportunity for agreement between the superpowers. The chapter thus builds on but does not repeat the work of other
scholars who have explored the same sources.[4]

The Background of Biological Disarmament

The control of biological (and chemical) weapons has a long history both
before and after the extensive use of CW in World War I.[5] In Chapter
16 Nicholas Sims discusses the role of the 1925 Geneva Protocol. After
World War II the topic of general and complete disarmament was taken
up by the newly founded United Nations. Throughout the postwar period
the Cold War dominated the political agenda, and nuclear weapons dominated the arms control and disarmament agenda. General and complete
disarmament and control of weapons other than nuclear weapons were

never totally absent from consideration, however.[6] In the CBW realm the Geneva Protocol was widely viewed as an inadequate instrument, yet strategies for addressing its inadequacies differed greatly. Some policy discussions demonstrated a reluctance to undermine the Protocol further by drawing attention to its weaknesses.

The extraordinary achievement of banning an entire class of weapons became possible only through the convergence of several phenomena in quite different spheres. The demonstration of the power of nuclear weapons at Hiroshima and Nagasaki and the subsequent development of nuclear weapons by the US, USSR (1949), UK (1952), France (1960), and China (1964) strengthened the institutions and motivations for multilateral arms control and disarmament. Simultaneously, serious academics developed a sophisticated literature on arms control and disarmament.[7] Third, in spite of the Cold War, the countries of NATO and the Warsaw Pact demonstrated their ability, together with the nonaligned countries, to reach agreement on multilateral nuclear weapons treaties, most importantly the Partial Test Ban Treaty in 1963 (also known as the Limited Test Ban Treaty) and the Non-Proliferation Treaty of 1968. Fourth, the negotiations on these multilateral treaties revealed a chasm between the US and Soviet approaches to arms control that would affect arms negotiations for decades: the Western insistence on verification (President Ronald Reagan's famous synopsis: "Trust but verify") and the consistent Soviet rejection of on-site verification measures.[8] Fifth, nongovernmental organizations concerned with arms control and disarmament, such as the Pugwash Conferences on Science and World Affairs, promoted discussions of CBW by governmental and nongovernmental participants. Finally, the Western European Union encountered technical difficulties with the implementation of portions of the Brussels Treaty of 1954 relating to BW.

The conceptual origins of the BWC can be traced to the Modified Brussels Treaty of 1954, in which treaty members created an obligation for Germany not to produce BW.[9] The Western European Union Armaments Control Agency (ACA) had responsibility for ensuring that Germany lived up to this obligation. The concept of "nonproduction control" had its origins in the ACA's effort to fulfill its task. A second provision of the Brussels Treaty established "quantitative control" of the size of stocks of BW on the mainland of Western Europe. Thirteen years later these provisions had "not yet been activated" by the Council of the Western European Union because "the Council have so far not been able to reach agreement on detailed regulations for the control of biological weapons."[10]

In the early 1960s the issue of BW control and disarmament centered on the use of BW and the nonratification of the Geneva Protocol by the US. UK and US discussions of disarmament proposals for CBW in 1963 give an early indication that the US government rejected CBW disarmament proposals because they might establish a precedent for nuclear weapons disarmament. In discussions between UK and US officials regarding the 1925 Geneva Protocol or new declarations regarding chemical or biological weapons, the US rejected ideas about restrictions on use and about the reduction or elimination of CBW not for any substantive reasons relating to CBW, but because of their possible effect on nuclear policy. Reporting to the Foreign Office in London, the UK embassy in Washington stated that the US would not "subscribe to any new declaration on chemical and biological weapons. In their view the principle involved is the same as that for declarations about the use of nuclear weapons, which are included in the category of weapons of mass destruction. Therefore a declaration about chemical and biological weapons would prejudice the principle which we have all maintained about the use of nuclear weapons."[11] The principle about the use of nuclear weapons was that NATO maintained the right to use nuclear weapons against a Soviet conventional attack in Europe in order to deter Soviet aggression.

The effect of nuclear policy on CBW policy was not limited to the US government. In March 1963 the UK delegation to the Eighteen Nation Committee on Disarmament (ENDC) wrote to the Foreign Office regarding the "alternative possibilities of a UK proposal for a reaffirmation of the 1925 Geneva Convention, or for the establishment of a study group on reductions of chemical and biological weapons." The minister of state considered that neither of these alternatives was advisable, because "it is quite likely that a Western proposal for a ban on the use of chemical and biological weapons would be exploited to our detriment in the Conference by the Soviet bloc as a precedent for a similar ban on the use of nuclear weapons: and on this issue the Soviet attitude would attract the sympathy of a number of the unaligned delegates."[12]

In 1963 and 1964 the US was carrying out studies of the problems associated with CBW. While both the US and the UK recognized the need to make progress in this area, at the end of January 1964 one UK official characterized it as "a matter on which there is no particular urgency."[13] The Americans repeatedly indicated that they wished to delay discussions of CBW issues until their studies were completed. In late September 1964 the UK embassy queried Washington officials on the status of the CBW studies. Washington officials told their UK colleagues that "the American studies were continuing, but were going slowly. For one thing, other matters had to be given priority."[14] The UK generally agreed that awaiting the conclusions of the studies was appropriate, but eventually the British began to display some impatience with the Americans.

In mid-1964 an official from the UK Foreign Office candidly admitted that "the primary purpose of a proposal for an agreement prohibiting the use of CBW would be to embarrass the Russians at Geneva. The West would try to expose Russian preparations in this field as far as would be compatible with security considerations. We would not be trying to force a change in Russian military doctrine . . . Our objective would, in short, be propagandistic rather than substantive." Problems with verification measures of a ban on CBW were clearly recognized as well:

> effective verification of a prohibition agreement may not be possible. This is because CW and BW agents have their source in toxicological research, in medicine, veterinary medicine and agriculture. Even if the facilities engaged in these activities in the U.K. could be listed and made available for inspection, it is not really conceivable that the Russians would be prepared to do the same. A proposal for an agreement that included provisions for verification would probably be too easily recognized as a propaganda move, leaving the alternative of an agreement without verification provisions.

The Foreign Office also outlined the argument for an agreement without verification provisions: "since there would be no dismantling of existing preparations or defenses, no harm to national security would result from the absence of verification arrangements."[15]

During 1964 the UK also began to consider separating BW from CW. The Ministry of Defence (MOD) argued that there was a case for distinguishing between the two on the basis of their tactical and strategic differences. "Potentially, given the appropriate delivery system, BW has the same strategic implications as nuclear warfare—widespread (national) devastation of the population, livestock and crops [using] very relatively small quantities, e.g., Botulism, Anthrax and other bacteria and virus strains. On the other hand CW would involve a wide variety of battlefield tactical delivery weapons with mainly local effects."[16] As we shall see, other, more complex arguments for the separation CW and BW were developed later.

As early as August 1964 the UK acknowledged differences between its own and US interests in CBW control. American interests were greater, according to the Foreign Office, because of the US offensive programs. Moreover, the UK acknowledged that the US was "engaged in a form of CBW, admittedly against plant life, in Vietnam" (see Chapter 10).[17] Thus, the propaganda value of any CBW proposals could have backfired on the West.

In March 1965, however, discussions shifted from plants to people as the *New York Times* revealed that the US was using nonlethal antipersonnel gas weapons (the tear gas CS) in Vietnam. National and international criticism of the US use of gas weapons followed.[18] The US defended its use of tear gas in Vietnam, insisting that because the military employed a riot-control agent the Geneva Protocol did not apply.[19] The war in Vietnam and US use of gas weapons continued to color discussions of BW disarmament proposals through the completion of the BWC in 1972.

In the mid-1960s disarmament issues were discussed in both the First Committee of the UN General Assembly, which met in New York, and the ENDC in Geneva.[20] On 7 November 1966 Hungary submitted a draft resolution to the First Committee seemingly directed squarely at the US use of gas in Vietnam.[21] The initial draft demanded "strict and absolute compliance by all States" with the Geneva Protocol, condemned "any actions aimed at the use of chemical and bacteriological weapons," and declared that CBW use "for the purpose of destroying human beings and the means of their existence" constituted an international crime.[22] This draft resolution generated considerable discussion within the First Committee. A group of African countries proposed amendments to the resolution that simultaneously strengthened the preambular paragraphs and softened the language of the operative clauses.[23] The US responded to the tabling of this draft resolution with a lengthy statement defending its use of tear gas and herbicides and declaring that the draft resolution was "motivated purely by propagandistic ends."[24] A group of Western nations submitted amendments to the resolution that further diluted its ability to cast blame on the US for its actions in Vietnam or for its status as a nonparty to the Geneva Protocol. The amended resolution noted that the ENDC had the task of "seeking an agreement for a cessation of development and production of chemical and bacteriological weapons" and called for strict observance of the principles and objectives of the Geneva Protocol. The resolution passed on a close vote, with Western states supporting the resolution, Eastern states abstaining, and neutral and nonaligned states split.[25]

In late 1965 the government of Malta disclosed to the UK that it intended to propose an extension of the 1925 Geneva Protocol "to cover all biological methods of warfare and armed conflicts other than war."[26] The UK urged Malta to show its proposals to the US, which was unlikely to react favorably because of Vietnam.[27] Nevertheless, in 1967 Malta introduced a draft resolution in the First Committee suggesting that the Geneva Protocol be revised, updated, or replaced. In addition, the draft resolution requested the secretary general to prepare a report on the effects of CBW, to be submitted to the ENDC.[28] In presenting this draft resolution the Maltese representative made a strong case for why the Protocol should be reviewed. He argued that it covered only some CW, that its scope included tear gas but not herbicides, and that whereas bacteriological weapons were banned, viral and fungal weapons were not. More generally, he concluded that the Protocol was "hopelessly out of date and should be either radically revised or a new international agreement negotiated."[29] He did allow that the Protocol "may be useful as a point of reference for the beating of propaganda drums."[30]

The Maltese interpretation of the Geneva Protocol was not widely shared—at least on the public platforms of the UN. Hungary, in particular, not only reintroduced a version of the resolution it had submitted the previous year, but also publicly challenged many of the assertions made by the Maltese representative. The Hungarian representative argued that the Protocol "unmistakably refers to all kinds of chemical and biological weapons . . . It is beyond dispute that this wording accurately covers not only any one of the existing chemical and bacteriological weapons, but also those that are now being developed." He argued that the Protocol did not need revision or replacement, "but to give a strong effect to its prohibitive clauses."[31]

The USSR rejected the Maltese arguments even more emphatically: "arguments that the Geneva Protocol is limited in content and does not cover all forms of chemical and bacteriological warfare are very dangerous and unfounded." Furthermore, the Soviet representative claimed that "a policy of replacement or revision [of the Protocol] would lead only to the undermining of most important and fundamental restraint on chemical and bacteriological warfare."[32]

The Netherlands stepped into the fray and proposed amendments to Malta's draft resolution that constituted a compromise. The Dutch agreed with the Maltese that the Geneva Protocol was outdated but supported the Hungarian position on its expression of value, noting that it "remains the only instrument of its kind" and that it exercised some "restraining influences" on the use of CBW.[33] The proposed amendments reaffirmed the Hungarian resolution on the Protocol passed the previous year. Ultimately both Hungary and Malta withdrew their draft resolutions after US opposition to a compromise draft.[34] The study by the secretary general of the effects of CBW, however, did take place. The report was concluded in 1969 and was influential in the subsequent negotiations.[35]

The influence of the Geneva Protocol in BW disarmament discussions should not be underestimated. The debates over the resolutions described above centered on three issues: first, the ban on first use of CBW was accepted as international law; second, of the major powers only Japan and the US were not parties to the Protocol; and third, the US was using chemicals in Vietnam—a fact interpreted by some, but not all, as a violation of the Protocol. As late as 1969 the US mission in Geneva, in arguing for US ratification of the Protocol, put it this way:

all discussions regarding CBW, in one sense or another, have to take into account fact that US has not . . . ratified Protocol, while virtually all other countries participating in discussion have accepted these obligations . . . if US should decide that its interests permit serious consideration of UK initiative for BW Convention, US would probably not . . . be able breathe life into British initiative and interest others in its negotiation, unless we had moved to ratify Protocol. Opposition to UK initiative derives in part from suspicion that US, as non-party to Protocol, would exploit study of such a BW measure either to draw attention from US non-adherence to protocol or to demonstrate that Geneva Protocol is not . . . a satisfactory instrument.[36]

The 1968 UK Working Paper on Microbiological Warfare

The UK delegation to the ENDC tabled a working paper on microbiological warfare on 6 August 1968 that described the principal elements of a new convention:

- A common understanding that any use of microbiological warfare and in any circumstances was contrary to international law and a crime against humanity and therefore a prohibition of use
- A ban on the production of agents for hostile purposes while recognizing the necessity of production of agents for peaceful purposes
- A ban on the production of ancillary equipment
- An obligation to destroy stocks of agents or equipment
- A ban on research aimed at production of prohibited agents and equipment
- Provisions for access by authorities to "all research which might give rise to allegations" of noncompliance
- Openness of relevant research to international investigation and public scrutiny

Regarding verification measures the working paper in tentative language recommended a "competent body of experts, established under the auspices of the United Nations," to investigate allegations of breaches of the convention and a commitment by parties to cooperate with any investigation. Finally, the working paper also contained tentatively phrased recommendations on entry into force and actions to counter the use or threatened use of BW.

The momentum behind the UK initiative can be said to have its roots in a series of papers from the Defence Research Policy Committee of the MOD in 1961.[37] These papers included separate operational assessments of CW and BW and recommendations for UK policy. In addition to ubiquitous recommendations for further study, a summary paper of the assessments prepared for the Chiefs of Staff concluded that there was no need for the UK to have either a CW or BW strategic offensive capability, because nuclear weapons "are far more effective and economical for the attack of civil populations." Nevertheless the summary acknowledged that such weapons might have some strategic use, in a sense keeping options open. Similarly, the paper ruled out any tactical potential of BW in *military* situations but allowed that "there might be uses in circumstances where immediate results are not required," thus suggesting a clandestine role for BW. Furthermore, it concluded that Russia was not "a suitable target system for BW attacks." In contrast, BW "could be a serious threat to the UK civil population."[38] These elements—UK vulnerability to BW, an assessment of limited strategic and limited military tactical value of offensive BW, and the unsuitability of Russia as a target of BW—continued to be reasons motivating the UK BW control initiative throughout the 1960s.

The 1968 UK working paper came at a busy time for the ENDC. The first priority for most delegations was a comprehensive test ban, and second, a treaty prohibiting nuclear or other WMD on the seabed. With these two draft treaties already on the agenda, some delegations expected the UK to follow through on its 1968 paper and introduce a draft BW treaty. However, London described the new Nixon administration as wanting to avoid CW issues, and the administration regarded CBW as a low priority: "the new people have so much on their plate already that they are unlikely to have a CBW policy before the summer session."[39] Indeed the policy review would not be completed until the fall of 1969. Although there are indications that neither the US nor the USSR was particularly supportive of the UK initiative, it was not a high enough US priority for Washington to bother to restrain the UK.

The UK argument for a prompt ban on the use and possession of BW contained four elements: (1) BW "are regarded with general abhorrence, possibly more so than any other means of waging war"; (2) "it seems unlikely that development or use of biological weapons is, at the moment, regarded by any state as essential to its security"; (3) new technological developments could lead to BW becoming an integral part of some states' armaments; and (4) it would be easier to achieve a ban before such armament took place than after.[40] The statement that development or use of BW was not regarded by any state as essential to its security was subsequently challenged by MOD personnel, who argued that the US did regard possession of BW—in order to retaliate in kind if necessary—as essential to its deterrent policy.[41]

On the key question of why to separate chemical and biological weapons the arguments involved mainly political pragmatism. Because biological methods were at an earlier stage of development, prohibiting BW should be easier. Because conditions were favorable, the paper argued, states might "simply throw away the chance of getting an effective and total prohibition" of BW if they tried to tackle the more difficult CW issues. Finally, getting agreement on the prohibition of BW would "create a favorable climate" for negotiations on a CW ban. Optimistically, the paper argued that "much of the preparatory work . . . for instance, work on methods of investigating complaints of infringements of any Convention—will be of great use when the problem of chemical weapons is tackled."[42] With hindsight the arguments for separating CW from BW were politically expedient while not necessarily substantively compelling. Nevertheless, the political difficulty of tackling CW, given the differences between East and West during the Cold War concerning verification, and given the US use of anticrop and antipersonnel chemicals in Vietnam, ensured that the political arguments won the day.

The dilemma posed by the Geneva Protocol—that drawing attention to its weaknesses could serve to exacerbate them—remained an important issue. The UK argued that the ban could be drafted so as to strengthen the Protocol through both preambular and operative clauses, which the subsequent UK draft did emphasize. The reason for what came to be known as the general-purpose criterion was succinct: "the trouble with listing exactly what is prohibited is that any agents of biological warfare subsequently developed remain outside the prohibition."[43]

The response to the issue of the difficulties of verification is worth quoting at length. It demonstrates the great differences between what the UK envisioned and what was actually achieved. The UK stated at the outset that "we do not think that verification, in the sense in which the term is normally used in disarmament negotiations, is feasible in the field of biological warfare." Yet Whitehall rejected this as a reason to eschew a ban on BW and explained:

> we think that the right solution is to ban the development and production of biological weapons, and to create the strongest possible deterrents against infringements of the ban. The objective should be to set up an effective procedure for investigating complaints that the Convention had been infringed, rather than to establish safeguards of the kind provided for in the Non-Proliferation Treaty. The procedure for investigating complaints would have two distinct aspects:
>   (i) machinery for receiving complaints and initiating an investigation;
>   (ii) machinery for carrying out the actual work of investigation;

These two aspects could be dealt with be a single body, but need not be
. . . We believe that it is of paramount importance that complaints be investigated very quickly in order to establish the facts. Given the impossi-
bility of elaborate verification procedures, only machinery which can go
into action quickly can provide an adequate restraint on aggressive activities. Because of this need for speed, the machinery should be as far as
possible automatic.

The paper went on to suggest two possibilities for a body to carry out the
actual work of investigations: "a Standing Committee or Panel of International Scientific Experts" or "a team of trained experts provided by the
defense authorities of a neutral country."[44] Each of these possibilities had
advantages and disadvantages. By the time the UK tabled its Draft Convention in July 1969, these ideas on the solution to the verification problem had changed considerably.

Internal disagreements within the UK government affected the language of the Draft Convention. Indeed, a confidential memo within the
MOD commenting on a draft BW paper from the Foreign and Commonwealth Office stated that the paper "does nothing to dispel the impression
. . . that this whole initiative by the UK is pretty half-baked."[45] This comment is but one of several comments by MOD personnel on a 12 May
1969 draft of the UK initiative. None of the comments was positive. For
instance, D. C. R. Heyhoe in mid-June of 1969 argued that, "while recognizing that a Convention must contain some provision forbidding possession, we consider that it is preferable not to make this as specific as the
present draft suggests."[46]

MOD personnel consistently voiced disagreements over the complaints
procedures elaborated above. In particular they emphasized the distinction between complaints procedures that would deal with the use of BW
and procedures to investigate violations of the ban on development and
production. In response to a request for a study related to verification of
production and possession, R. Holmes stated that such a study was unnecessary because investigation of a complaint of illicit production would
require "complete intrusion." Elaborating, he said that investigation of
an allegation that a plant was producing BW "could only be established
by on-site inspection including sampling and analysis of material in preparation. Any inspection would have to be virtually unannounced." The

MOD therefore concluded that "processes of verification (of production) are not possible."[47] This conclusion seems to have been a political one, based on what was thought to be politically possible in the USSR, the UK, or its allies, not on what was scientifically possible.

Earlier in 1969, the UK delegation to the ENDC pressed the government for a more modest proposal than the one outlined in the working paper, to focus on investigating use of BW only. The delegation did not believe that China, the USSR, or the US would accept an unverified ban on development and production: "It seems most unlikely that the Russians, with China and the U.S.A. in mind, would put their name to an agreement to cease production and development and to destroy stocks, even on what would amount to an unverifiable basis; the Americans would be even less likely to agree to an unverifiable undertaking of this kind."[48]

However, the Foreign and Commonwealth Office prevailed, and introduced a Draft Convention despite opposition from the MOD, skepticism regarding the substance of a ban on development, and pessimism regarding the political prospects for success. The draft banned all use of BW, including use in retaliation; it banned the development and production of BW; and it introduced procedures to investigate alleged use, development, and production.

The UK Draft Convention of July 1969

The Biological Warfare Draft Convention tabled by the UK delegation at the CCD in Geneva in July 1969 differed significantly from the BWC that opened for signature in 1972. The draft had seven preambular paragraphs, four of which pertained to the Geneva Protocol. The first two paragraphs acknowledged the contributions of the 1925 Protocol to mitigating the horrors of war. The third preambular paragraph recalled two UN General Assembly resolutions calling for observance of the Protocol, and the sixth expressed the desire to reinforce the Protocol. The remaining three preambular paragraphs concerned BW. The fourth paragraph expressed the belief that "biological discoveries should be used only for the betterment of human life and that their use for hostile purposes would be repugnant to the conscience of mankind." The sixth recognized the risk of BW, and the final preambular paragraph declared the belief that "provision should be made for the prohibition of recourse to biological methods of warfare in any circumstances."

Article I of the UK draft created an obligation "never in any circumstances" to engage in biological warfare or to use biological agents for hostile purposes. It explicitly outlawed hostile use of BW against humans, other animals, or crops.

Article II extended the prohibition on use to possession and research with the following language:

Each of the Parties to the Convention undertakes

(a) not to produce, or otherwise acquire, or assist in or permit the production or acquisition of

(i) microbial or other biological agents of types and in quantities that have no independent peaceful justification for prophylactic or other purposes;

(ii) ancillary equipment or vectors the purpose of which is to facilitate the use of such agents for hostile purposes;

(b) not to conduct, assist, or permit research aimed at production of the kind prohibited in sub-paragraph (a) of this Article; and

(c) to destroy, or divert to peaceful purposes, within three months after the Convention comes into force for that Party, any stocks in its possession of such agents or ancillary equipment or vectors as have been produced or otherwise acquired for hostile purposes.

Article III described the procedures for complaints of violation of the Convention. Allegations of use were to be taken to the UN secretary general along with a recommendation for an investigation, the report of which would be submitted to the Security Council. Allegations of breaches of the Convention that did not involve use were to be taken to the Security Council with a request that the complaint be investigated. Each of the parties was obliged to cooperate with the secretary general in any investigation.

Article IV affirmed parties' intentions to come to the assistance of any party that was the target of any biological methods of warfare. Article V bound parties to pursue negotiations to "strengthen the existing constraints on the use of chemical methods of warfare." Article VI reiterated that nothing in the Convention "shall be construed as in any way limiting or derogating from obligations" under the Geneva Protocol. The remaining articles included provisions for amendment, signature, ratification, entry into force, unlimited duration, and withdrawal.

Article III, which described two routes for complaints of noncompliance—one with the secretary general, the other with the Security Council—generated considerable internal debate. In reference to the Security Council's investigation of allegations of development or production: "the Ministry of Defense would have liked to omit altogether the phrase 'and request that the complaint be investigated' . . . on the grounds that no adequate means existed for investigating such complaints and that to included provision for such 'investigation' might be both misleading and detrimental to our own interests." Nevertheless, the Disarmament Department prevailed with its argument: "We contended that cases might well arise in which it was possible to investigate complaints that biological agents were being produced or stockpiled for hostile purposes . . . and that parties must be given the right to request that their complaints be investigated, even though the Security Council might not accede to the request. Otherwise the whole Convention would be gravely weakened."[49] The UK delegation at the ENDC in Geneva, commenting on the same article, noted that "even if we are not prepared to admit it, there would be little similarity in the procedure and prospects of success in the two cases."[50] Bifurcating the investigating procedures for complaints about the use of BW and about complaints of illicit development and production would have great significance as the UK Draft Convention was transformed through negotiations.

The ENDC delegation was also anxious to have the investigating machinery described in greater detail than in either the Draft Convention or the accompanying draft Security Council resolution. The Disarmament Department believed that it was premature to make a proposal on the composition of the investigating body. Although the ENDC delegation favored an international committee, the Disarmament Department had doubts about the "ability of such a committee to tackle the job with the necessary cohesion, efficiency and speed." The Disarmament Department described what it hoped would happen regarding investigating machinery for complaints:

> many states are likely to want to know what the investigating machinery is before they sign, or at any rate before they ratify, the Convention.

My hope is that the General Assembly will be prepared to adopt the Convention without knowing exactly what the investigating machinery would consist of; that the Secretary-General, in accordance with the Security Council Resolution, would thereafter establish a working group of some kind to consider the form of the investigating body; that he would then make appropriate recommendations; and that when these were approved a sufficient number of states would ratify the Convention to bring it into force.[51]

The UK did not seem particularly optimistic about the chances of success for its Draft Convention. The UK delegation in Geneva noted that although they had pointed out to the Russians that "the real danger comes from medium and small powers . . . this is not one we can press too hard." The nonaligned countries were not enthusiastic about a draft BW convention because of "their feeling that we are trying to divert attention from nuclear disarmament." The delegation concluded that "we are unlikely to win enough support from the non-aligned even to discomfort the Americans and Russians."[52]

US Policy Considerations

A top-secret memorandum of conversation between US Arms Control and Disarmament Agency (ACDA) officials and the counselor at the Canadian embassy in 1967 hinted at US interest in CBW disarmament. CBW was a concern of the US because of possible proliferation to non-nuclear countries, which might find the CBW alternatives to nuclear weapons "cheaper and easier."[53] In August 1969 the CIA issued an intelligence report, "Disarmament: Chemical-Biological Warfare Controls and Prospects for Improvement." The report noted once again that CBW had taken a backseat to other arms control initiatives, this time to nuclear weapons and the seabeds. It contained an assessment of the UK BW draft treaty, noting that "deference to the US no doubt played a part in Britain's decision not to treat CW in its initiative, since debate on the tear gas issue . . . is potentially embarrassing to the US." Indeed, the separation of CW and BW in the Draft Convention, Soviet and other delegations' reactions to the UK draft, and complications surrounding US use of tear gas in Vietnam warranted considerable discussion in the report. The report noted that the Soviets had expressed reservations about the verification aspects of the UK draft, particularly the role of the UN secretary general in investigating alleged use. The CIA report noted that the Soviets historically strove to "limit the powers of that office," preferring arrangements to go through the Security Council, presumably because of its veto power in that body. The report concluded that "the UK initiative on BW in the ENDC will probably founder, whatever its merits," because "the Soviets and their allies seem intent upon avoiding substantive consideration of it in Geneva in favor of taking the problem to the [General] Assembly" in New York, where the US expected the discussion to focus on US use of tear gas in Vietnam.[54]

In November 1969 a report to the US National Security Council outlined the advantages and disadvantages of what were seen as the important BW policy issues.[55] The primary issue was whether or not the US should support the UK draft treaty that had been tabled that year: Should the US restrict its BW programs to research, development, training, and education, or should it maintain and develop stockpiles of lethal, incapacitating, or anticrop agents? The report considered the pros and cons of lethal and incapacitating agents separately, but it omitted detailed discussion of anticrop agents. It did not consider restricting programs to research only. Nor did it reach any recommendations, instead only listing the advantages and disadvantages of each policy option. Nevertheless, the analysis reveals some of the thinking behind the ultimate US support for the UK Draft Convention. One major advantage was that it would be in the US interest to have other countries forgo a BW capability once the US had decided to do so. In addition, the provisions in the UK draft for "investigations of allegations of violations . . . provide some measure of security." Because the UK draft had some merit in inhibiting other states, in regard to security the US would be no worse off, and might be better off. The list of disadvantages centered on the effects of accepting an unverifiable treaty provision in other arms control contexts.

The Soviet Draft Conventions of September 1969 and March 1971

In September 1969 the Soviet Union tabled a Draft Convention prohibiting the development and production of both chemical and biological weapons. Because the Soviet draft combined CW and BW, excluded an explicit prohibition against use, and had been tabled at the UN General Assembly instead of at the CCD in Geneva, the "accepted forum for disarmament negotiations," the UK viewed the Soviet initiative as "a propaganda gesture."[56]

A few days after the Soviet Union introduced the Draft Convention, a confidential telegram described a luncheon conversation between Roland Timerbaev, of the Soviet Ministry of Foreign Affairs, and Alan Neidle, of the US delegation to the CCD in Geneva, on the topic of the recently tabled Soviet Draft Convention. Neidle expressed the opinion that the Soviet Draft Convention "looked like [a] superficial product not designed to promote serious negotiations." Timerbaev's response, however, indicated that the Draft Convention was not strictly a propaganda ploy. Timerbaev reportedly had preferred tabling the draft in Geneva; moreover, he had emphasized that the draft "may be viewed as bridge between propaganda of past and serious willingness to discuss actual arms control." The Soviets were willing to discuss limitations on development and production of CBW, Timerbaev said, but the Soviet government had to "go through certain procedures . . . which might have a propaganda appearance." The US delegation in Geneva was inclined to believe Timerbaev because of the "long history of Sovs entering into arms control negotiations (NPT, LTBT) via portal of odoriferous propaganda initiatives." The account of the conversation also emphasized the benefits of US ratification of the Geneva Protocol. Timerbaev stated that the Soviet Draft Convention was not a take-it-or-leave-it proposition and that "it would be necessary to give serious consideration to UK suggestions regarding complaints procedure."[57]

Less than two months later the US announced its decisions to unilaterally abandon its offensive BW program, to destroy its BW stocks, to place a moratorium on CW production, and to submit the 1925 Geneva Protocol to the US Senate for ratification. The internal government discussions leading to the US decision to unilaterally give up its offensive BW program have already been recounted in detail by Jonathan Tucker (see also Chapter 2).[58] In brief, early in the Nixon administration National Security Advisor Henry Kissinger launched an in-depth interagency review of CBW policies. Three interdepartmental groups conducted analytical work: one from the intelligence community to assess foreign capabilities in BW, a second to examine military options, and a third to explore diplomatic options. In addition, because of the technical nature of the topic, nongovernmental scientific experts provided advice. Despite opposition from the Joint Chiefs of Staff, who favored retention of an offensive program, the secretary of defense joined the other civilian principals in recommending unilateral renunciation of BW.[59] Tucker concluded that both military and political considerations influenced Nixon's decision. Militarily the weapons "had limited tactical utility . . . and did not constitute a reliable and effective strategic deterrent." Politically, the weapons had no powerful constituencies, it was important to discourage other countries from acquiring BW, and by renouncing them Nixon could deflect criticism about the use of chemicals in Vietnam.[60]

Following the US unilateral renunciation of BW in November 1969 and of toxins in February 1970, little progress was made on BW arms control. Efforts at the CCD were concentrated on convincing other delegations of the advisability of separating negotiations on the control of CW and BW.[61] Then, on 30 March 1971 the Soviet Union unexpectedly tabled a second Draft Convention at the CCD, dropping its long-standing opposition to separating CW and BW.

The new Soviet draft, tabled on behalf of nine Warsaw Pact states, differed from the UK draft in several crucial respects. First, it ignored Article I of the UK draft, which obligated parties never, in any circumstances, to use biological methods of warfare. Second, it omitted the UK draft's prohibition of research aimed at offensive production. Third, it required all complaints concerning a breach of obligations to go to the Security Council.

The Soviet draft included a few features absent from the UK draft. It required states to undertake legislative and administrative measures for prohibiting BW. It also created an obligation for states to facilitate the exchange of equipment, materials, and information for the use of biological agents and toxins for peaceful purposes.[62]

An internal UK document raised a number of concerns about the differences between the UK and Soviet Draft Conventions. One was that since the Soviet draft did not specifically outlaw use, the USSR might wish to retain the right to retaliate with BW. Another was that since the Soviet draft prohibited the production of "weapons" whereas the UK draft prohibited "methods of warfare," "the Soviet formulation would leave them free to manufacture, stockpile, etc. the *component parts* of the prohibited weapons," which could quickly and easily be turned into weapons.[63]

A third substantive criticism of the Soviet draft was that the Soviet method of consultation and cooperation to solve problems of implementation—which ultimately found its way into the BWC—was not a "realistic proposal to deter would-be violators." The UK document summed up its criticisms of the consultation and cooperation provisions of the Soviet draft in this way: "Consultation and cooperation may be all that is required in some arms control measures, e.g. on the sea bed, where States are free to observe other States' activities. But more than this is required when it is a question of a State's activities within its national territories."[64] The complaints procedure and the security assurances in the UK draft provided a more powerful deterrent.

Several states responded negatively to the attempt in the new Soviet Draft Convention to conform with the UK and US desires to separate chemical and biological weapons. Sweden, Mexico, Morocco, and others expressed both dismay, and doubt that effective controls on CW would now ever be achieved. The neutral and nonaligned countries, however, recognized that critical momentum was behind the initiative as soon as the Soviet Union and its allies tabled a Draft Convention similar to the UK draft supported by the US.

Following consultations with its allies in April (mere weeks after the new Soviet Draft Convention was tabled), the US decided to negotiate on the basis of the Soviet text. A National Security memorandum prepared for President Nixon focused on a few issues. First, the US planned to "firmly reject . . . the USSR's view that the 1925 Geneva Protocol prohibits the use of tear gas and herbicides in war." Second, the US proposed that the differences between the UK and Soviet drafts concerning the use of BW be handled by a preambular paragraph rather than by an operative provision as in the UK draft: "since a prohibition of possession or production of these weapons would accomplish the purpose of prohibiting use, an operative provision banning use is not considered necessary." Third, without an operative clause specifically banning use, the complaints procedure advocated by the UK for the secretary general to investigate allegations of use did not warrant any mention. However, the memorandum discussed the advantages to the US of the absence of on-site verification measures in both drafts: "Because on-site verification could not possibly be effective without also being extraordinarily intrusive, to us as well as to the USSR, and since biological weapons have questionable utility, all agencies consider the compliant procedures the only attainable system and hence adequate for these particular weapons."[65] The US State Department explained its position in much greater detail in April 1971 following the president's approval of the policy described above. Although the procedure involving the secretary general "could be useful in determining the facts in dispute, we are prepared to join in tabling a new draft without it because it is not essential to our security interests."[66]

Consultations between the UK and the US regarding the Soviet Draft Convention occurred on 4 May 1971. The UK still believed that a complaint procedure in case of BW use should be included in the Convention. It was suggested at this meeting that the UK and US continue to work on the problem and let the Soviets know that they might want to return to this point later in the negotiations. The US believed that the "only chance of achieving some provision would be at [a] later stage if non-aligned del[egation]s pressed for it." The US and UK believed that the Soviets had made a "major concession" in separating CW from BW. Suddenly, after years of lackadaisical attention to BW, "prompt negotiation of a first agreed draft with the USSR" was thought to be advantageous to nail down this concession. The next day the UK representative to the talks recommended that the "US and UK . . . give up idea of joint approach to [Soviets] and instead pursue separate, but mutually reinforcing, approaches." The separation of approaches allowed the UK to continue to press for a complaints procedure in case of use.[67]

During the summer of 1971 negotiations led to a generally agreed-upon draft text. In recommending approval of the treaty the most important issue for National Security Advisor Kissinger was a "political hooker . . . there are more references and a closer tie into chemical weapons." A memo to Kissinger detailed the references to CW in the then most recent draft and declared that "the references to CW have been the price for general support of a BW ban" from the nonaligned nations and the Soviets. The alternative to accepting the references to CW was to carry on negotiations with no guarantee of a more favorable outcome. The US delegation had added language that would "protect our position that any CW agreement would require more effective verification."[68] These discussions did not touch on the elimination of research from the UK Draft

Convention. Earlier discussions of the UK draft within the US government sought to clarify that research on measures to defend against BW would still be permissible under the treaty. Nevertheless, the idea of banning offensive research seemed to die a quiet death.

A confidential paper prepared for Kissinger in September 1971 described in detail, article by article, and commented upon all the proposed changes to the UK and Soviet drafts that led to a joint US-Soviet Draft in September 1971. Many of the changes had been made in response to requests by the nonaligned nations. Other proposals by the nonaligned were noted as unacceptable. The article affirming the objective of effective prohibition of CW was described as "the most difficult to negotiate."[69]

The Politics of Implementation

The final agreed-upon text of the BWC left a number of parties dissatisfied. France, for instance, did not become an original party to the treaty because it lacked effective verification. A few years later, in 1979, an outbreak of anthrax in the Soviet city of Sverdlovsk (see Chapter 13) highlighted the Convention's lack of on-site measures to address allegations of noncompliance. At the First Review Conference, in 1980, on the basis of the Sverdlovsk outbreak the US alleged Soviet violation of the treaty.

At each review conference the parties have sought to strengthen the ability of States Parties to verify and increase confidence in other states' compliance. At the First Review Conference a group of states led by Sweden attempted, unsuccessfully, to amend Articles V and VI of the Convention. They proposed that a Consultative Committee be established to conduct fact-finding, with on-site inspections as one of the tools at its disposal. Soviet opposition to amending the Convention at a review conference killed the Swedish proposal.[70] Commenting on the effect of the Sverdlovsk incident on the First Review Conference, Julian Perry Robinson observed:

> The Sverdlovsk allegation very much affected the content of the Final Declaration on the thorny issue of the Consultative Committee. On the one hand it illustrated most graphically the need for some form of international verification procedure. On the other hand it suggested that the

USSR would be the subject of the first complaint to be brought before the committee, and few states were happy to contemplate the political furor that would ensue, and the attendant threat to the BWC's continuation . . . [so it] paradoxically strengthened the position of the USSR.[71]

At the Second Review Conference, in 1986, delegations sought to strengthen the Convention through politically binding declarations of relevant activities, often referred to as confidence-building measures (CBMs). The 1986 CBMs were augmented with additional declaration obligations at the Third Review Conference, in 1991. Yet as early as 1991, states' compliance with even these minimal obligations proved disappointing.[72]

The Third Review Conference authorized the formation of a committee of experts, known as VEREX, to examine the technical feasibility of verification measures. After examining many on-site measures the experts concluded that some would make the treaty a more effective instrument.[73] At a 1994 Special Conference to discuss the VEREX findings, the States Parties created a mandate for an Ad Hoc Group (AHG) to draft proposals for a legally binding protocol to strengthen the effectiveness and improve the implementation of the Convention.

By April 2001, as a result of deliberations from January 1995 through February 2001, the AHG had produced a lengthy negotiating text. However, many substantive issues remained unresolved, as compromises were required that normally are not made until the diplomatic endgame. To provoke entry into that phase of negotiations, the group's chairman, Ambassador Tibor Tóth of Hungary, presented a "composite text" in which all outstanding issues were resolved in ways that he considered might be acceptable to all the drafting nations. The principal elements of the composite text protocol to strengthen the BWC would have required declaration of relevant facilities, established an organization to observe activities at declared facilities, and allowed States Parties to call for on-site investigations of alleged violations of the treaty. The protocol would have facilitated international cooperation in the peaceful uses of biology and biotechnology while monitoring transfers of biological agents, toxins, and relevant equipment.

The text garnered a groundswell of support from European countries, Canada, Australia, and key members of the nonaligned and eastern European groups, including South Africa, Brazil, Peru, Poland, and Croatia. Nevertheless, in July 2001 the US rejected the composite text and sought to end the mandate of the AHG. The US delegation offered a number of alternative measures, including stronger national penal legislation, better domestic control of pathogens, improved ability of the UN secretary general to investigate alleged use of BW or suspicious disease outbreaks, stricter biosafety standards, and better disease surveillance. There was no proposal to deal with suspected state programs of development and production of BW.[74]

The Fifth Review Conference began only a few months after US rejection of the fruits of the AHG negotiations. The conference averted the disastrous outcome of no final agreement only by taking the highly unusual step of suspending deliberations for a year. The suspension was called when the US insisted on termination of the mandate of the AHG in the face of fierce opposition. When the Review Conference resumed in November 2002, it agreed to yearly expert and political meetings on specific topics, with no mention of the AHG or its mandate, the US having backed away from its 2001 demand.[75] Thus, a 10-year international effort to address the weaknesses of the treaty has ground to an uncertain halt. While the Convention parties meet and debate the merits of the US proposals, the fate of the AHG and its mandate remain in limbo until the Sixth Review Conference, in 2006. The mandate remains in existence, with no action scheduled to fulfill its requirements.

Conclusion

Almost from start to finish the discussions of a ban on the development and production of BW within the governments most important to progress in the negotiations were affected by other, higher priorities. Those priorities included nuclear and CW policy, the use of chemical agents by the US in Vietnam, and the obligations and interpretations of the Geneva Protocol.

The UK initiated a propaganda exercise despite internal opposition, particularly within the MOD. The MOD did not press its concerns, because it did not think that the UK BW initiative would meet with substantive success. Meanwhile, US willingness to accept a BW Convention stripped of all but minimal attention to verification issues seemed to emanate from an assessment that BW were not particularly attractive weapons to the US, and that interest in BW by smaller states should be discouraged.

Although internal UK and US documents show intense attention to the BW disarmament issue at times, it was not a high priority for either government. Substantive aspects of the treaty, particularly the explicit banning of use and the complaints procedures for investigating use, were allowed to languish or were traded away in the desire to achieve some kind of disarmament agreement on weapons that both governments agreed were not essential to their security.

The absence of any framework of declarations and on-site measures to address compliance concerns has haunted the implementation of the Convention. Suspicions of Soviet noncompliance arose early on, and revelations about the extent of its massive treaty violations (see Chapter 6) have underscored the weaknesses of the instrument. Yet the complete rejection of the AHG by the US and the weakness of alternative national, bilateral, and multilateral measures proposed by the Bush administration have left the Convention in a familiar limbo. The treaty remains a necessary and profound articulation of the global norm against the possession and use of biological and toxin weapons, with no international machinery to investigate or to sanction alleged violations of that norm.

US conduct at the resumed Fifth Review Conference demonstrated that the pattern of assigning a low priority to BW control issues persists. Although the US government considered the AHG effectively dead, it did not push for formal termination of its mandate, instead acquiescing in the UK government's appeal to continue multilateral activities. Had it not been for the US need for UK support in Iraq, according to a US government official, the US government would have continued to press for a formal dissolution of the AHG.[76] Thus BW control continues to take a backseat to other international issues deemed more urgent or more important. Raising the political profile of BW control initiatives is likely to remain an imperative for advocates of international cooperation for a truly effective BW disarmament regime.

CHAPTER 16

Legal Constraints on
Biological Weapons

NICHOLAS A. SIMS

This chapter traces developments in legal constraints on biological weapons (BW) since 1945 and identifies legal issues of continuing relevance in biological disarmament diplomacy. It is a story of expansion both in constraints and in the number of states legally bound by
those constraints. However, it is not a story of uninterrupted progress.
Obstacles have had to be surmounted along the way, and there remains
much work to be done in consolidating and universalizing the legal con
straints on BW, in order to reinforce the humanitarian, ethical, political,
and technological constraints that also constitute bulwarks against the
threat of their use.

The chapter begins with a thematic outline of the two main categories
of legal constraints and their convergence. It continues with a historical
account woven around these two categories of constraint. It concludes
with an examination of outstanding issues.

Categories of Legal Constraint

Legal constraints on BW fall neatly into two main categories: constraints
on use and constraints on possession.

Constraints on Use

Constraints on use of BW are much longer established in international
law than constraints on possession. The Geneva Protocol of 17 June 1925
prohibited, as between the High Contracting Parties to that agreement,

the use in war of "bacteriological methods of warfare." Through the 20th century an originally somewhat inchoate norm of customary international law evolved into a rule that at least the *first* use of BW—not only those bacterial in origin, but BW of all types—and eventually *any* use of BW were prohibited *erga omnes.*

Uncertainty persisted longest over the legal permissibility of the use of BW in retaliation. Some States Parties to the Geneva Protocol, including several of the militarily most powerful states, attached to their instruments of ratification or accession to the Protocol a reservation pertaining to retaliation in the event of an enemy state's breaking its obligations under the Protocol. Typically such reservations purported to allow retaliatory use of either biological or chemical weapons against a State Party that had resorted to the use of either kind of weapon. Most of these reservations extended this right of retaliation to instances of attack by allies. Other reservations were less specific but by invoking a "condition of reciprocity" seemed to envisage a similar possibility of retaliation. The combined effect of all such reservations (shown by a superscript [R] in the lists below) was to lend the Geneva Protocol the appearance of a "no-first-use declaration" rather than a solemn renunciation by treaty of any use in war of chemical or biological weapons.

The Protocol began to shed this appearance as the number of reservations pertaining to retaliation diminished in the last decades of the 20th century. At the same time, even reservations that were not formally withdrawn came to be seen as incompatible, both with later treaty obligations that must override them, and with the customary norm, which had finally evolved into a rule against any use of BW by any state, regardless of its treaty status.

There remains a strong case for completing this process of withdrawal of the remaining reservations, if only as a measure for achieving consistency of obligation, notably for States Parties to the Biological Weapons Convention (BWC) of 10 April 1972. The Irish government's 1972 declaration withdrawing its Geneva Protocol reservation of 1930 before the Convention was opened for signature led the way and remains the classic statement: "Ireland considers that the Convention could be undermined if reservations made by the parties to the 1925 Geneva Protocol were allowed to stand, as the prohibition of possession is incompatible with the right to retaliate. As the Convention purports to strengthen the Geneva Protocol, there should be an absolute and universal prohibition of the use of the weapons in question."[1]

Other states, parties to both the Geneva Protocol and the BWC, still need to act to bring themselves into line with the Irish declaration of 7 February 1972. Finally, clarification is needed of the legal position of all states, whether parties to the Geneva Protocol or not, preferably through an authoritative statement of the law as prohibiting any use of BW by any state in any circumstances whatever.

Constraints on Possession

Constraints on possession of BW were nonexistent for most states until almost 50 years into the life of the Geneva Protocol.

For many years after 1925, negotiations, including a proposed prohibition on possession of BW, took place only in the context of unsuccessfully promoted plans for general and complete disarmament, and then only sporadically. Possession remained generally unconstrained until 26 March 1975, when the BWC entered into force for its 46 original States Parties (joined subsequently by 108 others). Only a small number—7 at most—of "former enemy states" and their associates were already prohibited from possessing, *inter alia*, BW by individual peace treaties or other postwar settlements negotiated between 1947 and 1955 with the victors of World War II. For the great majority of states, with no treaty obligation to observe and no norm of customary international law against the possession of BW equivalent in legal authority to the norm against their use, it remained lawful to possess them, if they chose to, until at least 1975.

In the event, most states chose not to avail themselves of this possibility. The few states that did were faced from 1975 with a treaty obligation to abandon possession of BW which, even if they did not themselves accept it, attracted so many other adherents that a customary norm against possession might emerge on the evidence of state practice and eventually *opinio juris*.

Since its entry into force in 1975, the BWC has been the principal legal constraint on possession.

"Possession" as a Convenient Portmanteau

To speak of "constraints on possession" as the counterpart to "constraints on use" is a matter of convenience, not of law. Its employment in this chapter may merit explanation.

"Possession" is not a word which occurs in the BWC list of prohibitions. (Its only occurrence anywhere in the text of the Convention is in the clause "which are in its possession or under its jurisdiction or control" in Article II.) But it is a convenient portmanteau for the nine activities that the BWC does explicitly constrain, either absolutely or by subjecting them to the "4P formula" ("of types and in quantities that have no justification for prophylactic, protective, or other peaceful purposes").

Stockpiling, acquisition, and *retention* are all forms of possession of the prohibited objects, which for simplicity may be abbreviated to BW. *Development* and *production* are activities antecedent to possession, and it can be argued that possession is strongly implied at least by *production;* for this implication not to follow, the improbable assumption would have to be made that what is produced is immediately destroyed.

Transfer implies possession because if there is no prior possession of the prohibited objects there is nothing to transfer.

The remaining prohibitions are on what might be called "ancillary" activities: *assistance, encouragement,* and *inducement.* For here it is assistance toward possession of BW, encouragement to possess, and inducement to possess that are prohibited.

Five of these activities are specified in Article I and four in Article III. In the text of these two articles printed below, they are again identified by the use of *italics:*

Article I

Each State Party to this Convention undertakes never in any circumstances to *develop, produce, stockpile* or otherwise *acquire* or *retain:*

(1) Microbial or other biological agents, or toxins whatever their origin or method of production, of types and in quantities that have no justification for prophylactic, protective or other peaceful purposes;

(2) Weapons, equipment or means of delivery designed to use such agents or toxins for hostile purposes or in armed conflict.

Article III

Each State Party to this Convention undertakes not to *transfer* to any recipient whatsoever, directly or indirectly, and not in any way to *assist, encourage*, or *induce* any State, group of States or international organizations to manufacture or otherwise acquire any of the agents, toxins, weapons, equipment or means of delivery specified in article I of this Convention.

Convergence of the Two Categories of Constraints

The relationship between the two categories of constraints has historically been an uneasy one. So long as the Geneva Protocol was regarded as merely a no-first-use declaration, with its reservations allowing retaliation to be threatened and thereby serving to bolster a tacit deterrence (whether in World War II or in the Cold War that followed), it was upon deterrence that the constraint on use principally relied. The availability of BW for use in retaliation logically ruled out a constraint on possession, because such a constraint would, on that logic, have weakened deterrence and hence the constraint on use. It was only when this tacit reliance on deterrence with a retaliatory BW capability lost whatever attraction it had once held, and universal biological disarmament became the goal of international policy and diplomatic endeavor instead, that a constraint on possession could logically be admitted to complement the constraint on use. The two constraints were henceforth on a convergence course, but their convergence is not yet perfect.

As the constraint on possession developed, especially through the expansion of BWC adherence and through the review process for the BWC, a need was recognized for the constraint on use to be strengthened. Unfortunately, the BWC does not refer explicitly to use in its operative part, although it was clear from its final preambular paragraphs that one of the purposes of the BWC was "to exclude completely the possibility of bacteriological (biological) agents and toxins being used as weapons" because "such use would be repugnant to the conscience of mankind." It was also widely recognized that any use of the prohibited objects by a State Party to the Convention must logically presuppose a prior breach of one or more of the prohibitions on development, production, stockpiling, acquisition, and retention. Nevertheless, an explicit ban on use was omitted from the Convention, an omission that left the constraint on use in need of strengthening in parallel with the constraint on possession.

This strengthening of the constraint on use was sought in three distinct ways: by the withdrawal (noted above) of Geneva Protocol reservations pertaining to retaliation; by the deliberate insertion of three "constraints on use" references in the Final Declaration of the Fourth (1996) BWC Review Conference;[2] and by an unsuccessful Iranian proposal to amend the Convention by adding the word "use" to its title and to the list of prohibitions in Article I. This last proposal was not widely welcomed, because although its substance was acceptable the procedure of formal amendment was thought too risky to be invoked on any occasion or on any issue; but the first two ways remain valid and need to be continued in a cumulative process centering on the Final Declaration of the Sixth (2006) BWC Review Conference and on any gathering of States Parties to the Geneva Protocol that may be convened, in order to achieve a more nearly perfect convergence of the two constraints.

The convergence of the two constraints can also be seen as a coming together of the law of war and the law of disarmament. In an eloquent concluding passage to his SIPRI study, *CBW and the Law of War,* Anders Boserup wrote, just after the BWC had been opened for signature: "At some point in the development of the law there is a recognition of the absolutely inhumane and uncivilized character of the use of BW under any circumstances and whatever the pretext. At this point, the logic of developments within the law of war breaks the confines of that field. The transition from a conditional prohibition to an absolute prohibition opens the way to the field of the law of disarmament: *renunciations of use fuse with renunciations of possession*" (emphasis added).[3]

Historical Perspective, 1945–2005

At its simplest, the legal status of BW since 1945 falls neatly into two halves. Although this statement will have to be qualified as the historical part of this chapter progresses, it remains essentially correct as defining the two distinct periods in which BW were very differently constrained.

From 1945 to 1975 the only legal constraint on most states with regard to BW concerned their use. Certain states, defeated in World War II, were subject to further constraints; but most states were not. They remained free to develop, produce, acquire, retain, and transfer BW. Even the constraint on use was far from absolute; but it existed, within limits, while other BW constraints for the great majority of states did not exist at all.

From 1975 to 2005 a large number of states, rising from 46 to 154, subjected themselves to additional treaty constraints on their freedom to develop, produce, stockpile, acquire, retain, and transfer BW. The question remains open whether the minority of states, which have stayed outside all treaty limitations on BW activities by being parties neither to the Geneva Protocol nor to the BWC, are nonetheless constrained by international customary law.

The midpoint in this 60-years saga was the entry into force, on 26 March 1975, of the BWC. This multilateral disarmament treaty is not the only post-1925 source of legal constraints on states in respect of BW, but it is the central one, and from it has flowed a treaty regime at the heart of the collective effort to combat the threat of BW. It has been complemented by the Chemical Weapons Convention (CWC) of 1993, in force since 29 April 1997, in respect of its prohibitions on all types of CW, including toxins as a category bridging the two Conventions; and by national measures, including legislation to give domestic legal effect to the BWC and CWC prohibitions. It may yet be complemented by further national measures, and prospectively by new international legal instruments, such as a CBW Criminalization Convention, which remain to be negotiated.

1945–1975: Constraints on Use

The main constraint on use is found in the following excerpt from the Geneva Protocol of 17 June 1925:

THE UNDERSIGNED PLENIPOTENTIARIES, in the name of their respective Governments:

Whereas the use in war of asphyxiating, poisonous, or other gases, and of all analogous liquids, materials, or devices, has been justly condemned by the general opinion of the civilized world; and

Whereas the prohibition of such use has been declared in Treaties to which the majority of Powers of the world are Parties; and

To the end that this prohibition shall be universally accepted as a part of International Law, binding alike the conscience and the practice of nations;

DECLARE:

That the High Contracting Parties, so far as they are not already Parties to Treaties prohibiting such use, accept this prohibition, *agree to extend this prohibition to the use of bacteriological methods of warfare,* and agree to be bound as between themselves according to the terms of this declaration. (Emphasis added)

1945   In 1945, 43 states were parties to the Protocol, 20 of them with reservations (shown by (R)) allowing BW retaliation: Australia(R), Austria, Belgium(R), Bulgaria(R), Canada(R), Chile(R), China(R), Czechoslovakia(R), Denmark, Egypt, Estonia(R), Ethiopia, Finland, France(R), Germany, Greece, India(R), Iran, Iraq(R), Ireland(R), Italy, Latvia, Liberia, Lithuania, Luxembourg, Mexico, Netherlands (reservation only in respect of chemical warfare), New Zealand(R), Norway, Paraguay, Poland, Portugal(R), Romania(R), South Africa(R), Spain(R), Sweden, Switzerland, Thailand, Turkey, USSR(R), UK(R), Venezuela, and Yugoslavia(R).

Six states had signed but not ratified the Protocol. These were Brazil, El Salvador, Japan, Nicaragua, US, and Uruguay. They were accordingly under the signatory obligation to refrain from acts that would defeat its object and purpose, at least until they made clear their intention not to proceed to ratification.

1945–1965   The Geneva Protocol acquired six more States Parties in the next 20 years: Hungary (1952), Ceylon (now Sri Lanka; 1954), Pakistan(R) (1960), Tanganyika (now Tanzania; 1963), Rwanda (1964), and Uganda (1965).

Accessions failed to keep pace with the number of newly decolonized states. By 1960 more states were outside the Protocol than inside. Within the UN, 55 percent of member states were parties to the Protocol in 1945, but only 43 percent in 1960 and 41 percent in 1965.

1965–1975   Then a raising of consciousness occurred. CBW came back onto the agenda of the UN General Assembly in 1966 and into the diplomacy of the Eighteen Nation Committee on Disarmament in 1968 (see Chapter 15). It also became a major issue for the International Commit-

tee of the Red Cross (ICRC) and for disarmament nongovernmental organizations (NGOs), which appealed to governments to take more resolute
and effective action against the threat of CBW.

Accession to or ratification of the Geneva Protocol was a central, and
relatively uncontentious, element in every resolution and appeal to states
under the heading of CBW. Adherence to the Geneva Protocol accordingly began to pick up speed in the years 1966–1969. The roster rose
from 49 at the end of 1965 to 68 at the end of 1969, with Cuba, Gambia,
Holy See, Cyprus, and Maldives (1966); Monaco, Niger, Sierra Leone,
Ghana, Tunisia, Madagascar, and Iceland (1967); Nigeria[R], Mongolia[R],
and Syria (1968); Israel[R], Lebanon, Nepal, and Argentina (1969).

Then came the 45th anniversary of the Protocol and the 25th anniversary of the UN. On 16 December 1969, by GA Resolution 2603B (XXIV),
the General Assembly:

> *Conscious* of the need to maintain inviolate the Geneva Protocol and to
> ensure its universal applicability . . . *Invites* all States which have not yet
> done so to accede to or ratify the Geneva Protocol in the course of 1970
> in commemoration of the 45th anniversary of its signing and the 25th
> anniversary of the UN.

A major letter-writing campaign in support of the resolution was undertaken from Geneva by the ICRC and the Special NGO Committee on
Disarmament. The result was spectacular. The Geneva Protocol roster
increased from 68 to 91 in 1970 and 1971. This expansion included ratification by Japan on 21 May 1970 and Brazil on 28 August 1970 and 21
instruments of accession or succession: Central African Republic, Côte
d'Ivoire, Dominican Republic, Ecuador, Jamaica, Kenya, Malawi, Malay-
sia, Malta, Mauritius, Morocco, Panama, and Trinidad & Tobago, (1970);
Arab Republic of (North) Yemen, Indonesia, Kuwait[R], Libya[R], Saudi
Arabia, Togo, Tonga, and Upper Volta (now Burkina Faso) (1971). As a
proportion of all the states in the world, this represented an increase from
50 percent to 65 percent in just two years. Almost two-thirds of all states
were now formally bound by the Protocol.

The years 1972 added Lesotho, and 1973 Fiji[R] and the Philippines. The
next year saw no new States Parties, but 1975 brought a historic event in
the life of the Protocol: on 10 April 1975 the US became the 95th State
Party by ratifying its signature of 50 years before.

1945–1975: Constraints on Possession

Throughout the period 1945–1975 constraints on possession of BW had been markedly less secure and extensive than the constraints on use, which by 1975 extended to 95 states by treaty (the Geneva Protocol), albeit only *inter se* and subject in many cases to reservations pertaining to retaliation, and probably to the remainder by custom, although whether the norm governing biological warfare prohibited *all* use or only *first* use was still uncertain.

Constraints on possession applied to "former enemy states" and their associates as almost incidental provisions of the military limitations imposed in peace treaties and other postwar settlements.

Peace treaties with Bulgaria, Finland, Hungary, Italy, and Romania were negotiated at the Paris Conference of Foreign Ministers and signed on 10 February 1947. Each contained an identical provision: the country was not to "manufacture or possess, either publicly or privately, any war material different in type from, or exceeding in quantity, that required for the forces permitted" with "war material" defined in annexes to the treaties, including "VI. Asphyxiating, lethal, toxic or incapacitating substances intended for war purposes, or manufactured in excess of civilian requirements." Nowhere else, in the annexes or in the main text of the treaties, was there any reference to chemical or biological weapons.

Julian Perry Robinson, from whose research this section has been drawn, commented in 1971, "It is not clear from the texts of the treaties whether the five countries were to be entirely debarred from possessing CB weapons, or whether some sort of quantitative prohibition was contemplated."[4]

In the case of Austria, not a "former enemy state" but officially the first state victim of Nazi aggression, the State Treaty for the Re-establishment of an Independent and Democratic Austria was negotiated by the four occupying powers of 1945–1955 (France, USSR, UK, US) and signed in Moscow on 15 May 1955. "Its military clauses," wrote Robinson, "contained a more explicit prohibition of CB weapons than did the earlier peace treaties."

In the absence of a peace treaty with the Federal Republic of Germany, the constraints imposed by the victors of World War II encompassed, but went far beyond, the mere prohibition of possession of nuclear, biological, and chemical weapons. Industry was to be stringently regulated, controlled, and inspected to ensure that no clandestine diversion from peaceful to military production occurred.

In sum, the effect of the peace treaties was to impose constraints on Bulgaria, Finland, Hungary, Italy, and Romania that were of uncertain application to BW; but the effect of the postwar settlements for Austria and (West) Germany respectively was to impose a clear prohibition on possession of *(inter alia)* BW, with in the former case a ban on "experimentation" as well as possession and construction, and in the latter case a prohibition on "manufacture" that left R&D formally unconstrained, but with a unique Western European Union regime of international verification introduced primarily to prevent any breach of this obligation through clandestine diversion from industry.[5]

As of 10 April 1972 the 85 original signatories to the BWC—joined by 14 more over the next three months—were under a signatory obligation to refrain from acts that would defeat the object and purpose of the Convention. Just which acts would cross the line was never defined, but it was only on 26 March 1975 with entry into force that the legal constraints on possession of BW and associated activities came fully into effect, initially for the 46 original parties and now for 154 States Parties to the BWC.

In France and the UK, national penal legislation had made a wide range of BWC-prohibited activities related to BW possession illegal and subject to criminal prosecution and penalties: in France from 9 June 1972 (by Law No. 72-467) and in the UK from 8 February 1974 (by the BW Act 1974). These laws came fully into effect immediately, and would have remained fully in effect even if the BWC had not entered into force, because in legal status they were wholly independent of the Convention. In the case of the UK the legislation was intended to enable the UK to ratify the Convention secure in the knowledge that it was already in a position to give domestic legal effect to the obligations it would be assuming upon the entry into force of the Convention. In the case of France, which had political objections to the BWC that would delay its accession until 1984, the legislation was all the more independent of the BWC, because in 1972 it appeared likely to be a permanent substitute for it.

1975–2005: Constraints on Use

For this second period of 30 years, there is relatively less to say about constraints on use but much more to say about constraints on possession, brought into full effect by the entry into force of the BWC on 26 March 1975.

Constraints on use remained rooted in the Geneva Protocol of 17 June 1925 and the related norm of customary international law.

The US ratification of the Protocol on 10 April 1975 was followed by a steady inflow of new States Parties, enlarging the roster from 95 to 132 by the end of the century:[6]

1976	Barbados, Qatar
1977	Jordan[R], Senegal, Uruguay
1978	Bhutan
1980	Papua New Guinea[R], Sudan, Vietnam[R]
1981	Solomon Islands[R]
1983	Cambodia, Guatemala
1985	Bolivia, Peru
1986	Afghanistan, Benin, People's Democratic Republic of (South) Yemen[R] (which united with the Yemen Arab Republic, already a State Party to the Protocol, on 22 May 1990)
1988	Antigua & Barbuda, Bahrain[R], Saint Lucia
1989	Albania, Bangladesh[R], Cameroon, Equatorial Guinea, Grenada, Guinea-Bissau, Democratic People's Republic of (North) Korea[R], Republic of (South) Korea[R], Laos, Saint Kitts & Nevis
1990	Angola[R], Nicaragua
1991	Cape Verde, Liechtenstein, Swaziland
1992	Algeria[R], Russia (in succession to the USSR, a party since 1928)[R]
1993	Czech Republic (in succession to Czechoslovakia, a party since 1938)
1997	Slovakia (also in succession to Czechoslovakia)
1999	Saint Vincent & the Grenadines

With the US ratifying in 1975, Uruguay in 1977, and Nicaragua in 1990, only one of the Protocol's original signatories—El Salvador—was left with its signature unratified and hence bound only by the obligation to refrain from acts that would defeat the object and purpose of the Protocol. The great majority of nonparties were now states that had never signed the Protocol; by 2004 they totaled about 60. Almost all were by then members of the UN, in which the proportion of member states that were parties to the Protocol had risen from 67 percent in 1971 to 77 percent in 1990 but then fell back to 70 percent in 2004.

WITHDRAWAL OF RESERVATIONS   Reservations pertaining to retaliation were withdrawn by Ireland in 1972, Barbados in 1976, Australia in 1986, New Zealand in 1989, Mongolia and Czechoslovakia in 1990, Romania, Bulgaria, and Chile in 1991, Spain in 1992, South Africa and France in 1996, Belgium in 1997, Estonia in 1999, and Russia in 2001. Canada and the UK modified their reservations in 1991 to assimilate themselves to the position—in which the reservation applied only to CW—occupied by the Netherlands and the US ever since they had ratified the Protocol in 1930 and 1975 respectively. The Republic of (South) Korea followed with a similar modification in 2002. Canada and the UK proceeded to full withdrawal of their remaining reservations after the 1997 entry into force of the Chemical Weapons Convention.

Appeals for withdrawal of the remaining reservations have been made by BWC States Parties in the Final Declarations of the Third (1991) and Fourth (1996) Review Conferences. The latter adopted two paragraphs, the first on the initiative of France and the Netherlands, the second on the initiative of Chile, Mexico, and Peru, which significantly strengthened this part of the Final Declaration from the 1991 text on which it was based and explained its reasoning (pioneered in the Irish statement of 1972 quoted above):

Article VIII

6. The Conference welcomes the actions which States Parties have taken to withdraw their reservations to the 1925 Geneva Protocol related to the Biological and Toxin Weapons Convention, and calls upon those States Parties that continue to maintain pertinent reservations to the 1925 Geneva Protocol to withdraw those reservations, and to notify the Depositary of the 1925 Geneva Protocol of their withdrawals without delay.

7. The Conference notes that reservations concerning retaliation, through the use of any of the objects prohibited by the Biological and Toxin Weapons Convention, even conditional, are totally incompatible

with the absolute and universal prohibition of the development, production, stockpiling, acquisition and retention of bacteriological (biological) and toxin weapons, with the aim to exclude completely and forever the possibility of their use.[7]

This clear declaration challenges states to act and not to hide behind the supposed supersession of the Protocol by the BWC as an excuse for inaction. In any case the BWC did not explicitly ban *use:* it was concerned with other activities, grouped around the *possession* of BW, and the ban on use had to be logically derived from the combined effect of the actual BWC prohibitions and belatedly emphasized in the Final Declaration of the Fourth Review Conference. In this context, there is all the more reason to regularize the Geneva Protocol status of BWC parties. Even if its import would be more political than legal, withdrawal of reservations still has an important part to play in achieving consistency of obligations and emphasizing the absolute character of the ban on use of BW.

Withdrawal of the remaining reservations would also help to clarify the extent of the norm against use in customary international law. As Adam Roberts and Richard Guelff observed, writing in 1999:

> Partly because of the large number of states bound by the 1925 Geneva Protocol and the repeated expressions of support for it in the UN General Assembly and Security Council, the prohibitions embodied in the Protocol are widely viewed as having become a part of customary international law. As customary international law, the Protocol would be applicable to all states and not merely those which have become formally bound by ratification, accession or succession. However, some suggest that the controversy over the Protocol's interpretation, as well as the character of reservations, have reduced the Protocol's usefulness as a guide to customary international law in this area. The weight of opinion has long been that at least the first use of lethal chemical and BW is prohibited by customary international law.[8]

What is needed now is an authoritative statement that the norm against use of BW in customary international law has evolved to become absolute. It would be much easier to generate such a statement if the remaining Geneva Protocol reservations were withdrawn.

1975–2005: Constraints on Possession

In the period 1975–2005 states were subject to constraints on possession of BW that flowed mainly but not exclusively from the BWC. Formally states may be grouped into four categories according to their BWC treaty status and other sources of international legal obligation:

- *States Parties to the BWC* (46 in 1975 at entry into force, 154 by 2005). These states were subject to the full range of constraints contained in the BWC (see below).
- *States signatories but not parties to the BWC* (64 in 1975, reduced to 16 by 2005). These states were subject only to the signatory obligation to refrain from acts that would defeat the object and purpose of the BWC, in the absence of any statement that they did not intend to proceed to ratification.
- *States neither parties nor signatories to the BWC* (approximately 22 in 2005). These states were subject only to the corresponding norm of customary international law, insofar as a norm had evolved that was coextensive with the BWC constraints.
- *States that were "former enemy states" or associated with them.* These states were subject to the provisions on BW in their respective peace treaties with the victors of World War II or other postwar settlements dating respectively from 1947 (Bulgaria, Finland, Hungary, Italy, and Romania) and 1955 (Austria and the Federal Republic of Germany). However, these provisions were in general less far-reaching and specific than the obligations flowing from the BWC, to which all the states in this category became parties between 1975 and 1983.

STATES PARTIES TO THE BWC   The following is a summary of the structure of obligations in the BWC, article by article:[9]

Article I   Never in any circumstances to develop, produce, stockpile, or otherwise acquire or retain: (1) microbial or other biological agents, or toxins whatever their origin or method of production, of types and in quantities that have no justification for prophylactic, protective, or other peaceful purposes; (2) weapons, equipment, or means of delivery designed to use such agents or toxins for hostile purposes or in armed conflict

Article II   To destroy them or divert them to peaceful purposes, as soon as possible but not later than nine months after entry into force

Article III   Not to transfer them to any recipient whatsoever, and not in any way to assist, encourage, or induce any state, group of states, or international organizations to manufacture or otherwise acquire them

Article IV   To take any necessary measures to prohibit and prevent the Article I activities within the territory of each State Party, under its jurisdiction, or under its control anywhere

Article V   To consult one another and cooperate in solving any problems that may arise

Article VI   To cooperate in carrying out any investigation that the Security Council may initiate following a complaint lodged with the Council under this article

Article VII   To provide or support assistance to any party to the Convention that so requests, if the Security Council decides that such party has been exposed to danger as a result of violation of the Convention

Article IX   To continue negotiations in good faith with a view to reaching early agreement on effective measures for an equivalent set of prohibitions on CW and for their destruction

Article X(1)   To facilitate, and participate in, the fullest possible exchange of equipment, materials, and scientific and technological information for the use of biological agents and toxins for peaceful purposes, and the application of scientific discoveries in the field of biology for prevention of disease or for other peaceful purposes; *(2)*   To implement the Convention in a manner designed to avoid hampering the economic or technological development of States Parties or international cooperation in the field of peaceful biological activities

Articles I and III contain the legal constraints on possession and related activities at the international level, and Article IV requires them to be given domestic legal effect by each State Party in such a way as effectively to prohibit and prevent any such activities occurring within its territory, under its jurisdiction, or under its control anywhere.

The 46 original parties, for which the BWC entered into force on 26 March 1975, were Afghanistan, Austria, Barbados, Brazil, Bulgaria, Byelorussian Soviet Socialist Republic (now Belarus), Canada, Costa

Rica, Cyprus, Czechoslovakia, Denmark, Dominican Republic, Ecuador, Fiji, Finland, German Democratic Republic (dissolved in 1990), Guatemala, Hungary, Iceland, India, Iran, Ireland, Kuwait, Laos, Lebanon, Mauritius, Mexico, Mongolia, New Zealand, Niger, Nigeria, Norway, Pakistan, Panama, Philippines, Poland, San Marino, Saudi Arabia, Senegal, Tunisia, Turkey, Ukrainian Soviet Socialist Republic (now Ukraine), Union of Soviet Socialist Republics (to the international obligations of which the Russian Federation succeeded on 26 December 1991), United Kingdom, United States, and Yugoslavia (now Serbia & Montenegro).

The following ratified or acceded subsequently, bringing the roster up to the total shown for each year:[10]

1975	Bolivia, Dahomey (now Benin), Ethiopia, Ghana, Greece, Italy, Jamaica, Jordan, Malta, Nicaragua, Portugal, Qatar, Rwanda, Singapore, South Africa, Thailand (62)
1976	Cuba, Guinea-Bissau, Kenya, Luxembourg, Paraguay, Sierra Leone, Sweden, Switzerland, Togo, Tonga (72)
1977	Australia, Cape Verde, Lesotho, Zaire (now Democratic Republic of Congo) (76)
1978	Bhutan, Congo, Dominica, Venezuela (80)
1979	Argentina, Belgium, Honduras, Romania, Sao Tome & Principe, Seychelles, Spain, People's Democratic Republic of (South) Yemen (united with North Yemen since 1990) (88)
1980	Chile, Papua New Guinea, Vietnam (91)
1981	Netherlands, Solomon Islands, Uruguay (94)
1982	Japan, Libya (96)
1983	Cambodia, Colombia, Federal Republic of Germany (99)
1984	China, France (101)
1985	Bangladesh, Peru (103)
1986	Bahamas, Belize, Grenada, Saint Lucia, Sri Lanka (108)
1987	Democratic People's Republic of (North) Korea, Republic of (South) Korea (110)
1988	Bahrain (111)
1989	Equatorial Guinea (112)
1990	Vanuatu, Zimbabwe (113, as German Democratic Republic ceased to exist)
1991	Brunei Darussalam, Burkina Faso, El Salvador, Gambia, Iraq, Liechtenstein, Malaysia, Saint Kitts & Nevis, Swaziland (122)
1992	Albania, Botswana, Indonesia, Oman, Slovenia, Uganda (128)

1993   Croatia, Czech Republic (in succession to Czechoslovakia),
        Estonia, Maldives, Slovakia (in succession to Czechoslovakia),
        Suriname (133)
1994   Armenia, Bosnia & Herzegovina (135)
1996   Georgia, Former Yugoslav Republic of Macedonia,
        Turkmenistan, Uzbekistan (139)
1997   Latvia (140)
1998   Lithuania (141)
1999   Monaco, Saint Vincent & the Grenadines (143)
2001   Algeria (144)
2002   Holy See, Morocco (146)
2003   Antigua & Barbuda, Mali, Palau, Sudan, Timor-Leste (151)
2004   Azerbaijan, Kyrgyzstan (153)
2005   Moldavia (154)

STATES SIGNATORIES BUT NOT PARTIES   States that signed the
BWC but did not ratify it numbered 18 for many years. Then in 2002 Morocco and in 2003 Mali ratified the BWC. The remaining 16 states in
this category are Burundi, Central African Republic, Côte d'Ivoire, Egypt,
Gabon, Guyana, Haiti, Liberia, Madagascar, Malawi, Myanmar (which
had signed as Burma), Nepal, Somalia, Syria, Tanzania, and United Arab
Emirates. For over 30 years these states have been subject to the signatory obligation to refrain from acts that would defeat the object and purpose of the Convention.

STATES NEITHER PARTIES NOR SIGNATORIES   There is no official
list of states in this category—understandably, because it would inevita-
bly run into controversy at the margins of statehood. Caution therefore
restricts us to the use of the UN membership list, without prejudice to
the statehood claims of nonmembers. There are 22 member states of the
UN that are neither parties nor signatories to the BWC. These are
Andorra, Angola, Cameroon, Chad, Comoros, Djibouti, Eritrea, Guinea,
Israel, Kazakhstan, Kiribati, Marshall Islands, Mauritania, Micronesia,
Mozambique, Namibia, Nauru, Samoa, Tajikistan, Trinidad & Tobago,
Tuvalu, and Zambia.

    Does customary international law constrain these states? Evidence for
the existence of a norm may be found in UN resolutions and other authoritative sources, but it would be stronger if state practice were consis-

tent. Noncompliance with BWC obligations (see Chapters 6, 8, and 9) has made it difficult to claim consistency of state practice as evidence for a customary norm.

Outstanding Issues

Geneva Protocol

The Geneva Protocol is notoriously imprecise; but its problems over inexact correspondence of terms in the adjectives (other/*similaires*) applied to gases/*gaz* in the two equally authentic languages only ever affected the CW prohibition.

But what of the difference between "bacteriological" and "biological"? Did "bacteriological methods of warfare" encompass methods that relied on viruses (which in 1925 were only just coming to be classified as a separate class of microorganism, rather than as very small bacteria) and other nonbacterial microorganisms?

In 1968 the UK professed to see in the supposed narrowness of "bacteriological" one of the reasons why a new instrument, strengthening the Geneva Protocol prohibition, was needed. Otherwise the deliberate spreading of viral or other nonbacterial infections might escape prohibition.[11] But other authorities were unimpressed by this apprehension. They found no loophole. They preferred to see the Protocol as applying to all "microbial and other biological agents" of warfare "which are intended to cause disease or death in man, animals or plants," and R. R. Baxter and Thomas Buergenthal noted that this interpretation had US government support.[12] George Bunn concluded from the negotiating history that "there . . . is no justification for limiting the scope of the ban on 'bacteriological warfare' because some new diseases have been discovered since 1925 which we do not classify as bacteriological."[13] This view was widely accepted. It was bolstered by evidence that the negotiators at Geneva in 1925 had used the two adjectives interchangeably. It was from the Military Committee of the 1925 conference which recommended "the prohibition of chemical and *biological* warfare" (emphasis added) that the "Polish addendum" prohibiting *bacteriological* methods of warfare was derived.[14]

Accordingly, Boserup could write in 1972: "Even though in its strict scientific meaning the term 'bacteriological' is narrower than the term 'biological,' it has always been accepted that in the *legal* context of the Protocol the two words are exact synonyms. In most texts the matter is taken for granted and the question not even raised."[15]

BWC

Most of the outstanding legal issues in the BWC have to do with the scope of definitions; and most of the definitional issues that have complicated interpretation of the BWC have arisen from Article I. Here they will be formulated as questions, to which answers are offered. The answers are necessarily tentative where no authoritative interpretation has been handed down; and this section ends, indeed, with a plea for the BWC States Parties to correct this situation by installing a mechanism for the resolution of such legal issues.

1. How comprehensive, in respect of developments in science and technology, is the formulation "microbial or other biological agents or toxins"?
Successive Review Conferences have tried to answer this question, but how they have answered it has varied according to the prevalent anxieties of the time.

In 1980 the First Review Conference simply declared: "The Conference believes that Article I has proved sufficiently comprehensive to have covered recent scientific and technological developments relevant to the Convention,"[16] without specifying whether this very general expression of confidence referred to the causative agents of four newly identified (between 1967 and 1976) diseases—Ebola virus, Lassa virus, *Legionella,* and Marburg virus—or to the "revolution in genetics" that had likewise been reported to the conference.

In 1986 the Second Review Conference declared: "The Conference reaffirms that the Convention unequivocally applies to all natural or artificially created microbial or other biological agents or toxins whatever their origin or method of production. Consequently, toxins (both proteinaceous and non-proteinaceous) of a microbial, animal or vegetable nature and their synthetically produced analogues are covered."[17]

In 1991 the Third Review Conference declared that the scope of the definition embraced "microbial or other biological agents or toxins harmful to plants and animals, as well as humans," and that "all microbial or other biological agents or toxins, naturally or artificially created or altered," were unequivocally covered.[18]

The words "or altered" extended the 1986 reaffirmation ("all natural or artificially created") to keep up with emerging concerns about genetic manipulation of bacteria, viruses, and toxins.

The same paragraph of the 1991 declaration included a more general response to some of the anxieties then current, in a sentence that emphasized the comprehensiveness of the formulation in Article I: "The Conference, conscious of apprehensions arising from relevant scientific and technological developments, *inter alia,* in the fields of microbiology, genetic engineering and biotechnology, and the possibilities of their use for purposes inconsistent with the objectives and the provisions of the Convention, reaffirms that the undertaking given by the States Parties in Article I applies to all such developments."[19]

In 1996 the Fourth Review Conference reaffirmed all the above-mentioned interpretative paragraphs of 1986 and 1991 and carried them forward in its declaration, together with new references to areas of scientific development that had acquired even greater salience since those earlier conferences: "molecular biology" and "any applications resulting from genome studies" were added to the 1991 trio of "microbiology, genetic engineering and biotechnology."[20]

In responding to new challenges from science and technology, which might, if not named and confronted, threaten the credibility of the BWC or alter the balance of incentives to compliance, Review Conferences are performing one of their essential functions, found in Article XII of the Convention itself: to "take into account any new scientific and technological developments relevant to the Convention."

Authoritative interpretation of the BWC proceeds by drawing out and making explicit the latent implications of the text. Its comprehensiveness was already signaled by the words "whatever their origin or method of production" in Article I. What States Parties were doing in the declarations of successive conferences from 1980 to 1996 was recording and extending, in a cumulative process, their shared understanding of these implications. The 1986 word "consequently" was well chosen. The States Parties were building up an authoritative but nonexhaustive list of examples that they specifically affirmed to be covered by Article I, without at any time allowing such a list to detract by omission from the comprehensive character of the prohibition as expressed in this formulation.

One further issue of scope, the "infestation" problem, will be considered separately below (at question 4).

2. How comprehensive, in respect of research, is the formulation "never in any circumstances to develop, produce, stockpile or otherwise acquire or retain"?
Here interpretation has proceeded much more slowly. Indeed, only one experimental activity has been specifically labeled as falling under this prohibition. In 1991 the Third Review Conference declared: "The Conference notes that experimentation involving open-air release of pathogens or toxins harmful to man, animals and plants that has no justification for prophylactic, protective or other peaceful purposes is inconsistent with the undertakings contained in Article I."[21]

What remains unclear is the boundary between research and development. Development, like production, stockpiling, acquisition, and retention is an activity prohibited unless justified for "prophylactic, protective or other peaceful purposes" (the 4P formula) with which the types and quantities of the agents or toxins involved are consistent. Research is not mentioned anywhere in the BWC.

In one view, research is unconstrained because it was finally omitted from the BWC: even "research aimed at production of the kind prohibited," which the UK Draft Convention of 1969 had, logically enough, envisaged the States Parties undertaking "not to conduct, assist or permit,"[22] escaped regulation, let alone prohibition.

In another view, research is constrained by the 4P formula, either because it is functionally indistinguishable from development in a seamless R&D process, or because research on microbial or other biological agents or toxins can be conducted only if those agents or toxins have first been acquired or retained; and acquisition or retention other than in types and quantities that are justified for prophylactic, protective, or other peaceful purposes would be a prior breach of Article I.

No authoritative statement has yet declared that research is constrained by the 4P formula, either via development or via acquisition/retention. It is sometimes informally suggested that "offensive" research may be regarded as incompatible with BWC obligations and "defensive" research as compatible. Yet the words "offensive" and "defensive" do not occur in the Convention, any more than "research" does. No Review Conference has yet pronounced on research, other than on experimentation involving open-air release of pathogens or toxins.

Silence on the legal status of research has left the scope of the formulation in this part of Article I uncertain. Any conclusion on legal status can be only tentative. An earlier study argued that "incorporation of research within the [BWC] treaty regime can proceed . . . through the final declarations of successive review conferences" and that the next one "might, for example, return to the language of the British draft conventions of 1969 and 1970 and state that the parties recognize an obligation 'not to conduct, assist or permit research aimed at production of the kind prohibited' under Article I." It "might also declare that research and development are so intricately related to each other that, in order for the ban on BTW [T for toxin] development to be upheld, it is necessary for research to be constrained by the same condition. Research would then require" justification by the 4P formula, and "the general purpose criterion would be understood as also encompassing research."[23]

3. Does the 4P formula apply to the "weapons, equipment or means of delivery" prohibited in the second part of Article I; and what is the legal effect of the qualifying phrase "designed to use such agents or toxins for hostile purposes or in armed conflict"?

The prohibition on "weapons, equipment or means of delivery" is qualified by a purpose criterion: "designed to use such agents or toxins for hostile purposes or in armed conflict." The words "such agents or toxins" refer back to the first part of Article I and thereby import the 4P formula, but only if the application of the word "such" is to the complete category "microbial or other biological agents, or toxins whatever their origin or method of production, of types and in quantities that have no justification for prophylactic, protective or other peaceful purposes." If, however, the reference of the word "such" is limited to the first part of the phrase only—"microbial or other biological agents, or toxins"—then the prohibition on weapons, equipment, and means of delivery is absolute, because the exceptions allowed by the 4P formula do not apply.

On the other hand, a new qualifying phrase does undoubtedly apply—

"designed to use [them] for hostile purposes or in armed conflict"—
which may be regarded as a functional equivalent of the (possibly) missing 4P formula. On this reasoning, just as under its first part Article I
allows States Parties to develop, produce, stockpile, acquire, or retain
agents or toxins provided they do so for prophylactic, protective, or other
peaceful purposes with which the types and quantities of agents or toxins
are consistent, so under the second limb Article I allows them to develop,
produce, stockpile, acquire, or retain weapons, equipment, and means of
delivery provided they are not designed to use agents or toxins for hostile
purposes or in armed conflict.

It was presumably on this reasoning that US government lawyers were
relying (those from the State Department reportedly dissenting) when
they apparently advised that US efforts to replicate key parts of a biological bomb designed in the former Soviet Union, in order to better understand the BW threat, would not be in breach of Article I. The absence of
an intention "to use such agents or toxins for hostile purposes or in
armed conflict" would on this argument (which depends crucially upon
the interpretation of "designed" to mean the same as "intended") render
the design of such a weapon compatible with US obligations under the
BWC.[24]

There remains the further objection that to advance such a permissive
interpretation of the design criterion is undesirably subjective, allowing
each State Party to decide for itself whether its intention, in designing a
biological bomb, is to hold it in readiness to use it for hostile purposes or
in armed conflict, or merely to acquire a better understanding of how
others might use it in such circumstances. There can be no assurance that
this latter intention will not change over time and carry the project over
the border of international legality (assuming that it was legal in the first
place). Because the assessment of intention is the State Party's own, it is
subject to no international scrutiny. It is hardly surprising if others' assessments differ and are seldom so charitable. Skepticism prevails. Perceptions of vulnerability translate easily into perceptions of threat. As
Malcolm Dando and Simon Whitby recommend, "This particular case
underlines also the essential need for mechanisms through which States
Parties can demonstrate their compliance with the Convention and
through which possible ambiguities, anomalies and uncertainties can be
addressed."[25]

4. Does the BWC prohibit agents of infestation?
This issue arose acutely on 30 June 1997, when Cuba formally invoked the BWC's Article V contingency mechanism, namely a Consultative Meeting open to all States Parties at expert level, over an allegation of infestation of crops.[26]

Cuba accused the US of deliberately spraying *Thrips palmi* pests from an aircraft overflying its western province of Matanzas so that the insects would devour its potato crop, thereby causing economic damage to Cuba. *Thrips* had been detected on 18 December 1996; the overflight had taken place, using a permitted air corridor, on 21 October. The US denied any connection between the two events.

Denmark and the Netherlands explicitly, and other States Parties tacitly, reserved their positions on the legal issue of the BWC's scope during the BWC Consultative Meeting process while addressing other issues in dispute between Cuba and the US over events between October and December 1996 and the origin of the *Thrips* infestation.[27]

The legal issue concerns the scope of Article I. When introducing the British Draft Convention in 1969, UK Minister of State Fred W. Mulley had explained its intended scope as follows: "The Convention is, of course, aimed at prohibiting the use for hostile purposes of disease-carrying microbes . . . However, it is possible to envisage the use in war of biological agents which are not microbes: hookworm, for instance, or the bilharzia worm, or even crop-destroying insects such as locusts or Colorado beetles."[28]

The British Draft Convention had accordingly given infestation equal prominence with infection in its proposed Article I: "Each of the Parties to the Convention undertakes never in any circumstances, by making use for hostile purposes of microbial or other biological agents causing death or disease by infection or infestation in man, other animals, or crops, to engage in biological methods of warfare."[29]

But the words "by infection or infestation" had fallen away during negotiation of the BWC, leaving uncertainty as to whether, in the definitive Article I, "microbial or other biological agents" included agents of infestation (such as Colorado beetles, locusts or, indeed, *Thrips palmi*) as well as agents of infection.

There has been no doubt since 1991 that "agents or toxins harmful to plants and animals, as well as humans" are covered by the definition in

Article I.[30] The only doubt is over the route by which harm is caused. Some who have examined the negotiating record of the BWC claim that agents of infestation were understood not to be covered by Article I as it was finally negotiated. However, it is hard to see how the actual text of Article I conveys this understanding. "The method of waging war was no longer specified, and thus it can only be assumed that actions preparatory to *both* infection and infestation are prohibited."[31] But is silence a sufficient basis for this assumption? Authoritative confirmation is needed. Only then will agents of infestation be conclusively covered by the BWC.

Given the prevailing uncertainty, it was prudent of those States Parties which took the view that the BWC does not cover agents of infestation to allow the Cuban allegation to be heard in 1997, on a "without prejudice" basis, rather than to try to have it excluded from the purview of the Article V contingency mechanism, which Cuba had invoked. Denmark, the Netherlands, and others who shared their doubts did not compromise their positions on the legal issue by taking part in the Consultative Meeting process. Much more harm would have been caused if they had tried to block it. Procedurally it was generally judged helpful to the BWC: a useful airing of views and pooling of scientific and technical input albeit with an inconclusive outcome.[32]

Conclusion

The existence of outstanding legal issues points to a need for BWC States Parties to resume the interrupted process of cumulative interpretation, achieved through recording extended understandings of the implications of the BWC in the declarations of their review conferences, on legal as well as scientific and other issues relevant to the Convention.

They would benefit from the expert advice of a legal advisory panel in the service of States Parties collectively, working alongside a scientific advisory panel. Such a panel could help with new issues as well as with those already identified. It could bring an international dimension to what has up to now been a solely national process of generating legal opinions, surely an unsatisfactory state of affairs for a multilateral treaty 30 years old with 154 States Parties, though, sadly, only one aspect of the wider institutional deficit that persists to the detriment of the Convention.

CHAPTER 17

Analysis and Implications

MALCOLM DANDO

GRAHAM PEARSON

LAJOS RÓZSA

JULIAN PERRY ROBINSON

MARK WHEELIS

In a little over 100 years we have moved from a situation of almost total ignorance to a significant understanding of pathogens and the diseases they, and the toxins they produce, cause in humans, animals, and plants. Our already considerable capabilities to use this knowledge for good or, regrettably, for ill are being profoundly enhanced by the ongoing genomics revolution in the life sciences. There must indeed be a risk that in the coming century such knowledge may be used for hostile purposes and in warfare.

Our objectives in undertaking this study, therefore, were, first, to provide as complete and accurate an history of BW since 1945 as is possible with the data available today, and second, to draw possible lessons from that history with regard to strengthening the long-standing total prohibition of biological weapons (BW).

The Central Issues

In considering the history of offensive BW programs since 1945 we address three central issues: (1) Why have states continued or begun programs for acquiring BW? (2) Why have states terminated BW programs? and (3) How have states demonstrated to other states that they have indeed terminated their BW programs? We consider these issues in two periods: before the entry into force of the Biological Weapons Convention (BWC), and since.

Before the Entry into Force of the BWC

In 1945, following the end of World War II, a number of states—Canada, the UK, the US, and the USSR—continued or, in the case of France, quickly restarted national programs to study or develop BW. They did so because of the seemingly devastating potential of the weapons, a potential recognized by their inclusion in the category of "weapons of mass destruction" that the Security Council of the United Nations was at that time defining. In most of these countries the priority assigned to BW during the late 1940s and early 1950s was comparable to that given to nuclear weapons. Although all had signed and, except for the US, had ratified the Geneva Protocol of 1925, which prohibited the use in war of asphyxiating, poisonous, or other gases, and of bacteriological means of warfare, all had entered reservations to the effect that they would not be bound by the prohibition in regard to states not party to the treaty, to an adversary that resorted to the prohibited weapons, or to an ally of such an adversary.

Within a decade the UK had begun to produce nuclear weapons and had reduced the priority accorded to its BW program, which was in effect abandoned altogether in the late 1950s. As for Canada, the NATO alliance and its security provisions seem to have served as sufficient reason for the country no longer to seek its own BW after 1956. France exploded its first nuclear device in 1960 and had also by then begun to shut down its BW program. Consequently, by the latter part of the 1960s only the USSR and the US of those five countries were continuing their offensive biological warfare programs with any vigor, alongside nuclear and chemical weapons (CW) programs.

Perceptions of the utility of BW had been changing during the 1950s and 1960s as nuclear weapons came increasingly to be regarded as an ultimate weapon that, in the hands of the two alliance leaders, could enhance the security of all member states of NATO and the Warsaw Pact. Concepts of "limited war" and—as always throughout the long history of biological warfare—sabotage and other covert operations now dominated the uses seen, at least in the West, for BW. Simultaneously, the technology of BW was advancing from the cluster bomb and other such agent-delivery systems envisaged at the end of World War II to aerosol-generating devices capable of enormously greater area effectiveness, comparable to that of nuclear, even thermonuclear, weapons. Increasing reliance on nuclear deterrence was thus removing the major incentive for states to acquire BW as an element in ensuring their national security at just the time that their potential utility for attacking large targets was becoming apparent.

It was into this security environment in the mid-1960s that the UK introduced its ideas for strengthening the 1925 Geneva Protocol in such a way as to prohibit CBW altogether. It became clear that at least the US was not then prepared to see CW included in such a disarmament treaty. As BW, unlike CW, had not been used extensively in war and consequently had not entered the mainstream of military doctrine and planning, there was less resistance to the idea that they should be totally prohibited. There was also an appreciation that there would be security benefits in the total prohibition of BW, as this would prevent the proliferation of weapons that had potential for effectiveness against large target areas, as well as for clandestine use, and that were potentially accessible to a much wider range of countries than could develop nuclear weapons.

The BWC was negotiated in 1969–1972 and opened for signature on 10 April 1972 in London, Moscow, and Washington, with the governments of the UK, the USSR, and the US as co-depositaries (the Russian Federation taking over this responsibility from the Soviet Union in 1992). It entered into force three years later, on 26 March 1975, with 46 original parties. The treaty for the first time totally prohibited an entire class of weapons.

Since the BWC

By 2005 the number of States Parties to the BWC had grown to 155 along with 16 signatory states. In addition, some states, mindful of the "never in any circumstances" provision of BWC Article I, had taken steps to lift the reservations they had entered when joining the 1925 Geneva Protocol, whereby they retained the right to retaliate in kind against biological (or chemical) attack. Although the US ratified the Geneva Protocol in 1975, it retains its reservation in regard to CW, and in December 2004 abstained from voting on UN General Assembly Resolution 59/70 calling upon all Protocol States Parties to lift their reservations.

BW programs that continued in States Parties to the BWC, such as the Soviet Union, or that were started subsequent to the entry into force of the BWC, such as in South Africa and Iraq, were consequently illegal and in breach of the solemn undertakings given by the countries concerned. The Soviet Union, succeeded by Russia, had ratified the BWC on 26 March 1975, and South Africa had done so on 3 November 1975. Iraq had signed (11 May 1972) though not yet ratified the BWC when it initiated its BW program. But in international law, the act of signing a treaty incurs the obligation to refrain from acts contrary to the object and purpose of the treaty, even if the signature remains unratified. Consequently Iraq, too, was in breach of the BWC in regard to its BW program. Iraq was also in flagrant breach of the Geneva Protocol, which it had joined in 1931, when it used CW against Iran in the 1980s.

In the case of the Soviet Union, it is now evident that its BW program was intensified following the entry into force of the BWC. Information is not yet available from Russian archives as to why this happened. Various ideas have been advanced, including Soviet disbelief that the US had terminated its own weapons program, and the Soviet perception that BW would acquire much added value in a world in which no other states possessed them. The program continued until at least 1992, when President Boris Yeltsin decreed that work on BW in Russia should stop. Although the UK and the US engaged with the USSR and then with Russia in a trilateral program intended to demonstrate that all the BW programs in Russia had indeed ended, the work proved inconclusive because of sustained refusal by Russia to allow UK and US access to its military biological facilities. This obstruction of transparency has resulted in suspicions that continue to this day.

The South African CBW program was restarted in the 1980s, when apartheid-era South Africa was becoming increasingly isolated internationally. The South African government was concerned that its forces in Angola might be attacked by such weapons, and its spokesmen have stated that it wished to explore the option of being able to retaliate in kind in accordance with South Africa's reservation to the Geneva Protocol (which was not lifted until 1996). South Africa had joined the Protocol in 1930, but Angola did not do so until October 1990. The biological component of the South African program was not clearly described either in open official documentation or in the declaration made by South Africa under the BWC confidence-building measures. Its objectives appear to have been more toward possible use in sabotage and assassination than as a potential weapon to be used in general war. A change in government together with joint UK-US démarches to South Africa led to the abandonment of what had been a well-hidden secret program known to relatively few in South Africa, which was in clear violation of South Africa's obligations since 1975 as a State Party to the BWC.

The Iraqi BW program is now known to have started in 1974, two years after Iraq had signed the BWC. It is clear that Iraq sought nuclear, biological, and chemical weapons in order to enhance its military capability and influence in the region. Iraq used CW, including nerve gases, extensively during its 1980–1988 war with Iran. As with past CW programs elsewhere in the world, the Iraqi one soon sought to develop toxins and pathogenic bacteria as a means of increasing the potency of the payloads carried by the weapons.

The termination of Iraq's BW program was demanded by the UN Security Council in its Resolution 687 (1991), after the response by the US-led coalition forces to the Iraqi attack on Kuwait. This resolution required not only that all Iraqi WMD be destroyed under the supervision of the United Nations Special Commission (UNSCOM) on Iraq, but also that Iraq ratify the BWC, which it duly did on 18 April 1991. The eradication of the Iraqi WMD programs and the ongoing monitoring and verification program to ensure that Iraq did not continue or resume WMD activities were to be carried out under the supervision of UNSCOM. It came to be widely believed that Iraq did not in fact abandon its WMD work but instead endeavored to conceal whatever it could of its past programs and to continue work to develop and acquire such weapons. It is evident that the Iraqi government took no steps to abandon its BW program or to build international confidence that it had renounced the program. Concerns and suspicions continued throughout the UNSCOM years and the rather different UNMOVIC years that succeeded them. Since the US-led invasion of Iraq in 2003, the Iraq Survey Group has undertaken detailed investigations from which it appears that Iraq was intent on preserving at least the capacity to restart an offensive program, but was not actively engaged in production or stockpiling of BW during this period.

Thus the straightforward answers to our three questions are not in dispute. The major states that ended on the winning side after World War II had developed BW programs because such weapons were seen to be potentially important militarily for retaliation in kind, and they continued or restarted them for the same reasons. The two states definitely known to have begun offensive BW programs later in the century also had military reasons for their programs. Programs were terminated because of regime change (South Africa and possibly Russia), because of imposed disarmament following defeat in war (Iraq), because BW were overshadowed by nuclear weapons (the UK, France, the US), or because major partners terminated their program (Canada). Finally, when states terminated their offensive programs, because they had been kept highly secret there was generally little attention given to providing information to convince other states that the offensive programs had indeed been terminated.

Overarching Themes

In the history of the period since 1945 there are several recurrent themes that help to elucidate the nature of BW programs: (1) changing perceptions of BW, and of their utility or lack of utility; (2) the limitations of intelligence; (3) the shifting balance between secrecy and transparency, suspicion and confidence; and (4) the influence of treaties and of international technological collaboration upon national BW programs.

Changing Perceptions of BW

By the end of World War II many countries had concluded that BW would be effective. The UK stockpile of cattle cakes inoculated with *Bacillus anthracis,* developed during World War II as an antianimal weapon, was destroyed after the war, and the subsequent high-priority programs in the UK, US, and Canada focused initially on weapons to deliver antipersonnel agents, such as anthrax spores, against military targets; but the development of antiagricultural BW, particularly means of attacking staple crops, also became important. As the nuclear weapons programs in the US and UK came to fruition, attention to BW began to be focused more on sabotage and special forces operations. Nevertheless, as the programs developed attention turned also toward large area coverage and on attaining a casualty-causing capability comparable to that of nuclear weapons.

With the increasing importance of nuclear weapons, BW were abandoned in the 1950s by the UK and incrementally by France in the 1960s. The nuclear capabilities of NATO led Canada likewise to abandon BW. The 1960s and early 1970s saw the US abandonment of its BW program (in 1969) and the negotiation of the BWC. However, following the entry into force of the BWC, the Soviet Union, one of the three co-depositaries, continued and accelerated its BW program for reasons that are still unclear.

The South African program in the 1980s was much smaller in scale, apparently focusing on sabotage and assassination applications. The Iraqi program had broader aims, with the production in the late 1980s of significant amounts of agent for delivery in missile warheads and aerial bombs. The anthrax letter attacks in the US in the autumn of 2001, which have produced so many concerns about bioterrorism attacks, are best regarded as being aimed at sabotage and creating alarm.

In the years after World War II the approach adopted, at least in the West, was to acquire BW to be used to retaliate in kind should such weapons be used against the possessor state. Consequently, the requirement was for a an offensive capability available for use on short notice. Following the entry into force of the BWC and its prohibition of the development, production, and stockpiling of BW, there are some indications—to some extent in Russia and also in Iraq—that with a change in the concept of use away from an ability to retaliate in kind to an ability to launch a first-strike BW attack, the focus was more on a capability to produce such weapons immediately before their use. Iraq had certainly followed this approach in regard to CW, thereby avoiding the need to produce agents of sufficient purity that they could be stored for years. It is also clear that a policy to produce and use BW without the need for a stockpile of agents or of filled munitions could make it much harder for an international inspectorate to detect noncompliant activities.

The Limitations of Intelligence

A fundamental characteristic of BW is the dual-use nature of both the agents and the equipment needed to produce the agents. As biological agents (microbes and toxins) occur in nature, the presence or absence of a particular agent or toxin cannot by itself be regarded as evidence of a prohibited activity. Nor does the presence or absence of equipment to produce the agents by itself indicate the presence or absence of a prohibited activity. As the language in the BWC makes it clear, the prohibition applies to agents "of types and in quantities that have no justification for prophylactic, protective or other peaceful purposes." Consequently, it is the quantity and type of agents—and the scale of the equipment used to produce the agent—that enter into the judgment as to whether a particular instance of possession of a biological agent is or is not possession of a BW agent. Underlying intention remains the determining consideration.

The information available from official sources about intelligence on other countries' interests in BW programs is limited, as such information is normally classified. It is evident that, given the dual-use nature of programs involving pathogenic microorganisms and toxins, it is difficult to be certain whether a program on which information becomes available is a national offensive program or merely an ongoing program for permitted purposes—prophylactic, protective, or other peaceful purposes. In developing policy on the basis of intelligence, it needs to be recognized throughout that intelligence analysts are expected to draw worst-case interpretations from the available wisps of information. Furthermore, because intelligence analysts are making their assessments against a background of knowledge of their own national programs in this area, an element of mirror-imaging is liable to enter the intelligence assessment. Although intelligence analysts are generally aware of the strengths and weaknesses of their assessments, it is by no means clear that the policymakers who make decisions based on those intelligence assessments are equally aware. Similarly, policymakers may err on the side of caution so that they cannot be accused of having failed to take steps to protect the security of their country.

It is, of course, clear, in the case of the 1979 accident at Sverdlovsk, that the US had information (not necessarily accurate) from both technical and human sources that convinced it that the Soviet Union had continued its offensive program after signing the BWC. Although it would appear that the US did not have a complete understanding of the scale and scope of the Soviet program until the defections at the end of the Cold War, it is clear that such a large-scale program could not be established by a major state in the future without substantial risk of detection. In contrast, the difficulties that Western intelligence had in understanding the Iraqi program suggests that smaller-scale but regionally significant programs may be much more difficult to assess with any great confidence. Unless means can be found to increase the transparency of states' activities to counter bioterrorism and BW in both civil and military sectors, it seems unlikely that the dilemmas of intelligence can be easily resolved.

Duality is also important in the context of the undertakings made by States Parties under Article III of the BWC "not to transfer to any recipient whatsoever, directly or indirectly, and not in any way to assist, encourage, or induce any State, group of States or international organizations to manufacture or otherwise acquire any of the agents, toxins, weapons, equipment or means of delivery specified in article I of this Convention." For this reason States Parties have increasingly required that all exports of biological agents and toxins and associated equipment, which might be misused to produce BW, be controlled and made subject to prior approval by the exporting government. In order to ensure that no novel agents or equipment are overlooked, States Parties have increasingly included "catch-all" clauses in their export control legislation and regulations, as well as including intangible technology within the scope of such legislation and regulations.

Secrecy versus Transparency

Secrecy engenders suspicions about the real purpose of activities being carried out by a state. Transparency contributes to building confidence that such activities are indeed permitted peaceful activities. It was largely for this reason that BWC States Parties agreed at the Second Review Conference in 1986 that all parties should make annual declarations under four confidence-building measures (CBMs) as a first step toward improving transparency. These CBMs were strengthened and extended at the Third Review Conference in 1991. It is, however, unfortunate that not all States Parties have made all the agreed annual declarations—only about a dozen have done so—and of those parties that have made one or more annual declarations, the information provided has been of uneven quality. Even so, it is important to recognize the contribution provided by such official declarations, for they may provide the only official statements on relevant national activities. A hopeful sign is that states such as Australia and the US have made their CBMs available on the Internet and thus opened them to public scrutiny. It is to be hoped that other states will quickly do the same.

Thus far there is little indication that States Parties other than the UK and the US have used declarations made under the CBMs as a basis for bilateral consultations on BWC-related matters of concern. The UK and the US did so during their démarches to South Africa already noted, as they had previously done following the démarches to the Soviet Union and then Russia that had furthered the trilateral process among them. There is substantial potential for States Parties to use their CBM declarations of past offensive programs to achieve as much transparency as possible, especially in regard to the use of facilities and capabilities subsequent to the abandonment of a national offensive program. Likewise, States Parties should use their CBM declarations of national biological defense programs and facilities to demonstrate that the capabilities and activities are indeed consistent with permitted protective purposes under the Convention. The omission of parts of national biodefense programs and activities from such declarations can only heighten suspicions and harm the building of confidence in compliance.

Although there are similarities between activities required for an offensive program and some activities required for a defensive program—for example, a defensive program needs to evaluate the hazard to which its forces or population may be exposed as a result of a biological attack—there would seem to be little need for practical demonstration on any significant scale of the feasibility of production, weaponization, or dispersion of agent. States Parties would be wise to review their national biological defense programs from the point of view of other states, which might form incorrect perceptions of their objectives—and, if so, to take steps aimed at providing sufficient transparency to reduce the likelihood of such incorrect perceptions.

The absence of any secretariat for the BWC and hence of any forum, other than the Review Conferences at five-year intervals, in which States Parties can collectively address issues relating to compliance concerns is a serious deficiency. A secretariat involving only a few people could encourage the timely submission of the annual CBM declarations by all States Parties, arrange for their translation into the UN official languages, and analyze the information they contain. Such an entity could provide a forum for individual States Parties to approach one another in timely fashion to seek clarification about ambiguities, uncertainties, and omissions, thereby increasing transparency and, over time, building confidence in compliance. For such a cooperative approach to be feasible, however, States Parties would have to take submission of full and accurate CBMs much more seriously than they have in the past.

Treaties and International Collaboration

What has been termed the "taboo" against the weaponization of poison and disease is long-standing, reaching far down the generations and across cultures. In a recent codification of the taboo, the 1925 Geneva Protocol, the prohibition applies both to asphyxiating, poisonous, and other gases and to use of bacteriological methods of warfare. This prohibition recognizes chemical and biological agents as forming a broad spectrum against which wide-ranging protective measures are required. There was therefore a particular logic in the view taken by those who pressed for a combined treaty to prohibit both CW and BW. Nevertheless, the BWC was negotiated to ban only BW, although it included a commitment to continue to work for a ban on CW, a commitment that came to fruition 20 years later in the Chemical Weapons Convention (CWC).

The BWC prohibits under any circumstances the development, production, stockpiling, or other acquisition or retention of "(1) Microbial or other biological agents, or toxins whatever their origin or method of production, of types and in quantities that have no justification for prophylactic, protective or other peaceful purposes." Successive Review Conferences have extended the understandings of States Parties that "the Convention unequivocally covers all microbial or other biological agents or toxins, naturally or artificially created or altered, as well as their components, whatever their origin or method of production, of types and in quantities that have no justification for prophylactic, protective or other peaceful purposes." The CWC prohibits the development, production, stockpiling, acquisition, retention, or use of CW, defined as including "toxic chemicals and their precursors, except where intended for purposes not prohibited under this Convention, as long as the types and quantities are consistent with such purposes." Toxic chemicals are defined as "any chemical which through its chemical action on life processes can cause death, temporary incapacitation or permanent harm to humans or animals. This includes all such chemicals, regardless of their origin or of their method of production, and regardless of whether they are produced in facilities, in munitions or elsewhere." Munitions or other devices designed to disseminate CW agents are also defined as chemical weapons.[1]

There is consequently today a comprehensive prohibition of both chemical and biological weapons, with the midspectrum area including toxins and bioregulators prohibited under both the BWC and the CWC. It is against this background that recent events—notably the Russian use of an anesthetic chemical during the Moscow theater siege in October 2002 and recent US efforts to develop "nonlethal" or "less than lethal" weapons—cause so much concern. The central prohibitions in the BWC and the CWC are enshrined in what is known as the "general-purpose criterion," which prohibits all biological agents unless "of types and in quantities" that are justified "for prophylactic, protective and other peaceful purposes" and all chemical agents unless "of types and in quantities" consistent with "purposes not prohibited under the Convention." Under the BWC, any program to develop "nonlethal" or "less than lethal" weapons using materials that fall within the extended understanding agreed by the States Parties quoted above is a violation of the BWC. The situation with regard to the CWC is more complex, as the CWC, unlike the BWC, includes "law enforcement including domestic riot control purposes" as a purpose not prohibited, although it obliges each State Party "not to use riot control agents as a method of warfare." Consequently, a program to develop "nonlethal" or "less than lethal" weapons exclusively for law enforcement, including riot-control purposes, is permitted under the CWC so long as such weapons are not used for warfare and so long as the chemicals concerned meet the CWC definition of a "riot control agent." However, a program to develop such weapons or agents for warfare is completely prohibited and in violation of the CWC.

International influences from 1945 to the present have clearly had a major impact on the nature of national offensive biological warfare programs. In the early years, the thrust was primarily toward a BW program that could be used in retaliation in kind should such weapons be used by another state. The reservations to the Geneva Protocol made by all the

States Parties that declared BW programs during the period from 1945 until the BWC in 1972—Canada, France, the USSR, and the UK—rendered such programs, so long as they were for retaliatory purposes only, not incompatible with the obligations of the Geneva Protocol. Although the US did not ratify its signature to the Geneva Protocol until 1975, after the entry into force of the BWC, it then did so with a similar reservation. The BWC, which totally prohibits the development, production, acquisition, and retention of biological and toxin weapons, rendered the reservations to the Geneva Protocol inconsistent with the obligations under the BWC. This situation led to the gradual withdrawal of these reservations—by Canada on 20 August 1991, by France on 25 November 1996, by Russia on 18 January 2001, and by the United Kingdom on 8 November 1991.

Implications

During the early 1980s CBW still did not lie within the mainstream of military theory and practice.[2] Constraints of international law and domestic political, military, and technological factors had prevented the effective assimilation of BW. Yet we now know that the Soviet Union undertook a thoroughgoing attempt to use the new revolution in the life sciences in its late Cold War offensive program.

The reasons for the Soviet decision to embark on this program remain unclear, but the historical record that we have examined here, and the earlier SIPRI study of the years before 1945, demonstrate that several factors enhance the likelihood that offensive BW programs will be initiated or maintained and that other factors deter initiation and development. The likely enhancing factors include:

- A regional or international security situation in which conflict has broken out or is feared and in which military forces are being strengthened
- Scientific and technological change that demonstrates that BW can be used effectively for hostile purposes in current conflicts
- Intelligence (threat analyses) that suggests that potential adversaries are involved in offensive biological warfare programs (even if they are not)
- Organizations set up to pursue offensive programs, and individuals seeking advancement within such programs

- Assimilation of weapons and equipment into operational military forces and development of doctrine for the use of such weapons
- Lack of transparency about defensive biological programs, leading to misperceptions of true intent
- Interaction of offensive and defensive programs, including misperceptions leading to action/reaction arms races

Factors that will tend to retard offensive programs include:

- Scientific and technical difficulties in the effective development, production, and use of biological and toxin weapons
- Prevention of the establishment and growth of organizations (and their individual experts) with offensive objectives
- General dislike among political leaders and military forces for such abhorrent weapons
- The prohibitory norm embodied in international arms control agreements if effectively implemented nationally and internationally
- Public support for the prohibitory norm and its further strengthening at international, national, and substate levels
- As much transparency as possible in biodefense in order to prevent misperceptions and inappropriate reactions

The balance between these sets of factors has varied over time and in different states and alliances. In particular, when cold or hot warfare breaks out and the threats to national security are seen to increase, it is far less likely that restraint will prevail.

A dangerous combination of events could be the discovery of means to produce effective BW more easily, coupled with military interest in developing such weapons.[3] We see such a combination currently with the interaction of discoveries in the biology of receptor systems and military desires for chemical "nonlethal" weapons for use in operations other than war. Should such combinations of events lead to the establishment and growth of national organizations having the objective of developing such new weapons, then when warfare does come—as in Vietnam in the 1960s, for example—the norm will be under great threat. Norms once violated on a large scale are not easily restored.

What do the lessons of history have to tell us about policy in regard to biological and toxin weapons and their effective prohibition in these early years of the 21st century? At the end of World War II the resulting bipolar power system and the new nuclear technology that had been shown to work effectively brought nuclear weapons to the fore. We are clearly now in a novel international security system in which the simple (nuclear) deterrence mechanisms of the past apply to a much lesser extent. In today's uncertain world, where adversaries are not easily identified and the scope and spread of the new biotechnology suggests that effective BW could be widely held, there is a danger that these weapons could move to the forefront of military thinking.[4]

A potential risk is surely the real (or misperceived as real) foundation of a new series of state-level offensive BW programs. Real or imagined, these programs might, as in the past, trigger an action/reaction process that would elevate the real or perceived level of threat and promote the further development of offensive BW programs. The net result would be that the norm would at best stagnate or even erode in peacetime and perhaps suffer irreparable damage in future conflicts.

A lesser problem at present is terrorism using BW. The weight of the evidence suggests that at the moment capabilities for causing mass human casualties could come only from leakage of agents or knowledge from state-level programs. Agricultural systems are more at risk today, but, undoubtedly, capabilities for harming humans will increase as the revolution in biology gathers pace, and this trend will require the development of new policies on biosecurity at the national level.

Policies for the 21st Century

The obvious conclusion that follows from the preceding sections is that the major objective of any policy designed to prevent the hostile use of biological agents in future decades must be concentrated now on the prevention of initiation and development of offensive BW programs by states. The question therefore is: What policies would best achieve this? Clearly, if any state (or substate organization) were actually to use biological or toxin agents, it would be necessary for the international community to react in the strongest possible manner in order to reestablish the norm. If the regime fails, and a program is detected, or BW used, the response must be swift and decisive. But while a clear international response is necessary, it is far from sufficient on its own.

It should be possible to make offensive programs difficult for proliferators (such as Iraq in the recent past) by careful control of key exports
of equipment and materials. Eventually, however, a determined state
could overcome such obstacles. Control of BW materials and equipment
is more difficult than control of essential fissile material for nuclear weapons. So although export controls are essential, they are not in themselves
sufficient.

As understanding of pathogenesis improves, it will also be possible to
improve our capabilities for biodefense—disease agent detection and disease prevention and treatment. This improved understanding, and defensive actions based upon it, would reduce the effectiveness of attacks and
thus act as a deterrent. But again, while helpful, improved defense will
always be difficult, and there will be increasing options for an attacker,
too, through modification of traditional agents or the development of advanced biological agents—including new synthetic agents.[5]

So international responses to use, export controls, and biodefense are
each part of the solution, but are not adequate in isolation. They indeed
have to be seen as parts of a web of deterrence.[6] Such policies have to be
bound together by the norm embodied in the Geneva Protocol, the BWC,
and the CWC. Moreover, it has to be accepted that these agreements,
which embody the norm, have significant weaknesses that will take time
and effort to rectify.

The Geneva Protocol, now recognized as being customary international law, prohibits only use of BW (and CW). The Chemical Weapons
Convention is the strongest element in the regime, but its verification
system needs some reorientation, and the challenge of new scientific and
technological developments needs to be faced up to squarely.[7] The BWC
is in need of the greatest attention.

The deficiencies in the compliance monitoring system of the BWC
were recognized from its inception. Efforts to strengthen the Convention
through confidence-building measures (annual data exchanges), negotiated in 1986 and extended in 1991, are widely viewed as not being as effective as they could be, and the decade-long effort of 1991–2001 to reach
agreement on a legally binding instrument also ended in failure.[8] Less
well known, but potentially as serious, is the lack of any international organization to care for and develop the treaty regime between its five-

yearly Review Conferences. The current inter-review conference process, agreed upon in the resumed 2001–02 Fifth Review Conference, may help to encourage the implementation of the Convention in some limited aspects in some states, but it does not address the two fundamental weaknesses of the BWC regime, which are the lack of any effective compliance verification system and the lack of an organization to sustain the Convention between its review conferences. If the legally binding instrument had been agreed, both of these deficiencies would have been corrected, as it would have necessitated the establishment of an organization comparable to the Organization for the Prohibition of Chemical Weapons (OPCW), which, in a division of labor with the States Parties, implements and oversees the CWC.[9]

The stumbling block to progress in strengthening the BWC is clearly the issue of verification. The question therefore is: Can a monitoring system involving declarations, and a process for validation of those declarations, be crafted that can create sufficient transparency to detect militarily significant, state-level, offensive programs without exposing crucial national security and commercial proprietary information to espionage? Most States Parties involved in the BWC-strengthening negotiations of the 1990s thought that a solution was possible. Indeed, many felt that a more intrusive solution would have been acceptable. A small number of states, however, were unenthusiastic about the course of the negotiations, and the US eventually refused to accept the proposed text—or any such text based on the agreed negotiation mandate. In its statement to the States Parties rejecting the proposed text, the US plainly offered as its overriding concerns the adequacy of the proposed compliance monitoring system and the safety of national security information.

Significantly, both the CWC and the BWC legally binding instrument were designed with a three-pillar system of declarations, visits, and challenge investigations that would make significant violations difficult to hide. Commentators have had difficulty in seeing how the system for the CWC was acceptable while that proposed for the BWC was not. The BWC design was based on much research into how visits and investigations could be carried out in numerous States Parties to both the CWC and BWC, and on a two-year-long investigation (VEREX) of what might be an effective BWC system. The BWC negotiators were therefore not trying to devise a system from scratch, but building on a great deal of relevant experience. The VEREX investigation by experts from BWC States Parties, for example, concluded in part in 1993 that "potential verification measures as identified and evaluated could be useful to varying degrees in enhancing confidence, through increased transparency, that States Parties were fulfilling their obligations under the BWC."[10]

Given the nature of some of the biodefense work carried out recently by the US, the suspicion has arisen that perhaps the US has decided upon a different policy—one that involves pushing at the boundaries in biodefense even at the risk of provoking increasing suspicions in other states. The historical record of misperceptions of the nature of biowarfare/biodefense programs over the last century suggests that it may well result in similar policies in other states and the initiation of precisely the action/reaction process that occurred in the past.[11]

A saner and safer policy would surely be to seek the greatest possible transparency in all activities in the life sciences in academia, industry, and government in all countries. Such transparency would help to diminish the possibility of misplaced threat perceptions and could be achieved without necessarily exposing crucial secrets—as has been demonstrated in the CWC inspection processes. Even in the present situation there is a clear-cut method by which this objective could be promoted, and that is for the States Parties to the BWC to make much more use of the mechanisms that already exist within the treaty regime. The CBMs agreed at the Second and Third Review Conferences, for example, include Measure F, "Declaration of Past Activities in Offensive and/or Defensive Biological Research and Development Programs." If States Parties were to use the opportunity to provide comprehensive information in this document, a great deal of suspicion might be reduced. For example, the US and the UK declarations went along with the spirit of the requirement and attempted to describe the whole of their offensive programs even though they needed to report only from 1946 onward. Other countries such as Russia, South Africa, and Iraq have yet to submit declarations of any adequacy under this heading. Similarly, it would be possible to attempt to resolve some, at least, of the past allegations if use were made of the opportunities provided for consultation and cooperation in Article V of the Convention itself. Yet in order achieve this end, much greater public and political support for enhancing the BWC will be required. The BWC—for so long the orphaned negotiation in the shadows—will have to be brought into the public spotlight.

The crucial question therefore is: How are this exposure and consequent action to be effectively brought about? It is all very well for a group of specialists to provide an account of biological warfare and its control from 1945 to the present day and to elucidate the relevant policy lessons—but how is effective political action to be generated? Certainly in order for the BWC to move up the political agenda the public must be better informed and concerned. However, whatever else is needed, one crucial ingredient is clear: people with scientific and medical expertise surely have a special responsibility to alert policymakers in governments around the world to the very real dangers of inaction in regard to the BWC. The World Health Organization report on CBW in the middle of the Cold War helped in the achievement of the BWC, and its recent updated second edition will help in giving credence to the escalating dangers. Furthermore, the International Committee of the Red Cross has clearly seen the dangers arising from the production of new biological weaponry and, referring back to its appeals against CW in the lead-up to agreement of the 1925 Geneva Protocol, has issued a new appeal, *Biotechnology, Weapons and Humanity,* which calls on states, specialists, and industry to work to control the dangers.[12]

March 2005 marked the 30th anniversary of the BWC's entry into force. Its Sixth Review Conference is scheduled for late 2006. The CWC will have its second five-yearly Review Conference in 2008. Thus in the first decade of the 21st century there are multiple opportunities to strengthen the norm against the hostile use of biological and chemical weapons before the outbreak of an arms race that could change the nature of warfare as profoundly as nuclear weapons did in the last century—and with unpredictable consequences.

Whether the lessons of history, as set out here, will lead to such progressive action remains to be seen.

Convention on the Prohibition of the Development, Production and Stockpiling of Bacteriological (Biological) and Toxin Weapons and on Their Destruction

Signed at London, Moscow, and Washington on 10 April 1972. Entered into force on 26 March 1975. Depositaries: UK, US, and Soviet (since 1991 Russian Federation) governments.

The States Parties to this Convention,

Determined to act with a view to achieving effective progress towards general and complete disarmament, including the prohibition and elimination of all types of weapons of mass destruction, and convinced that the prohibition of the development, production and stockpiling of chemical and bacteriological (biological) weapons and their elimination, through effective measures, will facilitate the achievement of general and complete disarmament under strict and effective international control,

Recognizing the important significance of the Protocol for the Prohibition of the Use in War of Asphyxiating, Poisonous or Other Gases, and of Bacteriological Methods of Warfare, signed at Geneva on June 17, 1925, and conscious also of the contribution which the said Protocol has already made, and continues to make, to mitigating the horrors of war,

Reaffirming their adherence to the principles and objectives of that Protocol and calling upon all States to comply strictly with them,

Recalling that the General Assembly of the United Nations has repeatedly condemned all actions contrary to the principles and objectives of the Geneva Protocol of June 17, 1925,

Desiring to contribute to the strengthening of confidence between peoples and the general improvement of the international atmosphere,

Desiring also to contribute to the realization of the purposes and principles of the United Nations,

Convinced of the importance and urgency of eliminating from the arsenals of States, through effective measures, such dangerous weapons of mass destruction as those using chemical or bacteriological (biological) agents,

Recognizing that an agreement on the prohibition of bacteriological (biological) and toxin weapons represents a first possible step towards the achievement of agreement on effective measures also for the prohibition of the development, production and stockpiling of chemical weapons, and determined to continue negotiations to that end,

Determined for the sake of all mankind, to exclude completely the possibility of bacteriological (biological) agents and toxins being used as weapons,

Convinced that such use would be repugnant to the conscience of mankind and that no effort should be spared to minimize this risk,
Have agreed as follows:

Article I

Each State Party to this Convention undertakes never in any circumstances to develop, produce, stockpile or otherwise acquire or retain:
   (1) Microbial or other biological agents, or toxins whatever their origin or method of production, of types and in quantities that have no justification for prophylactic, protective or other peaceful purposes;
   (2) Weapons, equipment or means of delivery designed to use such agents or toxins for hostile purposes or in armed conflict.

Article II

Each State Party to this Convention undertakes to destroy, or to divert to peaceful purposes, as soon as possible but not later than nine months after entry into force of the Convention, all agents, toxins, weapons, equipment and means of delivery specified in article I of the Convention, which are in its possession or under its jurisdiction or control. In implementing the provisions of this article all necessary safety precautions shall be observed to protect populations and the environment.

Article III

Each State Party to this Convention undertakes not to transfer to any recipient whatsoever, directly or indirectly, and not in any way to assist, encourage, or induce any State, group of States or international organizations to manufacture or otherwise acquire any of the agents, toxins, weapons, equipment or means of delivery specified in article I of this Convention.

Article IV

Each State Party to this Convention shall, in accordance with its constitutional processes, take any necessary measures to prohibit and prevent the development, production, stockpiling, acquisition, or retention of the agents, toxins, weapons, equipment and means of delivery specified in article I of the Convention, within the territory of such State, under its jurisdiction or under its control anywhere.

Article V

The States Parties to this Convention undertake to consult one another and to cooperate in solving any problems which may arise in relation to the objective of, or in the application of the provisions of, the Convention. Consultation and Cooperation pursuant to this article may also be undertaken through appropriate international procedures within the framework of the United Nations and in accordance with its Charter.

Article VI

(1) Any State Party to this convention which finds that any other State Party is acting in breach of obligations deriving from the provisions of the Convention may lodge a complaint with the Security Council of the United Nations. Such a complaint should include all possible evidence confirming its validity, as well as a request for its consideration by the Security Council.

(2) Each State Party to this Convention undertakes to cooperate in carrying out any investigation which the Security Council may initiate, in accordance with the provisions of the Charter of the United Nations, on the basis of the complaint received by the Council. The Security Council shall inform the States Parties to the Convention of the results of the investigation.

Article VII

Each State Party to this Convention undertakes to provide or support assistance, in accordance with the United Nations Charter, to any Party to the

Convention which so requests, if the Security Council decides that such Party has been exposed to danger as a result of violation of the Convention.

Article VIII

Nothing in this Convention shall be interpreted as in any way limiting or detracting from the obligations assumed by any State under the Protocol for the Prohibition of the Use in War of Asphyxiating, Poisonous or Other Gases, and of Bacteriological Methods of Warfare, signed at Geneva on June 17, 1925.

Article IX

Each State Party to this Convention affirms the recognized objective of effective prohibition of chemical weapons and, to this end, undertakes to continue negotiations in good faith with a view to reaching early agreement on effective measures for the prohibition of their development, production and stockpiling and for their destruction, and on appropriate measures concerning equipment and means of delivery specifically designed for the production or use of chemical agents for weapons purposes.

Article X

(1) The States Parties to this Convention undertake to facilitate, and have the right to participate in, the fullest possible exchange of equipment, materials and scientific and technological information for the use of bacteriological (biological) agents and toxins for peaceful purposes. Parties to the Convention in a position to do so shall also cooperate in contributing individually or together with other States or international organizations to the further development and application of scientific discoveries in the field of bacteriology (biology) for prevention of disease, or for other peaceful purposes.

(2) This Convention shall be implemented in a manner designed to avoid hampering the economic or technological development of States Parties to the Convention or international cooperation in the field of peaceful bacteriological (biological) activities, including the international exchange of bacteriological (biological) and toxins and equipment for the processing, use or production of bacteriological (biological) agents and toxins for peaceful purposes in accordance with the provisions of the Convention.

Article XI

Any State Party may propose amendments to this Convention. Amendments shall enter into force for each State Party accepting the amendments upon their acceptance by a majority of the States Parties to the Convention and thereafter for each remaining State Party on the date of acceptance by it.

Article XII

Five years after the entry into force of this Convention, or earlier if it is requested by a majority of Parties to the Convention by submitting a proposal to this effect to the Depositary Governments, a conference of States Parties to the Convention shall be held at Geneva, Switzerland, to review the operation of the Convention, with a view to assuring that the purposes of the preamble and the provisions of the Convention, including the provisions concerning negotiations on chemical weapons, are being realized. Such review shall take into account any new scientific and technological developments relevant to the Convention.

Article XIII

(1) This Convention shall be of unlimited duration.

(2) Each State Party to this Convention shall in exercising its national sovereignty have the right to withdraw from the Convention if it decides that extraordinary events, related to the subject matter of the Convention, have jeopardized the supreme interests of its country. It shall give notice of such withdrawal to all other States Parties to the Convention and to the United Nations Security Council three months in advance. Such notice shall include a statement of the extraordinary events it regards as having jeopardized its supreme interests.

Article XIV

(1) This Convention shall be open to all States for signature. Any State which does not sign the Convention before its entry into force in accordance with paragraph (3) of this Article may accede to it at any time.

(2) This Convention shall be subject to ratification by signatory States. Instruments of ratification and instruments of accession shall be deposited with the Governments of the United States of America, the United Kingdom of Great Britain and Northern Ireland and the Union of Soviet Socialist Republics, which are hereby designated the Depositary Governments.

(3) This Convention shall enter into force after the deposit of instruments of ratification by twenty-two Governments, including the Governments designated as Depositaries of the Convention.

(4) For States whose instruments of ratification or accession are deposited subsequent to the entry into force of this Convention, it shall enter into force on the date of the deposit of their instruments of ratification or accession.

(5) The Depositary Governments shall promptly inform all signatory and acceding States of the date of each signature, the date of deposit or each instrument of ratification or of accession and the date of entry into force of this Convention, and of the receipt of other notices.

(6) This Convention shall be registered by the Depositary Governments pursuant to Article 102 of the Charter of the United Nations.

Article XV

This Convention, the English, Russian, French, Spanish and Chinese texts of which are equally authentic, shall be deposited in the archives of the Depositary Governments. Duly certified copies of the Convention shall be transmitted by the Depositary Governments to the Governments of the signatory and acceding states.

Notes

1. Historical Context and Overview

1. SIPRI, *The Problem of Chemical and Biological Weapons*, vol. 1: *The Rise of CB Weapons* (1971), vol. 2: *CB Weapons Today* (1973); vol. 3: *CBW and the Law of War* (1973); vol. 4: *CB Disarmament Negotiations, 1920–1970* (1971); vol. 5: *The Prevention of CBW* (1971); vol. 6: *The Technical Aspects of Early Warning and Verification* (1975) (Stockholm: Almqvist & Wiksell); Erhard Geissler and J. E. van Courtland Moon, eds., *Biological and Toxin Weapons Research, Development and Use from the Middle Ages to 1945*, SIPRI Chemical and Biological Warfare Studies No. 18 (Oxford: Oxford University Press, 1999).

2. US Code, title 18, pt. I, chap. 10 (Biological Weapons), secs. 175–178.

2. The US Biological Weapons Program

Acknowledgments: Thanks to the expanding community of scholars in BW, to my colleagues and editors involved in the production of this book, and especially to Dr. Gregory Koblenz of MIT, who shared numerous documents with me; Ms. Sim Smiley, a Washington, D.C., freelance researcher who retrieved new sources for me; and my wife, Joan Mary van Courtland Moon, whose sense of the English language is always invaluable.

1. Susan J. Djonovich, ed., *United Nations Resolutions,* Series I: *Resolutions Adopted by the General Assembly,* vol. 1: *1946–1948* (Dobbs Ferry, N.Y.: Occana Publications, 1973), p. 53.

2. United Nations Report No. E 69 I 24, *Chemical and Bacteriological (Biological) Weapons and the Effects of Their Possible Use* (New York: Ballantine Books, 1970).

3. George W. Merck to Secretary of War Robert P. Patterson, 24 October 1945, RG 330, NARA.

4. Military Intelligence Division, War Department, "Biological Warfare, Activities and Capabilities of Foreign Nations," 30 March 1946, RG 319, Records of the Army Staff, NARA.

5. Tom Mangold and Jeff Goldberg, *Plague Wars: A True Story of Biological Warfare* (New York: St. Martin's, 1999), p. 61.
6. Background Briefing (Administration Policy Concerning Toxins), at the White House (Key Biscayne, Fla.), 14 February 1970, White House Central Files, Ronald Zeigler, White House Special Files, Numerical Subject File, Foreign Affairs and Defense, DOD Chemical and Biological Warfare, Nixon Presidential Materials Project, NARA.
7. JCS 1780, "Estimate of the Effects on the Nature of War of Future Technical Developments in Weapons," 20 May 1947, Combined Chiefs of Staff (hereafter CCS) 386.2 (10-3-48), RG 218, NARA.
8. Theodor Rosebury, *Peace or Pestilence: Biological Warfare and How to Avoid It* (New York: McGraw-Hill, 1949), p. 174.
9. Dorothy L. Miller, "Military Biology and Biological Warfare," Historical Office, Office of the Executive, Air Materiel Command, Wright-Patterson Air Force Base, September 1958, vol. 1, pp. 1–4; vol. 2, p. 184.
10. Ibid., vol. 1, pp. 5 (n. 9), 13.
11. Ibid., p. 6; vol. 2, pp. 186, 187.
12. James Forrestal, Secretary of Defense, to President Harry S. Truman, 16 March 1948, President's Secretary's Files, Subject File: James S. Forrestal, Special Letters folder, Papers of Harry S. Truman, Harry S. Truman Library, Independence, Mo.
13. NSC 62, General Omar Bradley, Chairman, JCS, to the Secretary of Defense, 18 January 1950, RG 273, NARA. It was still the policy of the US in the final year of the Korean War. See NSC 147, "Analysis of Possible Courses of Action in Korea," 2 April 1953, RG 273, NARA. Section "Under Operational Restrictions": "Chemical, biological and radiological weapons will *not* be used by the United States except in retaliation"; NSC 62, approved 17 February 1950.
14. "Report of the Secretary of Defense's Ad Hoc Committee on Chemical, Biological, and Radiological Warfare," 30 June 1950, RG 330, NARA. The reactions of the secretary of defense and the service secretaries are also found in RG 330. For President Truman's reaction, see the President to the Secretary of Defense: "I read it with a lot of interest"; same file.
15. JCS 1837/26, "Biological Warfare: Memorandum by the Joint Advanced Study Committee for the Joint Chiefs of Staff," 21 September 1951, CCS 385.2, sec. 13, RG 330, NARA; JCS 1837/36, "Statements of Policy and Directives on Biological Warfare," 3 July 1952, CCS 385.2, sec. 15, RG 218, NARA.
16. NSC 5602/1, "Basic National Security Policy," 15 March 1956, RG 273, NARA; printed in *Foreign Relations of the United States* (hereafter *FRUS*): *1955–1957*, vol. 19: *National Security Policy*, ed. John P. Glennon, William Klingman, David S. Patterson, and Ilana Stern (Washington, D.C.: USGPO, 1990), pp. 242–268; quotation on 246.
17. "Memorandum of Discussion at the 277th Meeting of the National Security

Council, Washington," 27 February 1956; in *FRUS: 1955–1957*, vol. 19, pp. 201–218; quotation on 207–208. This discussion of the penultimate draft, NSC 5602, led to modification of the final report, NSC 1506/1.

18. "Memorandum of Discussion at the 412th Meeting of the National Security Council," 9 July 1959; in *FRUS: 1958–1960*, vol. 3: *National Security Policy*, ed. Edward C. Keefer, David W. Mabon, and David S. Patterson (Washington, D.C.: USGPO, 1996), p. 249.

19. Discussion at the 435th Meeting of the NSC, 18 February 1960, Eisenhower Papers, 1953–1961, Ann Whitman file, Dwight D. Eisenhower Library, Abilene, Kansas. I am grateful to Julian Perry Robinson for providing this source.

20. JCS 1837/113, "Chemical and Biological Warfare: Report by the J-5 to the Joint Chiefs of Staff," 31 May 1960, CCS 3260, Central Decimal File, RG 218, NARA. Related documents dated 6 June 1960, 9 June 1960.

21. Edmund Russell, *War and Nature: Fighting Humans and Insects with Chemicals from World War I to Silent Spring* (Cambridge: Cambridge University Press, 2001), pp. 206–208.

22. "Basic National Security Policy," 22 June 1962, NSC Files, Subject: Basic National Security Policy, John F. Kennedy Library, Boston, Mass.

23. McGeorge Bundy to the Director, Arms Control and Disarmament Agency, Subject: Chemical and Biological Weapons, 5 November 1963, ACDA, Kennedy Library.

24. JCS to the Secretary of Defense, Subject: Joint Declaration on Disarmament, 9 August 1961, ibid.

25. "Joint Declaration on Disarmament: A Program for General and Complete Disarmament in a Peaceful World," 11 August 1961, ibid.

26. Memorandum for the File: Dr. Franklin A. Long, Subject: Initial Steps on the Problem of Biological Weapons Ban—Anti Disease Program, 8 June 1963, Entry 5041, RG 59, NARA.

27. CIA, Intelligence Memorandum No. 105, Subject: Soviet Capabilities to Wage Biological Warfare, 20 December 1963, CREST (CIA Records Search Tool).

28. The documents relating to the decision to initiate "nonlethal" chemical warfare in Vietnam are found in NAM 115, National Security Files: Meetings and Memoranda, Kennedy Library. The case for initiating the use of herbicides was argued at length in Roswell L. Gilpatric to the President, Subject: Defoliant Operations in Vietnam, 21 November 1961, ibid.

29. Letter, Office of the Assistant Secretary of Defense to Mr. Kastenmeier, 16 March 1963, Central Files, Kennedy Library.

30. CIA, "The Soviet Air Forces," 25 July 1949, NSC Files: Central Intelligence Files, Intelligence Memoranda, Truman Library.

31. I. L. Baldwin, Chairman Committee on Special BW Operations, Memorandum for the Research and Development Board of the National Military Establishment, Washington, D.C., 5 October 1948, Truman Library.

32. CIA, Intelligence Memorandum No. 105, Subject: Soviet Capabilities to Wage Biological Warfare, 20 December 1948, CREST; Carly P. Haskins to Secretary of Defense Louis Johnson, "Report of the Ad Hoc Committee on Biological Warfare," 11 July 1949, President's Secretary's Files, Subject File: Cabinet: Defense Reports, Truman Library.

33. Joint Intelligence Committee (hereafter JIC), "Strategic Vulnerability of the USSR to a Limited Air Attack," 3 November 1945; printed in *America's Plans for War against the Soviet Union: 1945–1950,* ed. Steven T. Ross and David Alan Rosenberg, vol. 1: *The Strategic Environment* (New York: Garland, 1990), app. B, pp. 9–10.

34. JIS 80/26, "Capabilities and Intentions of the USSR in the Postwar Period," 9 July 1946; ibid, pp. 101–110. For a subsequent estimate, see JIC 374/1, "Intelligence Estimate That War between Soviet and Non-Soviet Powers Breaks Out in 1956," 6 November 1946; ibid.

35. Joint Intelligence Group (hereafter JIG) 286/2, "Intelligence Estimate of Soviet Capabilities to Engage in Espionage, Subversion, and Sabotage in the Western Hemisphere," 17 September 1948; ibid., vol. 6: *Plan Frolic and American Resources.*

36. National Intelligence Estimate (hereafter NIE) 18, "The Probability of Soviet Employment of BW and CW in the Event of Attacks upon the United States," 10 January 1951, RG 263, NARA.

37. NIE 31, "Soviet Capabilities for Clandestine Attack against the US with Weapons of Mass Destruction and the Vulnerability of the US to Such Attacks (mid-1951 to mid-1952)," n.d. (cover bears the date 1951/09/04), ibid.

38. NIE 11-7-60, "Soviet Capabilities and Intentions with Respect to the Clandestine Introduction of Weapons of Mass Destruction into the United States," 17 May 1960, ibid.

39. Discussion at the 435th Meeting of the NSC, 18 February 1960, Eisenhower Papers, 1953–1961, Ann Whitman file, Eisenhower Library.

40. For the impact of this development, see NIE 11-8-59, "Soviet Capabilities for Strategic Attack through Mid-1964," 9 February 1960; NIE 11-8-60, "Soviet Capabilities for Long Range Attack through Mid-1965," 1 August 1960; NIE 11-9-61, "Soviet Capabilities for Long Range Attack," 7 June 1961, Entry 29, RG 263, NARA.

41. JCS, "Joint Strategic Objectives Plan for FY 1970–1974" (JOP-70); in *FRUS, 1964–1968,* vol. 10: *National Security Policy,* ed. David S. Patterson (Washington, D.C.: USGPO, 2002), pp. 112–141.

42. NIE 11-9-61, "Soviet Capabilities for Long Range Attack," 7 June 1961.

43. CIA, "The Soviet BW Program: Scientific Intelligence Research Aid," 24 April 1961, http://www.foia.cia.gov.

44. NIE 11-6-64, "Soviet Capabilities and Intentions with Respect to Biological Warfare," 26 August 1964, RG 263, NARA.

45. Backstopper for Briefings: Soviet BW/CW Capability, 13 February 1967, CREST.
46. Miller, "Military Biology and Biological Warfare," vol. 1, pp. 50–51.
47. Ibid., pp. 51, 8.
48. DOD Directive No. 5128.7, Subject: Responsibilities of the Assistant Secretary of Defense (Research and Development), 12 November 1953, National Security Archive, George Washington University.
49. DA SR 10-350-1, 15 September 1949, quoted in Historical Office, Office of the Chief Chemical Officer, "Summary History of Chemical Corps Activities, 9 September 1951 to 31 December 1952," pp. 51–52, RG 175, NARA.
50. Miller, "Military Biology and Biological Warfare," vol. 1, p. 53.
51. Lt. Col. Charles T. Anders, "The Army Technical Committee System," *Army Research and Development News Magazine,* October 1966.
52. Miller, "Military Biology and Biological Warfare," vol. 1, p. 58.
53. Ibid., p. 56.
54. Ibid., p. 9; vol. 2, p. 155.
55. *Hearings before the Select Committee to Study Governmental Operations with Respect to Intelligence Activities of the United States Senate, Ninety-fourth Congress, First Session,* vol. 1: *Unauthorized Storage of Toxic Agents,* 16 September 1975 (Washington, D.C.: USGPO, 1976), p. 6.
56. R&D Directive, "Army Responsibilities and Procedures for Coordinating of Research and Development on Biological and Chemical Warfare," 31 May 1955, Department of the Army, Office of the Chief of Staff, Records of the Chemical Corps Technical Committee, RG 175, NARA
57. Miller, "Military Biology and Biological Warfare," vol. 1, p. 52.
58. Ibid., vol. 2, pp. 156–157.
59. Ed Regis, *The Biology of Doom: The History of America's Secret Germ Warfare Project* (New York: Henry Holt, 2000), pp. 132–133.
60. Miller, "Military Biology and Biological Warfare," vol. 2, pp. 23–25, 157.
61. Dr. Henry Stubblefield, "Assessment of the BW Effort," 1959, p. 28, RG 175, NARA.
62. Miller, "Military Biology and Biological Warfare," vol. 2, p. 157.
63. Stubblefield, "Assessment," p. 27.
64. JCS 1837, "Biological Warfare," 20 February 1948, CCS 385.2 (12-17-43), sec. 5, RG 218, NARA.
65. JCS 1837/2, "Capabilities of Biological Warfare," 11 June 1948, CCS 385.2 (12-17-43), sec. 6; and JCS 1837/3, "Capabilities of Biological Warfare," 10 July 1948, CCS 385.2 (12-17-43), sec. 6; both ibid.
66. Stubblefield, "Assessment," p. 30.
67. Ibid., table 2, "Value of Chemical Corps Facilities and Appropriations for Construction and Improvement, FY 51, 52, 53," shows a total of $323,551,599.
68. Adapted from Item 2483, "Continuation of 271 Projects in the Chemical

Corps FY 1953 Program and Revisions Thereto," 28 April 1952, Office of the Chief Chemical Officer, CmlC Technical Committee, Army Chemical Center, Maryland, RG 175, NARA.

69. Item 3620, "Revised Numbers for Chemical Corps R&D Projects & Tasks," 10 September 1959; and Item 3758, "Chemical Corps FY 61 BW RD Program," 25 July 1960; both ibid.

70. DOD, Research and Development Board, Committee on Biological Warfare, Data on Selected Antipersonnel Biological Warfare Agents and Munitions, 9 February 1951, CD 285 (Bio. War), 1951, RG 330, NARA.

71. Item 2990, "Airborne Dispersing Equipment for BW and CW Munitions," 6 May 1954, Office of the Chief Chemical Officer, CmlC Technical Committee, Army Chemical Center, Maryland, RG 175, NARA.

72. Miller, "Military Biology and Biological Warfare," vol. 2, p. 84.

73. Ibid., pp. 82, 83, 85–86.

74. Ibid., vol. 1, pp. 19, 79, 80; vol. 2, pp. 102–104, 124, 110, 116–118.

75. Item 2935, "BW-CW Warheads for B-62," 10 February 1954; and Item 4052, "SD-2 (XAE-3) System, Future R&D Program," 15 June 1952, Headquarters, CmlC Research and Development Command, Records of the Chemical Corps Technical Committee, RG 175, NARA.

76. Item 2923, "Chemical Warheads for USAF [US Air Force] Guided Missiles," 3 August 1954, CmlC Technical Committee Office, ibid.

77. Mangold and Goldberg, *Plague Wars*, pp. 35–36.

78. Regis, *Biology of Doom*, p. 186.

79. Brian Balmer, *Britain and Biological Warfare: Expert Advice and Science Policy, 1930–65* (Basingstoke: Palgrave, 2001).

80. Regis, *Biology of Doom*, pp. 117–118; Mangold and Goldberg, *Plague Wars*, pp. 36–37; Special Report No. 142, "Biological Warfare Trials at San Francisco, California, 2027 September 1950," CmlC Biological Laboratories, 22 January 1951. I am indebted to Leonard A. Cole for a copy of this document. Cole, *Clouds of Secrecy: The Army's Germ Warfare Tests over Populated Areas* (Savage, Md.: Rowman and Littlefield, 1990), pp. 60–65, 80. The reports of the CmlC dealing with the 1950s city tests are "Behavior of Chemical Clouds within Cities," *Joint Quarterly Report No. 3*, January–March 1953; and "Behavior of Chemical Clouds within Cities," *Joint Quarterly Report No. 4*, April–June 1953, National Technical Information Service, US Department of Commerce.

81. Regis, *Biology of Doom*, pp. 132–137.

82. Ibid., pp. 154–155.

83. CmlC Biological Laboratories, Camp Detrick, Frederick, Md., "Seventh Annual Report of the Chemical Corps Laboratories (Fiscal Year 1953)," 1 July 1953, RG 175, NARA.

84. Stubblefield, "Assessment," pp. 28–29.

85. Item 3414, "Classification of Cluster, Biological Bomb, 750-lb, M39

(E133R2) & BOMB, Biological, 1/2-lb, M130 (E61R4) as Standard Types," 17 February 1958, Office of the Chief Chemical Officer, CmlC Technical Committee, Army Chemical Center, Maryland, RG 175, NARA.

86. Item 2963, "Establishment of Project 4-11-01-005, Vulnerability of Military Personnel to BW Attack," 3 November 1954, ibid.

87. Regis, *Biology of Doom*, pp. 168, 176.

88. CmlC Historical Office, "Summary of Major Events and Problems, US Army Chemical Corps, Fiscal Year 1959," January 1960, RG 175, NARA.

89. Chemical and Biological Defense Information Analysis Center, "Summary Report (Phase III): Chemical Weapons Exposure Study Task Force (CWEST) Event Database to Project Manager, Office of the Secretary of Defense," 29 April 1996, RG 330, NARA.

90. CmlC Historical Office, "Summary of Major Events and Problems: United States Army Chemical Corps, Fiscal Year 1958," March 1959, RG 175, NARA.

91. "Fact Sheets of the Special Assistant to the Undersecretary of Defense (Personnel and Readiness) for Gulf War Illnesses, Medical Readiness and Military Deployments and the Office of the Assistant Secretary of Defense (Health Affairs), Deployment Health Support Directorate, 1963–1973," in *Report to Congress on the Disclosure of Information to the Department of Veterans Affairs as Directed by PL 107-314;* http://deploymentlink.osd.mil/current_issues/shad/final_report/index.htm.

92. Regis, *Biology of Doom*, pp. 187–192.

93. Ibid., pp. 152–153, 196–197.

94. Fort Detrick, Department of the Army, "Study of the Vulnerability of Subway Passengers in New York City to Covert Attack with Biological Weapons," January 1968, Miscellaneous Publication 25; document provided by Gregory Koblenz.

95. Ibid.

96. Cole, *Clouds of Secrecy*, p. 66.

97. Fort Detrick, "Study of the Vulnerability of Subway Passengers in New York City." For coverage of these tests, see Regis, *Biology of Doom*, pp. 197–198; Cole, *Clouds of Secrecy*, pp. 65–71.

98. Miller, "Military Biology and Biological Warfare," vol. 1, pp. 35–36.

99. WSEG, Office of the Secretary of Defense, "An Evaluation of Offensive Biological Warfare Weapons Systems Employing Manned Aircraft, 15 July 1952," WSEG Report No. 8, Enclosure C, RG 330, NARA.

100. NSC 5515/1, "Study of Possible Hostile Soviet Actions," 1 April 1955; printed in Glennon et al., *FRUS: 1955–1957*, vol. 19, p. 74.

101. NSC 5602/1, "Basic National Security Policy," 15 March 1956; ibid., p. 258.

102. Director of Air Defense and Special Weapons, Atomic-CBR Division, "Strategic Appraisal on Use of BW/CW Weapons," 1 April 1960, Records of the Army Staff, RG 419, NARA.

103. "United States and Allied Capabilities for Limited Military Operations to 1 July 1962," 7 July 1960, Office of the Special Assistant for NSC Affairs, NSC Series, Policy Papers Subseries, Eisenhower Library.
104. Headquarters, Department of the Army, Office of the Deputy Chief of Staff for Military Operations, Director of CBR Operations, "Performance Characteristics of Some Chemical and Biological Weapon Systems," August 1962, Army Staff, General Correspondence, 1962, RG 319, NARA.
105. Headquarters, Department of the Army, *FM 3-10: Chemical and Biological Weapons Employment,* February 1962, p. 51. I am grateful to Jack McGeorge for providing me with a copy of this manual.
106. Private source. See also Judith Miller, Stephen Engelberg, and William Broad, *Germs: Biological Weapons and America's Secret War* (New York: Simon and Schuster, 2001), pp. 53–57.
107. "Report of the Secretary of Defense's Ad Hoc Committee on Chemical, Biological, and Radiological Warfare," 30 June 1950, RG 330, NARA.
108. Miller, "Military Biology and Biological Warfare," vol. 1, p. 12.
109. Ibid., p. 28.
110. Ibid., vol. 2, pp. 34–35, 38–44.
111. Ibid., vol. 1, p. 32.
112. Historical Office, Office of the Chief Chemical Officer, "Summary History of Chemical Corps Activities, 1951–1952," pp. 51–52.
113. Miller, "Military Biology and Biological Warfare," vol. 1, pp. 21–24, 71.
114. JCS 1920/5, "Long-Range Plans for War with the USSR: Development of a Joint Outline Plan for Use in the Event of War in 1957"; in Ross and Rosenberg, *America's Plans for War against the Soviet Union,* vol. 14: *Long Range Planning: Dropshot* (New York: Garland, 1989).
115. Miller, "Military Biology and Biological Warfare," vol. 2, p. 3.
116. Secretary of the Army Frank Pace Jr. to Secretary of Defense, Chemical and Biological Warfare Readiness, 23 April 1952, enclosing study titled "Report on Chemical and Biological Warfare Readiness: General Summary," CCS 385.2 (12-17-43), sec. 15, RG 218, NARA.
117. Miller, "Military Biology and Biological Warfare," vol. 2, pp. 5–6.
118. Ibid., pp. vii, 76.
119. Ibid., pp. 9–10.
120. CmlC Biological Laboratories, Camp Detrick, Frederick, Md., "Seventh Annual Report of the Chemical Corps Laboratories (Fiscal Year 1953)," 1 July 1953, NARA.
121. JSPC 954/29, "Chemical (Toxic) and Biological Warfare Readiness," 13 August 1953, CCS 385.2 (12-17-43), sec. 18, RG 218, NARA.
122. Herbert Scoville Jr. to Dr. Paul A. Weiss, Chairman, BW-CW Panel, President's Science Advisory Committee, Subject: Comments on Draft BW-CW Panel Report, 8 May 1958, Office of the Special Assistant for Science and Technology, Eisenhower Library.

123. Memorandum, G. B. Kistiakowsky to Dr. Killian, Subject: Biological Warfare, 30 July 1958, White House Office, Office of the Special Assistant for Science and Technology, ibid.

124. David Z. Beckler to J. R. Killian Jr., Subject: Technical Management Problems Related to Defense, 15 May 1959, White House Office, Special Assistant for Science and Technology, ibid.

125. Director of Air Defense and Special Weapons, Atomic-CBR Division, "Strategic Appraisal on Use of BW/CW Weapons," 1 April 1960, Records of the Army Staff, RG 419, NARA.

126. CmlC, "Summary of Major Events and Problems, FY 1961–1962," June 1962, National Security Archive.

127. Quoted in Regis, *Biology of Doom*, p. 185.

128. Ibid., pp. 185–186.

129. Memorandum, Secretary of Defense Laird to National Security Advisor Henry Kissinger, 30 April 1969, Nixon Presidential Materials, NSC Files, Subject Files: Chemical Biological Warfare (Toxins, etc), vol. 1, NARA; reproduced as document 1 in "Biowar: The Nixon Administration's Decision to End the US Biological Warfare Programs," ed. R. A. Wampler, vol. 3 of National Security Archive Electronic Briefing Book No. 58, 25 October 2001, updated 7 December 2001, http://www.gwu.edu/~nsarchiv/ NSAEBB. An excellent study of US biological disarmament is Jonathan B. Tucker, "A Farewell to Germs: The U.S. Renunciation of Biological and Toxin Weapons, 1969–1970," *International Security*, 27 (summer 2002), 107–148.

130. National Security Memorandum No. 59, Henry Kissinger to the Secretary of State, the Secretary of Defense, the Director of Central Intelligence, the Special Assistant to the President for Science and Technology and the Director US Arms Control and Disarmament Agency, Subject: US Policy of Chemical and Biological Warfare and Agents, 28 May 1969, Microfilm Collection, Harvard University.

131. National Security Decision Memorandum No. 35, "Issues for Decision, Policy on Biological Warfare (BW) and Chemical Warfare (CW) Policy Issues," n.d., Nixon Presidential Material Staff, NSC Institutional (H) Files, National Security Decision Memos, H-213, NARA.

132. Report to the NSC, "US Policy on Chemical and Biological Warfare and Agents," submitted by the Interdepartmental Political-Military Group in response to NSSM 59, 10 November 1969, in Wampler, "Biowar," docs. 6a and 6b.

133. "Statement on Chemical and Biological Defense Policies and Programs, 25 November 1969," in *Public Papers of the Presidents of the United States: Richard Nixon: Containing the Public Messages, Speeches, and Statements of the President: 1969* (Washington, D.C.: USGPO, 1971), doc. 461, pp. 968–969.

134. "Remarks upon Signing Instruments of Ratification of the Geneva Protocol

of 1925 and the Biological Weapons Convention, 22 January 1975," in *Public Papers of the Presidents of the United States: Gerald R. Ford: Containing the Public Messages, Speeches, and Statements of the President: 1975*, vol. 1: *January 1 to July 17, 1975* (Washington, D.C.: USGPO, 1977), doc. 37, pp. 72–73.

135. One of the best analyses is Julian Perry Robinson, "Origins of the Chemical Weapons Convention," in *Shadow and Substance: The Chemical Weapons Convention*, ed. Benoit Morel and Kyle Olson (Boulder: Westview, 1993), p. 48.

136. Mangold and Goldberg, *Plague Wars*, pp. 54–55; Tucker, "A Farewell to Germs," pp. 126–127.

137. Report to the NSC, "US Policy on Toxins," 21 January 1970, Entry 10, RG 273, NARA.

138. Department of the Army, "US Army Activities in the US Biological Warfare Program," 25 February 1977, vol. 1, pp. 54–55, RG 175, NARA.

139. *Hearings to Study Governmental Operations with Respect to Intelligence Activities*, vol. 1, pp. 66, 68.

140. National Security Decision Directive No. 11, "United States Chemical and Biological Weapons Arms Control Policy," 4 January 1982; printed in Christopher Simpson, ed., *National Security Directives of the Reagan and Bush Administrations: The Declassified History of U.S. Political and Military Policy, 1981–1991* (Boulder: Westview, 1991), pp. 85–87.

141. NIE 11-11-69, "Soviet Chemical and Biological Warfare Capabilities," 13 February 1969, RG 263, NARA.

142. DIA, "Soviet Genetic Engineering Status and Threat: A Comparative Analysis," 23 August 1976, National Security Archive.

143. Special NIE 11/50/37-82, "Use of Toxins and Other Lethal Chemicals in Southeast Asia and Afghanistan," 2 February 1982, RG 263, NARA.

144. Special NIE 11-17-83, "Implications of Soviet Use of Chemical and Toxin Weapons for US Security Interests," 15 September 1983, Entry 29, RG 263, NARA.

145. Message to the Congress Transmitting a Report and a Fact Sheet on Soviet Noncompliance with Arms Control Agreements, 23 January 1984; in *Public Papers of the Presidents of the United States: Ronald Reagan: 1984*, vol. 1: *January 1 to June 20, 1984* (Washington, D.C.: USGPO, 1986), p. 74.

146. Amy E. Smithson, *Toxic Archipelago: Preventing Proliferation from the Former Soviet Chemical and Biological Weapons Complexes* (Washington, D.C.: Henry L. Stimson Center, 1986), pp. ix, x, 9–10. For additional details on the Toxic Archipelago, see Ken Alibek with Stephen Handelman, *Biohazard: The Chilling True Story of the Largest Covert Biological Weapons Program in the World—Told from the Inside by the Man Who Ran It* (New York: Random House, 1999).

147. Smithson, *Toxic Archipelago*, pp. 8–9.

148. Extracts from the October 2002 NIE report can be found at http://www.washingtonpost.com/wp.

149. For the text of the announcement regarding the creation of the cancer cen-

ter at Detrick, see Office of the White House Press Secretary, Fact Sheet (Fort Detrick), 18 October 1971, White House Central Files, Ronald Zeigler, White House Special Files, Numerical Subject File, Foreign Affairs and Defense, DOD Chemical and Biological Warfare, Nixon Presidential Materials Project, NARA.

150. Executive Office of the President, Office of Science and Technology, Press Release: Second Draft, 26 January 1971, Nixon Presidential Materials Project, White House Special Files, White House Files—Biological Warfare; Chemical Warfare, Nixon Presidential Materials Project, NARA.

151. See, for example, Miller, Engelberg, and Broad, *Germs.* Opinions differ sharply among BW experts regarding the legitimacy under the BWC of these activities.

152. Lt. George Korch Presentation: "Leading Edge of Biodefense: The National Biodefense Analysis and Countermeasures Center," Jackson Naval Air Station, 9 February 2004; David Ruppe, "Proposed U.S. Biological Research Could Challenge Treaty Restrictions, Experts Charge," Global Security Newswire, 30 June 2004; Milton Leitenberg, *The Problem of Biological Weapons* (Stockholm: Swedish National Defense College, 2004), pp. 155–206; Milton Leitenberg, Ambassador James Leonard, and Dr. Richard Spertzel, "Biodefense Crossing the Line," *Politics and the Life Sciences,* 22, no. 2 (2004); Mark Wheelis and Malcolm Dando, "Back to Bioweapons?" *Bulletin of the Atomic Scientists,* 59 (January/February 2003), 40–46.

3. The UK Biological Weapons Program

Acknowledgments: Thanks to Gradon Carter (who commented on a draft chapter) and Helen Wickham (for research assistance).

1. Gradon Carter, *Chemical and Biological Defence at Porton Down, 1916–2000* (London: TSO, 2000); P. M. Hammond and G. B. Carter, *From Biological Warfare to Healthcare: Porton Down, 1940–2000* (Basingstoke: Palgrave, 2002).

2. G. B. Carter and Graham S. Pearson, "North Atlantic Chemical and Biological Research Collaboration: 1916–1995," *Journal of Strategic Studies,* 19 (1996), 74–103.

3. Brian Balmer, "The Drift of Biological Weapons Policy in the UK, 1945–1965," *Journal of Strategic Studies,* 20, no. 4 (1997), 115–145; idem, *Britain and Biological Warfare: Expert Advice and Science Policy, 1930–65* (Basingstoke: Palgrave, 2001).

4. Brian Balmer, "Killing 'Without the Distressing Preliminaries': Scientists' Own Defence of the British Biological Warfare Programme," *Minerva,* 40 (2002), 57–75.

5. Balmer, *Britain and Biological Warfare;* M. Hugh-Jones, "Wickham Steed and German Biological Warfare Research," *Intelligence and National Security,* 7, no. 4 (1992), 379–402.

6. Hammond and Carter, *From Biological Warfare to Healthcare;* Balmer, *Britain*

and Biological Warfare; Carter, Chemical and Biological Defence at Porton Down;
G. B. Carter and Graham S. Pearson, "British Biological Warfare and Biological Defence, 1925–1945," in Biological and Toxin Weapons Research, Development and Use from the Middle Ages to 1945: A Critical Comparative Analysis, ed.
Erhard Geissler and J. E. van Courtland Moon, SIPRI Chemical and Biological Warfare Studies No. 18 (Oxford: Oxford University Press, 1999).

7. PRO, WO 188/667, BW(46)17, ISSBW, Future Development of Biological
Warfare Research, Memorandum by Dr. Henderson, 30 April 1946.

8. PRO, CAB 121/103, Report from Chiefs of Staff to the Defence Committee,
Biological Warfare, Annex BW(45)22, 14 September 1945.

9. Ibid., DO(45)15, Defence Committee, Biological Warfare, Item 5, 5 October
1945.

10. PRO, DEFE 10/26, DRP(50)53, DRPC, BW Policy, Note by the Chairman,
BWS, 11 May 1950.

11. J. Agar and B. Balmer, "British Scientists and the Cold War: The Defence
Policy Research Committee and Information Networks, 1947–1963," Historical Studies in the Physical Sciences, 28 (1998), 209–252.

12. PRO, DEFE 10/19, DRPC, Final Version of Paper on Future of Defence Research Policy, 30 July 1947.

13. PRO, DEFE 10/18, DRP(47)53, DRPC, Future Defence Policy, 1 May 1947.

14. DRPC, Final Version of Paper on Future of Defence Research Policy, 30 July
1947.

15. PRO, WO 188/660, ISSBW, Shortage of Scientific Staff for Research in Biological Warfare, 30 January 1947.

16. DRPC, Final Version of Paper on Future of Defence Research Policy, 30 July
1947.

17. PRO, WO 188/660, Chiefs of Staff Committee, BWS, Note by Ministry of
Supply, 17 November 1947.

18. PRO, AIR 20/8727, OR/1006, Air Staff Requirement for a Biological Bomb,
November 1946.

19. PRO, AIR 20/8733, An Assessment of the Potential Value of Biological Warfare (Draft), 19 January 1947.

20. Ibid.

21. PRO, AIR 20/8727, OR/1065, Air Staff Requirement, 12 November 1947.

22. Ibid., Notes on DOR(A)'s Air Staff Requirement for a Strategic Toxic
Weapon, 27 October 1947.

23. PRO, AIR 20/8733, Paper on the Theoretical Effectiveness of BW Weapons
against Potential Enemy Targets in Comparison with the Effect of Atomic
Bombs on the Same Targets, n.d. (with 1948 papers).

24. PRO, AIR 20/11355, 1949 Report on Biological Warfare, Section V: The
Practical Requirements for Offence and Defence, Air Ministry Draft Contribution, November 1949.

25. M. Goodman, "British Intelligence and the Soviet Atomic Bomb, 1945–
1950," Journal of Strategic Studies, 26, no. 2 (2003), 120–151.

26. PRO, WO 188/663, BW(49), 5th Meeting, BWS, 31 October 1949. The statement is not clear considering that the US level of production of weapons in 1945 was zero.

27. Ibid.

28. PRO, WO 188/663, BW(49)41 (final), Research and Development Policy in Relation to Intelligence and Russian Development, Report to the Defence Research Policy Committee, 29 November 1949.

29. PRO, CAB 131/9, DO(50)45, 7 June 1950.

30. The term *Red Admiral* appears rarely in the open literature. It is most directly attributed to the bomb project in a 1952 DRPC Review of R&D mentioned in PRO, AIR 20/8727, OR/1065, Minute from Dr. O. H. Wansbrough-Jones, Ministry of Supply to "W," 20 June 1952.

31. PRO, WO 188/661, BW(48)12, BWS, Reconnaissance for Operation Harness, 9 April 1948.

32. PRO, WO 188/663, BW/4/2, Naval Report on Operation "Harness" by Captain G. S. Tuck, June 1949.

33. PRO, DEFE 10/263, BW(49)15, Operation Harness 1947–1949, Scientific Report, 20 July 1949.

34. Ibid.; PRO, WO 188/663, BW(49)33 (final), BWS, BW Trials at Sea—Operation Harness, Report to the Chiefs of Staff and the Defence Research Policy Committee, 18 August 1949.

35. PRO, WO 188/663, BW(49)28, BWS, Operation Harness, 20 June 1949.

36. PRO, WO 195/12213, BW(53)2, Operation Cauldron 1952, Scientific Report and Naval Report, April 1953.

37. Ibid.

38. Brian Balmer, "How Does an Accident Become an Experiment? Secret Science and the Exposure of the Public to Biological Warfare Agents," *Science as Culture*, 13, no. 2 (2004), 197–228.

39. PRO, WO 188/668, AC12384/BRBM29, BRAB 29th Meeting, 12 June 1953.

40. Ibid., AC12459/BRB114, Operation Hesperus, Summary of Results, 14 October 1953.

41. Ibid., AC12526/BRB118, BRAB, Report by the Chairman on the Work of the Board during the Year 1953, 27 October 1953.

42. PRO, DEFE 55/256, MRD R13, Operation Ozone, Scientific Report and Naval Report, 22 December 1954.

43. Ibid.

44. PRO, WO 188/666, BW(54)13, Trials with BW Agents in the Bahamas, 11 August 1954.

45. PRO, WO 188/668, AC12890/BRBM32, BRAB 32nd Meeting, 15 June 1954.

46. PRO, WO 188/670, AC13332/BRB.M34, BRAB 34th Meeting, 4 June 1955.

47. Ibid., AC13178/BRB.M33, BRAB 33rd Meeting, 10 February 1955.

48. Ibid., AC13524/BRB.M35, BRAB 35th Meeting, 5 December 1955. *B.*

globigii is currently known as *Bacillus subtilis*. It was chosen as a simulant of anthrax primarily for its ability to form spores.

49. Ibid., AC13700/BRB.M36, BRAB 36th Meeting, 5 April 1956.

50. PRO, WO 195/13780, AC13784/BRB144, SAC, Notes on a visit to the Microbiological Research Establishment Porton on Thursday 5th July 1956, 15 August 1956.

51. J. Baylis and A. MacMillan, "The British Global Strategy Paper of 1952," *Journal of Strategic Studies*, 16, no. 2 (1993), 200–226.

52. R. Ovendale, ed., *British Defence Policy since 1945* (Manchester: Manchester University Press, 1994).

53. PRO, WO 188/705, CS(M) (no title), 12 March 1952.

54. PRO, PREM 11/49, COS(52)362, Report by the Chiefs of Staff on Defence Policy and Global Strategy to be communicated to the Governments of the old Commonwealth Countries, and to the United States Joint Chiefs of Staff, 15 July 1952.

55. PRO, WO 188/705, Biological and Chemical Warfare Research and Development Policy, Report by the Chiefs of Staff, Draft, n.d. (1952).

56. The principal director of scientific research was the channel for advising BRAB on official policy; PRO, WO 188/668, AC12127/BRBM27, BRAB 27th Meeting, 6 December 1952.

57. BRAB 27th Meeting, 6 December 1952.

58. PRO, WO 188/668, AC12227/BRBM27, BRAB 28th Meeting, 10 February 1953.

59. Ibid., AC12552/BRB120, Directive by the Minister of Supply, On Biological Warfare Research and Development, 19 November 1953.

60. PRO, ADM 1/27325, D(53)44, Biological Warfare Research and Development Policy, Memorandum by the Minister of Defence, 12 October 1953.

61. Ibid., C(53)224, BW Trials in the Bahamas, From Deputy Chief of the Naval Staff, 7 August 1953.

62. Biological Warfare Research and Development Policy, Memorandum by the Minister of Defence, 12 October 1953.

63. PRO, WO 188/668, AC12595/BRBM30, BRAB 30th Meeting, 6 November 1953.

64. Directive by the Minister of Supply, On Biological Warfare Research and Development, 19 November 1953.

65. Ibid.

66. PRO, DEFE 10/33, DRPC, Review of Defence R&D, 10 March 1954. The review was largely dismissed by the Chiefs of Staff for reasons that had little to do with chemical and biological warfare and were more concerned with a long-term strategy that by 1957 replaced the earlier concept of preparedness—and the arrival of the hydrogen bomb. See PRO, AVIA 54/1749, MOD (Ministry of Defence), DRPC Review, 1954.

67. DRPC, Review of Defence R&D, 10 March 1954.

68. PRO, DEFE 10/33, DRP/P(54)40, DRPC, Report by the Committee on the Review of the R&D Programmes, 18 November 1954.

69. PRO, WO 286/78, DRP/P(54)40, Extract from 1954 Review of R&D Programme.

70. PRO, AIR 20/8727, CMS.2484/54, OR/1065, Air Staff Target, Biological Warfare Agents and Weapons, and the Associated Problems of Their Storage and Handling under Storage Conditions, 16 July 1954.

71. Ibid.

72. PRO, WO 195/13798, AC13802, SAC Executive Officer's Report for the 134th Meeting of Council to be held on 25th October 1956, 15 October 1956.

73. L. Arnold, *Britain and the H-Bomb* (Basingstoke: Palgrave, 2001).

74. PRO, WO 286/78, Memo from Chief Scientist (Ministry of Supply) to DGSR(M), 7 November 1956.

75. PRO, DEFE 10/281, DRPS/M(56)28, Defence Research Policy Staff, Meeting, 2 November 1956.

76. PRO, WO 286/78, Extract from DRP/M(56)12, DRPC Meeting, 6 November 1956.

77. PRO, CAB 131/17, DC(56)6, Defence Committee Minutes, 10 July 1956. See Graham S. Pearson, "Farewell to Arms," *Chemistry in Britain*, 31 (October 1995), 782–786; Carter and Pearson, "North Atlantic Chemical and Biological Research Collaboration."

78. PRO, DEFE 10/281, DRPS/P(56)49, Defence Research Policy Staff, Offensive Biological Weapons, Note by DRPS (Air), 24 October 1956.

79. The full list in the directive was "research and development in this field should be designed primarily to determine (a) methods to protect civilians, service personnel, livestock, crops, water supplies and food stocks against infection or contamination; (b) methods for the treatment of human and animal casualties and for decontamination of persons and premises; (c) the practical potentialities of biological methods of warfare and, in particular, their relative effectiveness and cost as compared with other methods of attack; (d) suitable biological agents for use in different roles; (e) processes and techniques for the bulk production and storage of such agents and for the filling of weapons; (f) suitable forms of weapons for the delivery and distribution of such agents upon targets of various kinds."

80. PRO, WO 286/78, RDB/P(56)82, Ministry of Supply, Research and Development Board, Requirement for Further BW Sea Trials in 1957/58 (Note by C. M. and Chief Scientist), 24 August 1956.

81. Ibid.

82. PRO, WO 188/670, AC13968/BRB.M37, BRAB, Minutes 37th Meeting, 18 and 19 January 1957.

83. Ibid., AC14088/BRB.M39, BRAB, Minutes 39th Meeting, 8 and 9 May 1957.

84. PRO, WO 195/14170, AC14176/CDB237, Chemical Defence Advisory Board, Annual Review of the work of the Board for 1957, 4 November 1957.
85. PRO, WO 195/14822, AC14831A, BW Potential 1959, 26 November 1959 (revised 21 February 1960).
86. PRO, WO 195/14064, AC14069.OEC.176, Ptn/Tu.1208/2129/57, Offensive Evaluation Committee, Study of the Possible Attack of Large Areas with BW Agents, n.d. (1957).
87. Spencer R. Weart, *Nuclear Fear: A History of Images* (Cambridge, Mass.: Harvard University Press, 1988); J. Hughes, "The Strath Report: Britain Confronts the H-Bomb, 1954–1955," *History and Technology,* 19, no. 3 (2004), 257–275.
88. A previous series of trials with zinc cadmium sulfide took place in various parts of southern England from 1953 through 1955. These sprayed material from a ground-based source for distances between 25 and 50 miles. The trials were carried out by the CDEE, which had the necessary staff and knowledge. Although these earlier trials were connected with what would be called the large area concept, the term itself did not enter advisory and policy discussion until 1957. See R. Evans, "Germ Warfare Cloud Floated over Shire Counties," *The Guardian,* 2 November 1999, p. 4.
89. PRO, WO 195/14405, AC14413/BRBM41, BRAB 41st Meeting, 18 July 1958.
90. Ibid., Minute 286.
91. PRO, DEFE 41/156, JP(58)65 (final), Chiefs of Staff Committee, Joint Planning Staff, Biological and Chemical Warfare, Report by the Joint Planning Staff, 30 July 1958. The report is cited as DRP/P(58)29.
92. Ibid.
93. PRO, DEFE 10/356, DRP/P(59)6, Ministry of Defence, DRPC, Biological and Chemical Warfare, Note by the DRP Staff, 28 January 1959.
94. PRO, WO 195/14905, AC14914/CDB271 OECM5, Offensive Evaluation Committee 5th Meeting, 15 January 1960.
95. PRO, WO 195/14745, AC14754/BRBM43, BRAB 43rd Meeting, 28 July 1959.
96. PRO, WO 195/14811, AC14820/BRB172, BRAB Report, 1959.
97. PRO, WO 195/15014, SAC89/CDB281 OECM6, Offensive Evaluation Committee 6th Meeting, 15 July 1960.
98. PRO, WO 195/15142, Porton Note 203, Large Area Coverage by Aerosol Clouds Generated at Sea, 22 March 1961.
99. PRO, DEFE 13/440, Confidential Annex to COS(60) 59th Meeting, 27 September 1960.
100. PRO, WO 195/15168, SAC243/BRBM49, BRAB 49th Meeting, 15 April 1961; PRO, AIR 8/1936, COS.204/15/2/61, United Kingdom BW and CW Weapons Release of Information, 20 February 1961.

101. PRO, WO 195/14995, SAC70/BRBM48, BRAB 48th Meeting, 16 July 1960, Minute 324.

102. Membership: Sir Harry Melville, Dr. F. J. Wilkins, Professor E. R. H. Jones, Professor E. T. C. Spooner, Major General J. R. C. Hamilton, Professor P. B. Medawar.

103. PRO, WO 195/15287, SAC362/M51, BRAB 51st Meeting, 6 December 1961.

104. PRO, DEFE 10/490, DRP/P(62)33, DRPC, Chemical and Biological Warfare, Note by the Joint Secretaries, 10 May 1962.

105. S. Greenwood, *Britain and the Cold War, 1945–1991* (Basingstoke: Palgrave, 2000), p. 153.

106. T. C. Salmon, "Britain's Nuclear Deterrent Force: Changing Environment," in *The Defence Equation: British Military Systems—Policy, Planning and Performance since 1945,* ed. M. Edmonds (London: Brasseys, 1986); L. Freedman, *Britain and Nuclear Weapons* (Basingstoke: Palgrave, 1980).

107. PRO, DEFE 11/660, COS(62) 69th Meeting, Chemical and Biological Warfare, 1 November 1962.

108. PRO, WO 11/660, JP(62)96 (final), Chiefs of Staff Committee, Joint Planning Committee, Chemical and Biological Warfare, 12 October 1962.

109. PRO, DEFE 11/660, COS(62), 69th Meeting, Chemical and Biological Warfare, Annex to Minute 1, Amendments to JP(62)96, 1 November 1962.

110. Ibid., JP(62)96 (final), Chemical and Biological Warfare, Report by the Joint Planning Staff, 12 October 1962.

111. Ibid.

112. PRO, DEFE 11/660, Cabinet Defence Committee, Biological and Chemical Warfare Policy, Memorandum by the Minister of Defence, 16 April 1963.

113. Ibid., JIC(63)28 (final), Cabinet Joint Intelligence Committee, The Soviet Chemical and Biological Warfare Threat, Report by the Joint Intelligence Committee, 26 March 1963.

114. PRO, DEFE 13/440, SZ/1177/62, Chemical and Biological Warfare Policy, Note to the Minister from G. Owen (signed on behalf of CSA), 20 December 1962.

115. PRO, DEFE 24/31, d/Ds 6/8, Briefs for New Ministers, Chemical and Biological Warfare, 7 October 1964.

116. PRO, CAB 131/28, D(63)3, Cabinet Defence Committee Meeting, 3 May 1963.

117. Brian Balmer, "Using the Population Body to Protect the National Body: Germ Warfare Tests in the United Kingdom after World War II," in *Useful Bodies: Humans in the Service of Twentieth Century Medicine,* ed. J. Goodman, A. McElligott, and L. Marks (Baltimore: Johns Hopkins University Press, 2003).

118. Balmer, *Britain and Biological Warfare.*

119. PRO, WO 195/15610, SAC685/BRBM54, BRAB 54th Meeting, 13 May 1963.

120. PRO, WO 188/670, AC13968/BRB.M37, BRAB, Minutes 37th Meeting, 18 and 19 January 1957.

121. OR/1065, Air Staff Target, Biological Warfare Agents and Weapons, and the Associated Problems of Their Storage and Handling under Storage Conditions, 16 July 1954.

122. The number (162) denotes a particular strain of the bacteria used at Porton.

123. K. P. Norris, G. J. Harper, and J. E. S. Greenstreet, "The Viability, Concentration and Immunological Properties of Airborne Bacteria Released from a Massive Line Source," MRE Field Trial Report No. 4, Harvard-Sussex Program Information Bank, SPRU, University of Sussex (hereafter HSP), September 1968.

124. Another four trials took place off Lyme Bay between 3 February and 26 April 1966.

125. "Comparison of the Viability of Escherichia coli in Airborne Particles and on Microthreads Exposed in the Field," MRE Field Trial Report No. 5, HSP Information Bank, n.d.

126. PRO, WO 195/15819, SAC893/BRBM56, BRAB 56th Meeting, 13 March 1964; PRO, WO195/15750, SAC824/BRB232, "A Method for Studying the Viability of Micro-organisms of any Particle Size Held in the 'Airborne' State in Any Environment for Any Length of Time," 14 February 1964.

127. PRO, WO 195/14405, AC14413/BRBM41, BRAB 41st Meeting, 18 July 1958, Minute 286.

128. PRO, DEFE 13/557, Secretary of State to Prime Minister, Biological and Chemical Warfare, 8 November 1965.

129. Ibid., CA (studies), A. H. Cottrell, "Research Programme on Chemical and Biological Warfare," 8 February 1967.

130. Ibid.

131. PRO, AIR 8/2391, Air Commodore A. C. L. Mackie (DASB) to Chief of Air Staff, Research Programme on Chemical and Biological Warfare, 22 February 1967.

132. PRO, DEFE 11/660, Defence Research Committee, Review of CW and BW Research, Note by the Defence Research Staff, 1 December 1966.

133. PRO, DEFE 13/557, To Minister of Defence from JP, Chemical Warfare and Biological Warfare—The Future of MRE and CDEE Porton, 22 August 1967.

134. PRO, DEFE 12/557, From Roy Mason, Minister of Defence for Equipment to CA(S), Chemical Warfare and Biological Warfare—Future of MRE and CDEE Porton, 29 August 1967.

135. Hammond and Carter, *From Biological Warfare to Healthcare.*

136. PRO, AIR 8/2391, From Chief of the Defence Staff to CA(Studies), Biological Warfare, 14 July 1967.

137. PRO, DEFE 13/557, Sir Solly Zuckerman to the Rt. Hon. Dennis Healey, 7 November 1967.

138. See Chapters 15 and 16; Susan Wright, "Geopolitical Origins," in *Biological Warfare and Disarmament: New Problems, New Perspectives,* ed. Wright (Lanham, Md.: Rowman and Littlefield, 2002).

139. PRO, DEFE 11/672, The Porton Establishments, Note by the Secretary of State (Revised Draft), 3 October 1968.

140. Ibid.

141. PRO, AIR 8/18791, The UK Military Requirements for Biological Defence Research and Development, Draft, Attached to: Future of MRE Porton, 11 May 1970.

142. There is very little documentation in the public archives regarding these trials. At the time of writing, the sources cited in this section had been declassified and were awaiting acceptance by the PRO. Copies of reports had also been made available in the House of Commons Library and Dorset County Library.

143. PRO, DEFE 11/660, Defence Research Committee, Review of CW and BW Research, Note by the Defence Research Staff, 1 December 1966.

144. US Department of Defense, 2003 Report to Congress, "Disclosure of Information on Project 112 to the Department of Veterans Affairs, as directed by PL107-314," 1 July 2003.

145. G. J. Harper, J. E. S. Greenstreet, and K. P. Norris, "The Survival of Airborne Bacteria in Naval Vessels: Tests with E. coli," MRE Field Trial Report No. 8, May 1969.

146. G. J. Harper, F. A. Dark, and J. E. S. Greenstreet, "The Penetration of an Airborne Simulant into HMS Andromeda," MRE Field Trial Report No. 11, October 1971.

147. Currently named *Francisella tularensis.*

148. G. J. Harper, F. A. Dark, J. E. S. Greenstreet, and F. P. Errington, "Ship Defence against Biological Operations. Navy Trial Varan," MRE Field Trial Report No. 14, January 1974.

149. BRAB 43rd Meeting, 28 July 1959.

150. G. J. Harper, J. E. S. Greenstreet, and F. P. Errington, "Decontamination and Cleansing in Biological Operations," MRE Field Trial No. 15, August 1974.

151. "Studies in the Protection Training Unit, Phoenix NBCD School (U), Navy trial Gondolier," MRE Field Trial Report No. 21, July 1976.

152. PRO, WO 195/15014, SAC89/CDB281 OECM6, Offensive Evaluation Committee 6th Meeting, 15 July 1960.

153. The trial report also contained recommendations on clothing to be worn and procedures for undressing and decontamination in defense against a biological attack.

154. Hammond and Carter, *From Biological Warfare to Healthcare.*

155. Ibid.

4. The Canadian Biological Weapons Program and the Tripartite Alliance

1. The literature on Canada's involvement with WMD and the Cold War is limited. Some of the more useful studies include William Barton, *Research, Development and Training in Chemical and Biological Defence within the Department of National Defence and the Canadian Forces* (Ottawa: Queen's Printer, 1989); John Bryden, *Deadly Allies: Canada's Secret War, 1937–1947* (Toronto: McClelland and Stewart, 1989); James Eayrs, *In Defence of Canada,* vol. 1: *Peacekeeping and Deterrence* (Toronto: University of Toronto Press, 1972); D. J. Goodspeed, *A History of the Defence Research Board of Canada* (Ottawa: Queen's Printer, 1958); Robin Ranger, "The Canadian Contribution to the Control of Chemical and Biological Warfare," CIIA Paper (Wellesley Paper 5/1976); Erika Simpson, *NATO and the Bomb: Canadian Defenders Confront Critics* (Montreal: McGill/Queen's University Press, 2001); Denis Smith, *Diplomacy of Fear: Canada and the Cold War, 1941–1948* (Toronto: University of Toronto Press, 1988); Reg Whitaker and Gary Marcuse, *Cold War Canada: The Making of a National Insecurity State, 1945–1957* (Toronto: University of Toronto Press, 1994).

2. The debate over offensive and defensive BW is of long standing, with the retaliatory or deterrent dimensions often being viewed as defensive. In Canada, even having a BW retaliatory capability was seriously considered only during the late 1960s.

3. Canada was not alone in lacking a carefully formulated CBW policy. The 1969 Nixon administration review, for instance, noted that "the United States has not had a fully developed national policy in either the CW or BW fields"; Report to National Security Council (NSC), US Policy on Chemical and Biological Warfare and Agents, Submitted by the Interdepartmental Political-Military Group in Response to NSSM 59, 10 November 1969, National Security Archive, George Washington University.

4. The Defence Research Board (1946–1974) was a unique civilian-dominated operation within the Department of National Defence. Its first director general, Dr. Omond Solandt, had direct access to the deputy minister and the Chiefs of Staff and was able to establish his own liaison with British and American defense research organizations, military and civilian. After it was abolished, its various functions were integrated within the DND.

5. Donald Avery, *The Science of War: Canadian Scientists and Allied Military Technology during the Second World War* (Toronto: University of Toronto Press, 1999).

6. Ibid., pp. 151–175, 235–255.

7. Closely related was the work of the Tripartite Technical Cooperation Program, established in 1957 at the behest of the Eisenhower administration

for collective defense and mutual help in both nuclear and nonnuclear weapons.

8. During the spring of 1947 a special committee was created to assist the DRB in its biological and chemical warfare planning. It was chaired by Otto Maass, scientific advisor to the Chiefs of the General Staff, who had directed Canada's World War II program; National Archives of Canada, Ottawa (hereafter NAC), National Research Council, vol. 6, file 3-12-M3-24, Maass Report, 4 September 1947.

9. PRO, WO 188/440, Report of Meeting, 16 August 1946, Gravelly Pt., Va. In 1946 there were 10 qualified scientists at Suffield and the Kingston Laboratory, while at Detrick there were 150 professional scientific positions, with another 300 technicians and support personnel.

10. Peter Hammond and Gradon Carter, *From Biological Warfare to Healthcare: Porton Down, 1940–2000* (Basingstoke: Palgrave, 2002); Brian Balmer, *Britain and Biological Warfare: Expert Advice and Science Policy, 1930–65* (Basingstoke: Palgrave, 2001), pp. 55–184.

11. PRO, WO 188/660, Report to the Defence Research Policy Committee by Air Marshall Sir Norman Bottomley, Chair of the Bological Warfare Sub-Committee, 28 November 1947.

12. PRO, WO 188/705, Report for the British Chiefs of Staff, 20 November 1947. Wood had previously worked closely with Lord Stamp and David Henderson in the 1945 *Brucella* field trials at Suffield.

13. Defence Research and Development Canada–Suffield, Closed Collection (hereafter DRDC-S), Declassified Records, Operation Harness, 1947–1949. Scientific Report Prepared by D. W. Henderson and J. D. Morton for the Chiefs of Staff Committee, Biological Warfare Sub-Committee, Ministry of Supply and Biological Research Advisory Board, 20 July 1949.

14. PRO, WO 188/705, Walter Lamb to David Henderson, 25 November 1949; ibid., Report of Biological Warfare Sub-Committee, 12 January 1951.

15. DRDC-S, Declassified Records, Test, 21 April 1946, Bot Tox; 6 Dec 1945-US, "The Casualty Producing Power of US-S when Suspended from the Bomb A/C L," 14 December 1945.

16. NAC, Defence Research Board Papers (hereafter DRBP), vol. 4133, file 4-953-43-1. At the fourth meeting of the BWRP (17 February 1948) there had been extensive discussion of the various means of defending against various types of biological and toxin agents, with the threat of airborne clouds of BW agents deemed the most serious, since "this means of attack has been demonstrated successfully."

17. DRBP, vol. 4313, file DRBS 4-935-43-2, Minutes of Fifth Meeting of BWRP, 20 January 1949; ibid., Minutes of Sixth Meeting of BWRP, 12 November 1949; ibid., Report, Guilford Reed, 27 June 1950.

18. DRBP, vol. 7328, file 83-84/167, DRBS 100/24/0 (declassified), "Biological

Warfare in the Northern Half of the Northern Hemisphere: Possible Dissemination of Disease via Biting Flies and Other Northern Insects," 22 July 1949.

19. In this research, Reed drew upon World War II Japanese BW experiments, material he obviously obtained from Detrick. In 1955 one of Reed's co-workers was given the task of coordinating insect vector BW research between DRBK and Detrick; DRDC-S, Declassified Records, Report No. 43, A. S. West and G. B. Reed, "Fleas as Vectors of Plague in Bacterial Warfare," 30 October 1954.

20. Ibid., Suffield Technical Paper No. 118, D. E. Davids and A. B. Lamb, "Collection of Bacterial Aerosols under Cold Weather Conditions," 28 June 1957.

21. DRBP, vol. 4220, file 700-900-267-1, Group Commander H. G. Richards, Canadian Joint Staff Mission, Washington, to Chief of the Air Staff, Ottawa, 25 October 1951; ibid., A. L. Wright, DRB, Canadian Joint Staff Mission, to Chairman, DRB, 7 December 1951; ibid., G. W. Rowley, Arctic Research, to Chairman DRB, 14 December 1951; ibid., Memorandum, Special Weapons Test Facilities: Suffield Experimental Station, October 1951.

22. Tom Mangold and Jeff Goldberg, *Plague Wars: A True Story of Biological Warfare* (New York: St. Martin's, 1999), pp. 28–38; PRO, WO 188/660, BRAB Meeting, 29 March 1952.

23. NAC, Brock Chisholm Papers, vol. 1, file 90-92 (1949), Address to the World Union of Peace Organizations, Switzerland, 9 September 1949; *Ottawa Journal,* 10 September 1949.

24. Many of these accusations have reappeared in Stephen Endicott and Edward Hagerman, *The United States and Biological Warfare: Secrets from the Early Cold War and Korea* (Bloomington: Indiana University Press, 1998).

25. NAC, Records of the Department of External Affairs (hereafter DEAR), vol. 5921, file 50208, Address by the Secretary of State for External Affairs, Lester B. Pearson, 25 May 1952; ibid., Canadian Ambassador in Washington to DEA, 20 May 1952.

26. DEAR, vol. 5921, file 50208, W. H. Brittain, Vice-Principal, Macdonald College, McGill University, to Lester B. Pearson, 28 May 1952. The other two members of the team were A. W. Baker, head of the Department of Entomology and Zoology, Ontario Agricultural College; and C. E. Atwood, assistant professor of zoology, University of Toronto. Guilford Reed also found "the supposed method of dispersal ridiculous"; West and Reed, "Fleas as Vectors of Plague in Bacterial Warfare."

27. DEAR, vol. 5920, file 50208-40, Reports from the Canadian Permanent Representative to the United Nations, New York, to DEA, 19 June 1952, 28 March 1953, 12 August 1953, 8 September 1953, 26 October 1953.

28. Whitaker and Marcuse, *Cold War Canada,* pp. 364–425; Robert Bothwell and William Kilbourn, *C. D. Howe, a Biography* (Toronto: McClelland and Stew-

art, 1979), pp. 224–300; B. D. Hunt, ed., *Canada's Defence: Perspectives on Policy in the Twentieth Century* (Toronto: Copp Clark Pitman, 1993), pp. 129–272. Secretary of State John Foster Dulles' so-called New Look strategy made it clear that the US would use nuclear weapons if either the USSR or China committed aggression.

29. Report to NSC, US Policy on Chemical and Biological Warfare and Agents, 10 November 1969. This presidential policy guideline was incorporated into the Basic National Security Policy issued on 5 August 1959 (and rescinded January 1963): "The United States will be prepared to use chemical and biological weapons to the extent that such use will enhance the military effectiveness of the armed forces. The decisions as to their use will be made by the President."

30. In 1951 the Ottawa-based BW research facilities were moved to the new facility at neighboring Shirley's Bay; in 1956 this operation was further expanded with the shift of the Kingston BW laboratory to this location; Jim Norman and Rita Crow, eds., "A History of the Defence Research Establishment Ottawa, 1941–1991," Ottawa, March 1992.

31. DRBP, vol. 4220, file 423-935-267, Colonel E. Staples, Canadian Army Technical Representative, US Army Chemical Corps, Maryland, 29 November 1954; ibid., Charles Mitchell, Chief, Department of Agriculture Animal Research Institute, 6 December 1954.

32. DRBP, vol. 4224, file DRBS 1820-11, Charles Mitchell, Dominion Animal Pathologist (Dept. of Agriculture), to Glen Gay, Special Weapons Research Section, DRB, 28 October 1948; ibid., Lt. Col. J. C. Bond, Canadian Technical Representative, Edgewood, to DRB, 1 December 1949.

33. Ibid., Mitchell to Solandt, 4 December 1950. The advantages of using strains of rinderpest virus for growth in eggs was that its virulence could be sufficiently attenuated to allow its use as a vaccine for cattle, with a high degree of immunity.

34. Ibid., Major General A. C. McAuliffe, Chief Chemical Officer, to Solandt, 9 March 1951; ibid., E. L. Davies to McAuliffe, 24 March 1951; ibid., Colonel Oram Woolpert, Director, Headquarters Camp Detrick, to Davies, 3 April 1951; ibid., H. M. Barrett (for chairman DRB) to S. C. Barry, Director, Production Service, Dept. of Agriculture, 21 March 1958.

35. DRBP, vol. 4224, file 4-935-43-2, Minutes of the Meeting of the BWRP, 23 October 1954; *Langley* (B.C.) *Advance News,* 27 March 2001.

36. DRBP, vol. 4224, file DRBS 171-80/B1, Advisory Committee on BW Research (successor to the BWRP), 2–3 November 1959.

37. DRDC-S, Declassified Records, Technical Cooperation Programme, Thirteenth Tripartite, Ottawa-Suffield, 15–26 September 1958. This transition from small-area field tests, using bombs and bomblets, to the large area concept, was described as having enormous potential, since then it would "cover areas of up to 10/6 square miles by emission of agents from a line

source hundreds of miles in length with subsequent drifting by the elongated cloud by the prevailing winds."

38. DRBP, vol. 1450, file 83/84-167, Minutes of SES [Scientific Establishment Suffield]-Dugway Conference, 21 May 1957; interview with Archie Pennie (superintendent of DRES, 1957–1963), Ottawa, June 2004.

39. US Army Activity in the US Biological Warfare Programs, vols. 1 and 2, February 1977.

40. The Project 112 tests have been described as a "Cold War-era chemical and biological warfare test program . . . initiated in 1962 out of concern for our nation's ability to protect and defend against these potential threats." Of the 134 tests planned, 46 were confirmed to have been conducted with fact sheeting describing the goals and results; Current Issues—Project SHAD Chart Printer Friendly: Scoop DoD Releases Orihect 112 Chem Weapons Fact Sheets, deploymentlink.osd.mil/ current_issues/shad_chart/shad.

41. The bitter confrontation over Canada's response to the Cuban Missile Crisis and its refusal to accept nuclear warheads for either the Bomarc B ground-to-air missile system or for Canadian air squadrons assigned to NATO are not discussed here. There is, however, extensive literature on the subject. See Jocelyn Ghent-Mallet and Don Munton, "Confronting Kennedy and the Missiles in Cuba, 1962," in *Coming of Age: Readings in Canadian History since World War II*, ed. Donald Avery and Roger Hall (Toronto: Harcourt Brace, 1996), pp. 319–342; Erika Simpson, *NATO and the Bomb;* and Denis Smith, *Rogue Tory: The Life and Legend of John G. Diefenbaker* (Toronto: MacFarlane, Walter and Ross, 1995).

42. *Canadian Medical Association Journal,* 87 (1 December 1962), 1142, 1156–60. A 1960 DRB report had concluded that Canada's ability to survive a sustained CBW attack, on the basis of 25 key defensive measures, was almost entirely unsatisfactory; DRBP, vol. 4226, file DRBS 2001-1, DRB Report No. 29, December 1960. With regard to bioterrorism, Suffield scientists were aware of the secret field trials that had been carried out in the London Underground's Northern Line during the late 1950s; Hammond and Carter, *From Biological Warfare to Healthcare*, p. 117.

43. DRBP, vol. 4226, file DRBS 2000-1, J. C. W. Scott, DRB Officer, Canadian Joint Staff Mission, London, 26 September 1962; ibid., G. R. Vavasour, DRB, Memorandum on BW Policy, 22 November 1962.

44. The original policy was included in COS Committee Paper 1/63, 1 May 1963. In a 1966 policy statement DRB planners predicted that during the next 10 years Canadian troops could expect to encounter CB weapons on the battlefield; DEAR, vol. 5, file 28-6-6, Statement, Brigadier General H. Tellier (Director General Plans Canadian Forces) to J. S. Nutt, Office of Politico-Military Affairs, DEA, 31 December 1968; DRBP, vol. 4226, file DRBS 2000, Memorandum, G. R. Vavasour, 13 March 1966.

45. NATO Archives, Brussels (hereafter NATO files), Civil Defence Committee

(AC/023), file 1-79, First Meeting of Senior Civil Emergency Planning Committee, 10 February 1956; ibid., D/499, Report of the Meeting of the Restricted Working Party on Protection against Chemical Warfare, 12 August 1965; ibid., SGM 117/62, NATO Military Committee, Standing Group CBR, 15 April 1962; ibid., Long-Term Scientific Studies for the Standing Group North Atlantic Treaty Organization, 1962.

46. It was also predicted that in the CBW field the Soviets would greatly expand their R&D "to acquire a clear military advantage over the West"; ibid., MCM 122/62, Memorandum for Members of the Military Committee, 22 October 1962; ibid., SHAPE 84/62-1450/20, Chemical and Biological Warfare Policy, 13 April 1962.

47. This report also analyzed the CBW capabilities of other Warsaw Pact countries—concluding that "Czechoslovakia, Poland, East Germany and Bulgaria are conducting BW research . . . [none] of the satellites have an offensive biological warfare capability"; ibid., SG 161/19, Report of Standing Group on CBR, "Soviet Bloc Strength and Capabilities: Part I: The Soviet Bloc Threat to NATO, 1963–1967" (meeting of 27 April 1965); ibid., SG 265, Final Report, sec. 11, Biological Warfare, 14 March 1966.

48. Ibid., DPC/D(67)23, "Report by the Military Committee to the Defence Planning Committee on Overall Strategic Concept for the Defence of the North Atlantic Treaty Organization Area," 11 May 1967. The committee concluded that the USSR would "continue to support their objectives from a position of impressive military strength based on nuclear, massive conventional, chemical and possibility biological capabilities"; Gregory Pedlow, ed., *NATO Strategy Documents, 1949–1969* (Brussels: NATO, 2000), pp. 353, 366; DEAR, CBW Disarmament, vol. 3, file 28-6-6, Memorandum, Assistant Deputy Minister External Affairs, 6 January 1969.

49. A detailed study of possible candidates for an ideal BW agent had been prepared by Suffield scientists before the April 1966 Tripartite meetings. After examining the respective advantages of bacterial, rickettsial, and viral agents, it was decided to concentrate on the last group of pathogens because of their superior aerosol stability and respiratory pathogenicity. Among the viruses, they eventually chose influenza A and B, although serious consideration was given to dengue fever of the abovirus group and to vaccinia (smallpox) of the poxvirus category. DRDC-S, Declassified Records, Canadian Position Paper to the Quadripartite CBR Conference, Standing Group on Biological Warfare, Suffield, 4–6 April 1966 (hereafter Canadian Position Paper); DRBS 2000-1, Alex Longair, Director of Atomic Research, DRB, Memorandum, 26 October 1960.

50. Canadian Position Paper. This report also indicated the extent to which DRES was involved with "incapacitating but non-lethal agents," both in terms of understanding their chemical properties and in terms of their impact. Of particular interest were the so-called hallucinogenic or psychologi-

cal compounds such as 3-quinuclidiny benzilate (BZ) and lysergic acid diethylamide (LSD). In 1963 Australia and New Zealand had been added to the Tripartite CBW organization, largely because of US strategic interests in Southeast Asia.

51. DRDC-S, Declassified Records, Suffield Memorandum No. 125/68, Brief to the DRB, 10 October 1968/21 December 1968; ibid., Planning of BW/CW Programme in DRB, 9–10 July 1968. In May 1970, fearing a possible "invasion," DND officials even considered cordoning off the base with combat troops; DEAR, file 28-6-6 (pt. 13), Memorandum for the Minister of Defence: Proposed Demonstration against CBW at Suffield, May 29–31.

52. DEAR, CBW Disarmament, vol. 3, file 28-6-6, Memorandum, Marcel Cadieux, Minister of Defence, 23 August 1968; DRDC-S, Declassified Records, Suffield Memorandum No. 16/68: Brief Summary of Activities at the Defence Research Establishment Suffield during 1967 by E. J. Bobyn, 2 February 1968; DEAR, vol. 5, file 28-6-6, Statement, Brigadier General H. Tellier (Director General Plans Canadian Forces) to J. S. Nutt, Office of Politico-Military Affairs, DEA, 31 December 1968.

53. DEAR, file 28-6-6 (pt. 13), H. B. Robinson, Memorandum to the Minister, 7 October 1968. The first of these meetings had been held in London in 1962, with the second in Washington in 1965.

54. DEAR, file 28-6-6 (pt. 5), Report of Joint Meeting, 18 November 1968; ibid., Under-Secretary DEA to Chief of Defence Staff, 29 January 1969.

55. PRO, DEFE 13/557, Ministry of Defence, Memorandum for Prime Minister Harold Wilson, 21 August 1967; DEFE 11/672, Chiefs of Staff Report, 3 October 1968.

56. There are numerous secondary sources dealing with Canadian-US relations during the Vietnam War. The following were of particular value for this study: J. L. Granatstein and Robert Bothwell, *Pirouette: Pierre Trudeau and Canadian Foreign Policy* (Toronto: University of Toronto Press, 1990); Ian Lumsden, ed., *Close the 49th Parallel: The Americanization of Canada* (Toronto: University of Toronto Press, 1970); and Victor Levant, *Quiet Complicity: Canadian Involvement in the Vietnam War* (Toronto: Between the Lines, 1986).

57. In July 1969 Trudeau had briefly described Canada's ongoing review of its CBW policies and its support of BW disarmament negotiations. He subsequently considered attending the spring 1970 meetings of the CCD "to deliver a speech on CBW"; *Debates of the House of Commons* (hereafter *HC Debates*), 3 July 1919; DEAR (pt. 13), Peter Walker, Memorandum, 24 July 1970.

58. DEAR, vol. 11534, file 28-6-6 (pt. 9), British High Commissioner, Aide Memoire, 9 October 1969. For a good analysis of the efforts of the British Foreign Office to "kick-start" discussions of a separate BW Convention, see Susan Wright, "Geopolitical Origins," in *Biological Warfare and Disarmament: New Problems/New Perspectives*, ed. Wright (Lanham, Md.: Rowman and Littlefield, 2002), pp. 313–342. The Pugwash Study Group on Biological War-

fare, at both its 1965 and 1967 meetings, had also condemned this form of warfare, as had the special 1969 study on CBW carried out by SIPRI; Matthew Meselson, "Behind the Nixon Policy for Chemical and Biological Warfare," *Bulletin of the Atomic Scientists,* 26 (January 1970), 23–34; Julian Perry Robinson, "Origins of the Chemical Weapons Convention," in *Shadows and Substance: The Chemical Weapons Convention,* ed. Benoit Morel and Kyle Olson (Boulder: Westview, 1993), p. 48.

59. DEAR, file 28-6-6 (pt. 6), Disarmament, Arms Control and Non-Proliferation Division (IDA) to Legal Division, 20 February 1969, 21 March 1969; NARA, Richard Nixon Papers, box 310, Morton Halperin, Memorandum for Henry Kissinger: US Policy, Programs and Issues on CBW, 28 August 1969; ibid., Lee Du Bridge, Science Advisor, Memorandum for Dr. Henry Kissinger, 22 October 1969. The Trudeau papers and those of foreign minister Mitchell Sharp are closed, and Cabinet documents for this period remain elusive.

60. By 12 November Nixon and his advisors had finalized their policy, and arrangements were made for a carefully orchestrated scenario for the morning of 25 November. During this interval, consultations took place with Japan and Brazil, two US allies that had not yet ratified the Protocol. There is no evidence that the Canadian government received prior notification, despite the fact that Trudeau and Nixon were on relatively good terms—at least compared to their later acrimonious relationship; NARA, Nixon Papers, NSC, box 310, Briefings for NSC meetings on CW/BW, 19 November 1969.

61. *HC Debates,* 25 November 1969; *Montreal Gazette,* 26 November 1969; DEAR, file 28-6-6 (pt. 6), Telegram of DEA to Washington Embassy, 28 November 1969; George Ignatieff, Statement to the UN General Assembly, 28 November 1969. Surprisingly, most questions in the House of Commons involved the movement of surplus US nerve gas through Pacific coastal waters, rather than the implications of Nixon's declaration.

62. *HC Debates,* 25 March 1970; George Ignatieff, *The Making of a Peacemonger* (Toronto: Penguin, 1987); Ottawa, DND Historical Directorate, 82/445, Annual Reports of DRES (1970, 1971, 1972, 1974).

63. DEAR, file 28-6-6 (pt. 17), IDA Report, 4 September 1972; ibid., Mitchell Sharp to Deputy Minister of Defence, 19 August 1975.

64. Department of Foreign Affairs (DFA), vol. 3, file 26-6-6 (BW), Office of the Advisor on Disarmament and Arms Control Affairs, IDA, Telegram, D. S. McPhail, Permanent Disarmament Ambassador in Geneva, 24 March 1980; ibid., US DIA Report, "Foreign Technology Weapons and Systems," 3 March 1980; Leonard Cole, "Sverdlovsk, Yellow Rain and Novel Soviet Bioweapons: Allegations and Responses," in *Preventing a Biological Arms Control Race,* ed. Susan Wright (Cambridge, Mass.: MIT Press, 1990), pp. 199–219.

65. Ottawa, "Study of the Possible Use of Chemical Warfare Agents in South-

east Asia" (1982), by Dr. H. B. Schiefer, Toxicology Group, University of Saskatchewan; G. Humphrey and J. Low, "An Epidemiological Investigation of Alleged CW/BW Incidents in S. E. Asia" (1982), Report of Preventive Medicine/Surgeon General, Department of National Defence; DEAR, vol. 3, file 26-6-6 (BW), Washington Embassy to Ottawa, 21 September 1981; ibid., A. Mathewson, Chief Policy Planning, DND, to R. P. Cameron, Assistant, IDA, 6 August 1982.

66. DFA, Review of Reports from the BTWC Review Conferences: 1986, 1991, 1996; DEA unclassified records (vols. 7–11), file 28-6-6 (BW).

67. Health Canada, Centre for Emergency Preparedness and Response, http//www-hc-sc-gc. Several other federal agencies were also given a mandate to deal with bioterrorism, coordinated by the Ad Hoc Committee of Ministers on Public Safety and Anti-Terrorism. In the fall of 2001 the federal government also established a five-year $170 million fund to improve Canada's ability to respond to chemical, biological, radiological, and nuclear terrorist attacks.

68. Exercise Global Mercury: Post-Exercise Report to GHSI Ministers, November 2003; information on GHSI can be found at www.hc-gc.gc.Q-148, US Department of Homeland Security, Top Officials (TOPOFF) Exercise Series, After Action Summary Report for Public Release, 19 December 2003; additional information at www.dhs.gov.

69. This statement was made by William Patrick, one of the key scientists and administrators connected with the US BW operation; Judith Miller, Stephen Engelberg, and William Broad, *Germs: Biological Weapons and America's Secret War* (New York: Simon and Schuster, 2001), p. 64. See also Ken Alibek with Stephen Handelman, *Biohazard: The Chilling True Story of the Largest Covert Biological Weapons Project in the World—Told from the Inside by the Man Who Ran It* (New York: Random House, 1999).

70. DEAR, vol. 11535, file 28-6-6 (pt. 13), A. K. Longair (for deputy chairman, scientific) to D. M. Cornett, DEA Disarmament Division, 2 October 1968.

71. See Bryden, *Deadly Allies*, pp. 120–133, 259–260. Of specific value was the DRES development of the Canadian Integrated Bio/Chemical Agent Detection System (CIBADS), used by the Canadian forces during the first Gulf War. After the 2001 letter-bomb attacks, DRES techniques for anthrax detection were adopted by the US Centers for Disease Control.

72. Hugh Gusterson, *Nuclear Rites: A Weapons Laboratory at the End of the Cold War* (Berkeley: University of California Press, 1998); Hammond and Carter, *From Biological Warfare to Healthcare*; University of Toronto Archives, Banting Wartime Diary, 10 May 1940.

5. The French Biological Weapons Program

Acknowledgments: I wish to thank Daniel Kiffer (Délégation aux Affaires Stratégiques du Ministre de la Défense) and Camille Grand (Office du

Ministre de la Défense) for their valuable assistance. Without their kindness and support, this chapter would probably not have been written. Thanks also to Hannah Dyson for translating this chapter from its original French.

1. André Corvisier, ed., *Histoire militaire de la France*, vol. 4: *From 1940 to the Present* (Paris: Presses Universitaires de France, 1997).
2. Service Historique de l'Armée de Terre (hereafter SHAT), Memorandum of information from the Scientific Bureau of the Army to the Army chief of staff, 27 February 1947.
3. SHAT, File addressed to the Army chief of staff, Proposals regarding a program of research into germ warfare, 21 March 1947.
4. SHAT, Minutes of 11 March 1947. In the absence of explicit references, it is impossible to determine whether this decision was handed down by the government, which seems probable, or to date precisely the circumstances and motivation that prevailed at this resumption of France's BW activities.
5. Ibid., p. 3.
6. SHAT, Memo from the Services des Poudres, 8 April 1948. For analysis of the French BW program between 1919 and 1940 see Olivier Lepick, "French Activities Related to Biological Warfare," in *Biological and Toxin Weapons Research, Development and Use from the Middle Ages to 1945: A Critical Comparative Analysis*, ed. Erhard Geissler and J. E. van Courtland Moon, SIPRI Chemical and Biological Warfare Studies No. 18 (Oxford: Oxford University Press, 1999), pp. 70–90.
7. Memo from the Services des Poudres, 8 April 1948.
8. SHAT, Note from Colonel Krebs to the lieutenant general chief of staff, 21 March 1947.
9. Memo from the Services des Poudres, 8 April 1948.
10. SHAT, Memo from the Section Armement et Etudes of the Staff of the Second Bureau, 29 July 1947.
11. SHAT, Dossier on biological warfare (overseas information).
12. SHAT, Report to the Army chief of staff regarding research on BW, 18 August 1947.
13. SHAT, Parameters for research into germ warfare taking into account data currently available, 1953–1954.
14. Memo from the Services des Poudres, 8 April 1948.
15. SHAT, STA, Summary of trials on the contamination of animals in enclosures using germ aerosols delivering intestinally pathogenic germs, 16 March 1948.
16. Lepick, "French Activities Related to Biological Warfare."
17. SHAT, Parameters for research into germ warfare taking into account data currently available, 1953–1954, Minutes of 19 February 1948.
18. Ibid., Minutes of 31 October 1950.
19. Ibid., Minutes of 1 November 1950.
20. Ibid., Minutes of 4 November 1950.
21. Ibid.

22. Ibid., Minutes of October 1951.
23. Ibid., Minutes of March 1953. Surgeon General Costedoat stated in connection with these trials, "no animal died of anthrax, and no enzootic broke out."
24. Ibid., trials of 25 and 27 February 1953.
25. Ibid., p. 7.
26. The Special Weapons Command was created in November 1951. Headed by General Ailleret, this body was responsible for matters relating to chemical, biological, and nuclear armaments; SHAT, Decision No. 15486/EMA/IOS, signed by the secretary of state for war (Pierre de Chevigné), 20 November 1951.
27. SHAT, General plan to establish special weapons units, 22 January 1952.
28. SHAT, Resolution creating the CEECB, signed by the secretary of state for war (Pierre de Chevigné) and the minister of defense (René Pleven), 26 August 1952.
29. SHAT, Resolution from the minister of defense concerning the creation of the CEECB, 26 August 1952.
30. SHAT, Parameters for research into germ warfare taking into account data currently available, 1953–1954.
31. SHAT, Minute Nos. 9 and 14/ASC/TS, 22 December 1955; Memo Nos. 6929 and 6930 EM/ARMET/S, 10 April 1956.
32. For further reading: Claude Carlier, "The Genesis of Atomic Weapons," in Corvisier, *Histoire militaire de la France*, vol. 4, pp. 349–356.
33. SHAT, Minutes of the CEECB, 6 July 1959.
34. Ibid., 20 November 1957.
35. SHAT, Minutes of the CEECB, 6 July 1959.
36. SHAT, Minutes of the CIEECB, 25 April 1960.
37. At the CIEECB meetings, members of the Service de Santé left the room when questions relating to offensive weapons were discussed.
38. Minutes of the CIEECB, 25 April 1960.
39. Ibid., 27 February 1961.
40. Within the DMA, the Direction des Recherches et des Moyens d'Essai (DRME) approved contracts with private laboratories. The Services des Poudres had the CEB, which controlled the program of research and study on attack, detection, and protection. In the field of manufacture, the CEB had responsibility for the making of toxic products and loading of ammunition.
41. The SBVA carried out microbiological research with regard to both attack and protection. It made use of the Laboratoire Militaire de Recherches Vétérinaires at Alfort (18 people) and the Centre Biologique d'Expérimentations at Tarbes (26 people).
42. SHAT, File no. 011, Organization of Biological and Chemical Activities, addressed to the armed forces minister, 27 January 1966.

43. SHAT, Minutes of the Section d'Etudes de Biologie, 7 October 1960.

44. SHAT, Minutes of the CIEECB, 3 November 1961.

45. SHAT, Section d'Etudes de Biologie et de Chimie, Service de Bactériologie, A-61 F-151, Report on a mission to the US under the MWDDEA (Mutual Weapons Development Data Exchange Agreement), 2–7 June 1969.

46. SHAT, Minutes of the CIEECB, 23 January 1962, chaired by General Thiry of the Air Division.

47. Ibid., 16 May 1963.

48. SHAT, Ministerial resolution No. 6234 DMA/ORG/TS, 12 April 1963.

49. SHAT, B.E. No. 1437/CIAS/2/S, Remarks on the directive enhancing the studies of the SGTEB, 10 December 1963.

50. SHAT, Minutes of the CIEECB, 23 January 1962.

51. SHAT, B.E. No. 1437/CIAS/2/S, Directive enhancing the studies of the SGTEB, 10 December 1963.

52. Minutes of the CIEECB, 23 January 1962.

53. SHAT, Armed Forces Ministry, CIAS, CIEECB, Minutes of the SGTEB, 1963–1966, Minutes of 14 January 1964, chaired by Veterinary Colonel Courrèges.

54. The existence of an undated memo regarding "Soviet potential in the field of biological warfare" attached to ibid. allows us to assess how much the French authorities knew of the Soviet Union's BW program: "Having regard to the information gathered over recent years, it appears that the Soviets have a complete program of defense against biological warfare. There is no definitive proof regarding an offensive weapons program, certain strong assumptions are made, and it appears that Professor N. N. Zhukov-Verezhnikov, Member of the Academy of Medicine, Stalin Prize 1950, is the government's technical advisor in these matters."

55. Ibid., Minutes of 14 January 1964.

56. SHAT, Minutes of the SGTEB, 20 December 1966.

57. SHAT, Remarks on the directive enhancing the studies of the SGTEB, B.E. No. 1437/CIAS/2/S, 10 December 1963.

58. SHAT, Letter from General Lavaud (ministerial delegate for armaments) to the Army chief of staff, 21 August 1964.

59. At the meeting of the SGTEB on 20 December 1966, the chairman explained that "the execution of a biological and chemical armaments program was not pursued as originally envisaged for financial reasons"; SHAT, Minutes of the SGTEB, 20 December 1966.

60. SHAT, Minutes of the SGTEB, 2 June 1964.

61. SHAT, Resolution from the ministerial delegate for armaments aimed at creating the Section d'Etudes de Biologie et de Chimie, 11 January 1965.

62. SHAT, Minutes of the SGTEB, 18 December 1964.

63. Resolution aimed at creating the Section d'Etudes de Biologie et de Chimie, 11 January 1965.

64. SHAT, file no. 011, Organization of Biological and Chemical Activities, addressed to the armed forces minister, 27 January 1966.
65. SHAT, file addressed to the armed forces minister, 29 March 1966.
66. SHAT, Memo from General Ailleret, No. 437/EMA/PROG.4/C.D, 20 July 1967.
67. SHAT, Minutes of the SGTEB, 20 December 1966.
68. Ibid.
69. SHAT, File from the Army Staff, Object: creation of a permanent consultative committee NBC, signed by General Cantarel, Army chief of staff, 6 February 1970.
70. Armament et riposte biologique et chimique du CINBC.
71. SHAT, Minutes of the Défense NBC, chaired by Lieutenant Colonel Fontanges, 6 February 1969.
72. In 1969 the Section d'Etudes de Biologie et de Chimie du Bouchet included only three officers (one part-time, one scientist, and one veterinarian), with two laboratory assistants and five junior personnel.
73. Minutes of the meeting of Défense NBC, 6 February 1969.
74. SHAT, Minutes of the meeting of Défense NBC, 20 March 1969.

6. The Soviet Biological Weapons Program

Acknowledgments: I wish to thank those who kindly reviewed previous versions of this chapter, including the editors, the other contributors, Benjamin Garrett, Richard Guthrie, Jens Kuhn, Milton Leitenberg, Julian Perry Robinson, and Roger Roffey. Any errors or omissions are mine. The views expressed are my own and do not necessarily reflect those of SIPRI.

1. Statement on 29 January 1992 by B. N. Yeltsin, President of the Russian Federation, on Russia's Policy in the Field of Arms Limitations and Reduction, in Letter Dated 30 January 1992 from the Representative of the Russian Federation Addressed to the President of the Conference on Disarmament Transmitting the Text of the Statement made on 29 January 1992 by B. N. Yeltsin, the President of the Russian Federation, on Russia's Policy in the Field of Arms Limitation and Reduction, Conference on Disarmament doc. CD/1123, 31 January 1992. Copies of CD documents are maintained by the UN in Geneva.
2. "On Ensuring Fulfillment of International Obligations in the Area of Biological Weapons," Russian Presidential Decree No. 390, 11 April 1992.
3. See, e.g., Ken Alibek with Stephen Handelman, *Biohazard: The Chilling True Story of the Largest Covert Biological Weapons Program in the World—Told from the Inside by the Man Who Ran It* (New York: Random House, 1999); I. V. Domaradskij and W. Orent, *Biowarrior: Inside the Soviet/Russian Biological War Machine* (Amherst, N.Y.: Prometheus Books, 2003); and Ivan V. Domaradskij, *Troublemaker, or the Story of an "Inconvenient" Man* (Moscow,

1995), unofficial translation; original, in Russian, available at http://
domaradsky.h1.ru/; Milton Leitenberg, "Biological Weapons and Arms
Control," *Contemporary Security Policy,* 17 (April 1996), 3–12; idem, *The Problem of Biological Weapons* (Stockholm: Swedish National Defense College,
2004); and Anthony Rimmington, "From Offence to Defence? Russia's Re-
form of Its Biological Weapons Complex and the Implications for Western
Security," *Journal of Slavic Military Studies,* 16 (March 2003), 1–43.

4. On World War II Soviet BW munitions, including aerosol disseminators and
   frangible air bombs, see Walter Hirsch, *Soviet BW and CW Capabilities ("The
   Hirsch Report")* (declassified) (Washington, D.C.: US Army Chemical Intelligence Branch, 15 May 1951), pp. 105–110. For an account of BW-tipped intercontinental ballistic missiles, see Tom Mangold and Jeff Goldberg, *Plague
   Wars: The Terrifying Reality of Biological Warfare* (New York: St. Martin's,
   1999), pp. 83–84.

5. At least part of the officer corps of the Radiological, Chemical and Biological
   Defense Troops (and its predecessors) is included in the estimates. The
   number of BW experts with key knowledge or expertise would have been
   smaller.

6. Alibek, *Biohazard;* Domaradskij and Orent, *Biowarrior;* and Domaradskij,
   Troublemaker.

7. Pasechnik was a professor of medical science, a member of the Ministry of
   Defense's Council on Toxicology, and held the honorary rank of general. He
   was born on 12 October 1937 and died on 21 November 2001; S. Cooper,
   "Life in the Pursuit of Death," *Seed,* January/February 2003, 67–72 and
   104–107. For background on Alibek and Pasechnik, see also R. Preston,
   "Annals of Biowarfare: The Bioweaponeers," *New Yorker,* 9 March 1998,
   52–65.

8. For a discussion of more recent vaccine development activities in Russia,
   see Kristina S. Westerdahl, *Building and Measuring Confidence: The Biological
   and Toxin Weapons Convention and Vaccine Production in Russia,* (FOI) Report
   No. FOI-R-0189 (Umeå: Swedish Defence Research Agency, 2001); and
   K. S. Westerdahl and Roger Roffey, "Vaccine Production in Russia: An Update," *Nature Medicine Vaccine Supplement,* 4 (May 1998), 506.

9. See, for example, V. N. Orlov et al., eds., *My Zashchitily Rossiyu: Istoricheskii
   Ocherk o Sozdanii i Deyatel'nosti Nauchno-Tekhnicheskogo Komiteta, Upravleniya
   Zakazov, Proizvodstva i Snabzheniya i Upravleniya Biologicheskogo Zashchity UNV
   RKhB Zashchity MO RF* (We defended Russia: Historical outline of the estab-
   lishment and activities of the Scientific-Technical Committee, the Director-
   ate of Orders, Production, and Supply, and the Directorate of Biological De-
   fense of the Directorate of Radiological, Chemical, and Biological Defense
   Forces Command of the Russian Federation Ministry of Defense) (Moscow:
   Ministry of Defense of the Russian Federation, 2000). This book appears to
   be based largely on Russian and Soviet sources.

10. E.g., A. V. Stepanov, L. I. Marinin, A. P. Pomerantsev, and N. A. Staritsin, "Development of Novel Vaccines against Anthrax in Man," *Journal of Biotechnology*, 44 (1996), 155–160 (translated from Russian); Elie Shlyakhov, Ethan Rubinstein, and Ilya Novikov, "Anthrax Post-Vaccinal Cell-Mediated Immunity in Humans: Kinetics Pattern," *Vaccine*, 15, no. 6/7 (1997), 631–636 (translated from Russian); and A. P. Pomerantsev, N. A. Staritsin, Yu. V. Mockov, and L. I. Marinin, "Expression of Cereolysine AB Genes in *Bacillus anthracis* Vaccine Strain Ensures Protection against Experimental Hemolytic Anthrax Infection," ibid., no. 17/18 (1997), 1846–59 (translated from Russian). See also V. G. Frolov and Yu. M. Gusev, "Stability of Marburg Virus to Lyophilization Process and Subsequent Storage at Different Temperatures," *Voprosy Virosologii*, 41 (November–December 1996), 275–277 (in Russian); and A. A. Bukreev et al., "A Promising Method for Preparative Production and Purification of Marburg Virus," ibid., 40 (July–August 1995), 161–165 (in Russian). The English-language version of this journal is *Problems of Virology*.

11. E.g., L. A. Belikov, *The Bacteriological Weapon and Methods of Protection from It* (Washington, D.C.: US Joint Publications Research Service, 1961) (translated from Russian); and *Organization for the Medical Protection of the Population during Mass Attack, Medical Service in Mass Attack* (Washington, D.C.: US Joint Publications Research Service, 8 September 1959; originally published in Kiev in 1957).

12. Ya. Fishman, "Rabota Bakteriologicheskoi Laboratorii VOKHIMU" (Work of the Bacteriological Laboratory of the Military-Chemical Directorate), fond 33987, op. 1, d. 657, 1.143–144, 10 February 1928, Russian State Military Archive, Moscow, quoted in translation in S. W. Stoecker, *Forging Stalin's Army: Marshal Tukhachevsky and the Politics of Military Innovation* (Boulder: Westview, 1998), ref. 69, p. 109.

13. Hirsch, *Soviet BW and CW Capabilities*, pp. 105–110.

14. M. Popovsky, *Manipulated Science: The Crisis of Science and Scientists in the Soviet Union Today* (Garden City, N.Y.: Doubleday, 1979), p. 72.

15. E. S. Levina, "Eksperimental'naya Biologiya v Sisteme Bezopasnosti Rossii Vtoroi Polovinoi XX Veka: Biologicheskoe Oruzhie ili Zdravookhranenie?" (Experimental biology in the system of Russian security of the second half of the twentieth century: Biological weapons or healthcare?), in *Nauka I Bezopasnost' Rossii: Istorichesko-Nauchnie, Metodologicheskie, Istorichesko-Tekhnicheskie Aspekty* (Science and security of Russia: Historico-scientific, methodological, historico-technical aspects), ed. R. S. Golovina et al. (Moscow: Nauka, 2000), p. 373.

16. The archives of the Academy of Sciences and the State Committee on Science and Technology have volumes containing such correspondence; Levina, "Eksperimental'naya Biologiya," p. 372.

17. Domaradskij and Orent, *Biowarrior*, p. 143; Levina, "Eksperimental'naya Biologiya," p. 372.

18. Domaradskij and Orent, *Biowarrior*, p. 143.

19. Ibid., p. 93. See also E. A. Stavskiy, N. B. Cherny, A. A. Chepurnov, and S. V. Netesov, "Anthology of Some Biosafety Aspects in Russia (up to 1960)," in *Anthology of Biosafety*, vol. 5: *BSL-4 Laboratories*, ed. J. Y. Richmond (Mundelein, Ill.: American Biological Safety Association, 2002), pp. 29–91. I thank Dr. Jens Kuhn for drawing my attention to this reference.

20. T. E. Popova, *Razvitie Biotekhnologii v SSSR* (The development of biotechnology in the USSR) (Moscow: Nauka, 1988), p. 76.

21. Levina, "Eksperimental'naya Biologiya," p. 377.

22. Domaradskij and Orent, *Biowarrior*, p. 168. Domaradsky says the council was "new" as of the winter of 1972; ibid., pp. 118–119.

23. Levina, "Eksperimental'naya Biologiya," pp. 376–377.

24. Domaradskij and Orent, *Biowarrior*, pp. 144–145, 155. The council was headed by V. A. Lebedinsky (a military officer, immunologist, and microbiologist), while Domaradsky served as deputy head; ibid., p. 145.

25. Voenno-Promyshlennoi Komissii Soveta Ministrov SSSR po Probleme Obespecheniya Sozdaniya Novykh Vidov Biologicheskogo Oruzhiya Fundamental'nymi Razrabotkami.

26. Levina, "Eksperimental'naya Biologiya," p. 373.

27. S. Pluzhnikov and A. Shvedov, "Ubiistva iz Probirki" (Murder from a test tube), *Sovershenno Sekretno*, no. 4 (1998), 12.

28. Cooper, "Life in the Pursuit of Death," p. 70.

29. Smirnov headed the Main Military-Sanitary Directorate of the Red Army from 1939 to 1946. The name of the directorate was changed in 1946 to the Main Military-Medical Directorate (still headed by Smirnov). From 1947 to 1952 Smirnov served as the Soviet minister of health. Beginning in 1953, he headed the MOD's military-biological research. From 1953 to 1985 he headed the Seventh Directorate of the General Staff of the Soviet military forces.

   The 15th Directorate was established in accordance with a 25 June 1973 decision of the Soviet Communist Party Central Committee (no. 444-138) and an 11 January 1973 Soviet MOD decree (no. 99). Professor V. N. Orlov holds a doctorate in military sciences. In 2000 he was deputy head of the Radiological, Chemical, and Biological Defense Forces for Armaments and Scientific Research Work; Orlov et al., *My Zashchitily Rossiyu*, p. 207.

30. Ibid., pp. 207–208.

31. "O Sozdanii Komiteta po Konventsial'nym Problemam Khimichskogo i Biologicheskogo Oruzhiya pri Prezidente Rossiiskoi Federatsii" (On the establishment of the Presidential Committee on Problems of Chemical and Biological Weapon Conventions), Russian Federation Presidential Decree No.

160, 19 February 1992. This body was also known as the Conventional Committee.

32. Russian Presidential Decree No. 314, 9 March 2004.

33. Domaradskij and Orent, *Biowarrior,* p. 189.

34. Hirsch, *Soviet BW and CW Capabilities,* p. 77.

35. W. E. Lexow and Julian Hoptman, "The Enigma of Soviet BW," *Studies in Intelligence,* 9 (1965), 15–20; declassified CIA publication.

36. D. C. Kelly, "The Trilateral Agreement: Lessons for Biological Weapons Verification," in *Verification Yearbook 2002,* ed. T. Findlay and O. Meier (London: VERTIC, 2002), p. 103; and Mangold and Goldberg, *Plague Wars,* p. 95.

37. See Roger Roffey and K. S. Westerdahl, *Conversion of Former Biological Weapons Facilities in Kazakhstan: A Visit to Stepnogorsk, July 2000,* (FOI) Report No. FOI-R-0082 (Umeå: Swedish Defence Research Agency, 2001); and S. B. Ouagrham and K. M. Vogel, *Conversion at Stepnogorsk: What the Future Holds for Former Bioweapons Facilities,* Cornell University Peace Studies Program, Occasional Paper No. 28 (Ithaca: Cornell University, 2003).

38. Orlov et al., *My Zashchitily Rossiyu,* pp. 212–213.

39. From the 1950s through the 1970s much of this and related work was carried out under the direction of Doctor of Technical Sciences A. M. Masaltsevy.

40. Orlov et al., *My Zashchitily Rossiyu,* pp. 213, 214.

41. Ibid., p. 213.

42. Cooper, "Life in the Pursuit of Death," p. 70.

43. Alibek, *Biohazard,* pp. 123–132.

44. Levina, "Eksperimental'naya Biologiya," p. 382.

45. Mangold and Goldberg, *Plague Wars,* p. 92. A preliminary listing of Biopreparat companies and institutions is provided by R. Roffey, W. Unge, J. Clevström, and K. S. Westerdahl, *Support to Threat Reduction of the Russian Biological Weapons Legacy: Conversion, Biodefence and the Role of Biopreparat,* (FOI) Report No. FOI-R-0841-SE (Umeå: Swedish Defence Research Agency, 2003), pp. 127–129.

46. Cooper, "Life in the Pursuit of Death," pp. 70, 72, 104.

47. "Introduction," Vector web page, http://www.vector.nsc.ru/index-e.htm. See also Sergey V. Netesov, "The Scientific and Production Association Vector: The Current Situation," in *Control of Dual-Threat Agents: The Vaccines for Peace Programme,* ed. Erhard Geissler and J. P. Woodall, SIPRI Chemical and Biological Warfare Studies No. 15 (Oxford University Press: Oxford, 1994), pp. 133–138. See also Jens H. Kuhn, "Experiences of the First Western Scientist with Permission to Work inside a Former Soviet Biowarfare Facility" (Ph.D. diss., Charité-University Medicine Berlin, 2004).

48. K. Alibek, "Statement of Dr. Kenneth Alibek, President, Advanced Biosytems, Inc., Former Deputy Chief, Civilian Branch, Soviet Offensive Biological Weapons Program," in *Russia, Iraq, and Other Potential Sources of An-*

thrax, Smallpox and Other Bioterrorist Weapons, Hearings before the Committee on International Relations, House of Representatives, 107th Congress, First Session, 5 December 2001, Serial no. 107-56 (Washington, D.C.: USGPO, 2001), p. 8.

49. See Roger Roffey, "Biological Weapons and Potential Indicators of Offensive Biological Weapon Activities," in *SIPRI Yearbook 2004: Armaments, Disarmament and International Security* (Oxford: Oxford University Press, 2004); and Milton Leitenberg, "Distinguishing Offensive from Defensive Biological Weapons Research," *Critical Reviews in Microbiology,* 29, no. 3 (2003), 223–257.

50. I. V. Darmov, I. P. Pogorel'sky, and V. N. Velikanov, "Nauchno-Issledovatel'skomu Institutu Mikrobiologii Ministerstva Oborony Rossiiskoi Federatsii—70 let" (To the Scientific-Research Institute of Microbiology of the Russian Federation Ministry of Defense at 70 years), *Voenno-Meditsinsky Zhurnal,* 8 (1989), 79–81. STI stands for Sanitarny-Tekhnichesky Institut (Sanitary-Technical Institute); ibid., p. 80; and Orlov et al., *My Zashchitily Rossiyu,* p. 210.

51. Col. Hoover, "Before the United States Department of Defense Armed Forces Epidemiological Board," *Capitol Hill Reporting,* 8 July 1994, 36, http://www.ha.osd.mil/afeb/meeting/Transcripts/Transcripts%20-%20July%208%201994.PDF; and personal communication from former Soviet Anti-Plague System official, April 2001.

52. Judith Miller, Stephen Engelberg, and William Broad, *Germs: Biological Weapons and America's Secret War* (New York: Simon and Schuster, 2001), p. 180. See also V. A. Lebedinsky, *Ingalyatsionny (Aerogenny) Metod Vaktsinatsy* (Inhalation [Aerogenic] method of vaccination) (Moscow, 1971).

53. Darmov, Pogorel'sky, and Velikanov, "Nauchno-Issledovatel'skomu Institutu Mikrobiologii Ministerstva," p. 81.

54. John Hart, "Preventing Health and Proliferation Problems Stemming from the Soviet BW Legacy in Central Asia," *ASA Newsletter,* no. 84 (12 June 2001), 10–11.

55. Orlov et al., *My Zashchitily Rossiyu,* pp. 215–216. For background on the Virology Center, see R. N. Lukin et al., eds., *Dostoiny Izvestnosti: 50 let Virusologicheskomu Tsentru Ministerstva Oborony* (Worthy of fame: 50 years of the Ministry of Defense Virology Center) (Sergiev Posad: Ves' Sergiev Posad, 2004); and John Hart, "A Historical Note: The 50th Anniversary of Russia's Virology Center of the Ministry of Defense," *ASA Newsletter,* no. 106 (28 February 2005), pp. 1, 19–22.

56. NOVA Online, Bioterror series, undated interview with Ken Alibek and William Patrick, http://www.pbs.org/cgi-bin/wgbh/printable.pl?http%3A%2F%2Fwww.pbs.org%2Fwgbh%2Fnova%2Fbioterror%2Fbiow_alibek.html.

57. Domaradskij and Orent, *Biowarrior,* pp. 205–206.

58. Miller, Engelberg, and Broad, *Germs,* pp. 301–303.

59. "Interview—Sergei Popov," *Journal of Homeland Security*, 1 November 2000, http://www.homelandsecurity.org/journal/Interviews/ displayInterview.asp?interview=3 (updated 19 November 2002).

60. See Miller, Engelberg, and Broad, *Germs*, p. 302.

61. Domaradskij and Orent, *Biowarrior*, pp. 218–220.

62. NOVA Online, Bioterror series, undated interview with Sergei Popov, http:/ /www.pbs.org/cgi-bin/wgbh/print- able.pl?http%3A%2F%2Fwww.pbs.org%2Fwgbh%2Fnova%2Fbioterror %2Fbiow_popov.html; and Miller, Engelberg, and Broad, *Germs*, p. 220.

63. Alibek, *Biohazard*, pp. 123–132.

64. J. B. Tucker and R. Zilinskas, eds., *The 1971 Smallpox Epidemic in Aralsk, Kazakhstan, and the Soviet Biological Warfare Program*, Occasional Paper No. 9 (Monterey, Calif.: Monterey Institute of International Studies, Center for Nonproliferation Studies, 2002), http://cns.miis.edu/pubs/opapers/op9/ op9.pdf.

65. U-2 overflights of island were conducted in 1957 and 1959; Lexow and Hoptman, "The Enigma of Soviet BW," pp. 17, 19.

66. Mangold and Goldberg, *Plague Wars*, pp. 93–94.

67. Lexow and Hoptman, "The Enigma of Soviet BW," p. 15.

68. Ibid., pp. 16–17. The US trust in the veracity of the BW-related information contained in the Hirsch Report was based on the fact that it was able to cor- roborate a great deal of the report's CW-related information by other means. Others have questioned the reliability of the report partly because much of the information was based on Hirsch's memory and because his brother was interned by the Soviets for a number of years following the war. Hirsch died before he could complete his report.

69. April 1998 interview with Melvin Laird as reported in Mangold and Goldberg, *Plague Wars*, p. 62. Corona satellite and U-2 imagery of Soviet BW infrastructure on Vozrozdeniye Island is available at Global Security.org, "Vozrozhdeniye Island, Renaissance/Rebirth Island," http://www.serendip- ity.li/cia/foia.html.

70. April 1998 interview with Melvin Laird as reported in Mangold and Goldberg, *Plague Wars*, pp. 62–63.

71. Ibid., pp. 64–65; Arkady N. Shevchenko, *Breaking with Moscow* (London: Jonathan Cape, 1985), pp. 173–174, 179.

72. W. Beecher, *Boston Globe*, 28 September 1975, quoted in Mangold and Goldberg, *Plague Wars*, p. 63. For an account of information disclosures see Milton Leitenberg, "Biological Weapons Arms Control," Center for Interna- tional and Security Studies at Maryland's Project on Rethinking Arms Con- trol (PRAC), PRAC paper No. 16, University of Maryland, May 1996, pp. 3– 16.

73. Mangold and Goldberg, *Plague Wars*, pp. 83–86.

74. Second Review Conference Final Declaration, BWC document BWC/

CONF.II/13/II, 8–26 September 1986, p. 6. All BWC documents are held by the UN in Geneva and New York.

75. Kelly, "The Trilateral Agreement," p. 102.

76. Ibid., p. 95.

77. Mangold and Goldberg, *Plague Wars*, pp. 130–131.

78. Joint Statement on Biological Weapons by the Governments of the United Kingdom, the United States, and the Russian Federation (10–11 September 1992) (Trilateral Agreement), 14 September 1992, available at http://projects.sipri.se/cbw/docs/cbw-trilateralagree.html. For an authoritative overview of the Trilateral process, see Kelly, "The Trilateral Agreement," pp. 93–109.

79. Joint Statement on Biological Weapons, paras. 4(a), 5(a).

80. Points for Introducing Joint Statement, September 1992, p. 1, declassified US government interagency talking points, NSC Collection, John A. Gordon Files, box 1, BW, September 1992, OA/ID CF 01653, George Bush Presidential Library, College Station, Texas.

81. Kelly, "The Trilateral Agreement," p. 99.

82. Mangold and Goldberg, *Plague Wars*, p. 197. Viruses are traditionally grown in unfertilized eggs. Some viruses may be grown, with varying levels of ease, in tissue cultures. For example, influenza viruses do not grow well in tissue cultures, while adenoviruses generally do. The ability to grow viruses in tissue cultures, however, is improving.

83. Ibid., pp. 198–199.

84. Allbek, "Statement," pp. 25–26.

85. Moodie, "Soviet Union, Russia, and BWC," p. 67.

86. B. Gertz, "Russia Told to Unveil All Biological Arms," *Washington Times*, 24 July 1992, p. A8.

87. Mangold and Goldberg, *Plague Wars*, p. 165.

88. Gertz, "Russia Told to Unveil Biological Arms"; J. Smith, *Washington Post*, 9 January 1992, as quoted in "Chronology of Russian Clarification of the BW Program—31 August," *Arms Control Reporter*, 701.B.98.

89. Mangold and Goldberg, *Plague Wars*, p. 166.

90. Smith, *Washington Post*, 9 January 1992.

91. Gertz, "Russia Told to Unveil Biological Arms."

92. Smith, *Washington Post*, 9 January 1992.

93. M. Moodie, "The Soviet Union, Russia, and the Biological and Toxin Weapons Convention," *Nonproliferation Review*, 8 (spring 2001), 64–65.

94. Smith, *Washington Post*, 9 January 1992.

95. Russian Federation, "Section F: Declaration of Past Activities in Offensive/Defensive Biological Research and Development Programs" (in Russian), UN doc. DDA/4-92/BWIII/ADD.2, 1992, p. 85.

96. V. V. Tomilin and R. V. Berezhnoi, "Razoblachenie Prestupnoi Deyatel'nosti Yaponskikh Militaristov po Podgotovke i Primeneniyu

Bakteriologicheskogo Oruzhiya" (Exposure of the criminal activity of the Japanese militarists in preparation for the use of bacteriological weapons), clipping given to the author, journal unknown, date unknown, pp. 26–29.

97. For a Soviet view of the Japanese BW program see M. Yu. Raginsky, S. Ya. Rozenblit, and L. N. Smirnov, *Bakteriologicheskaya Voina—Prestupnoe Orudie Imperialisticheskoi Agressii (Khabarovsky Protsess Yaponskhikh Voennikh Prestupnikov)* (Bacteriological war: The criminal weapon of imperial aggression [The Khabarovsk Proceedings of (against) Japanese Criminals]) (Moscow: USSR Academy of Sciences, 1950).

98. *Materials on the Trial of Former Servicemen of the Japanese Army Charged with Manufacturing and Employing Bacteriological Weapons* (Moscow: Foreign Languages Publishing House, 1950), p. 5.

99. Ibid., p. 6. Zhukov-Verezhnikov (b. 1908, d. 1981), a microbiologist and immunologist, became a full member of the USSR Academy of Sciences in 1948. From 1938 to 1947 he headed the Anti-Plague Department at the "Microb" Institute (in Saratov). He was also deputy director of the Rostov Anti-Plague Institute. In 1950 he was named vice president of the USSR Academy of Medical Sciences. From 1952 to 1954 he served as deputy minister of the Health Ministry; Levina, "Eksperimental'naya Biologiya," p. 392. For other biographical information, see "Anniversary Dates: Nikolai Nikolaivich Zhukov-Verezhnikov (on His 70th Birthday)," *Zhurnal Mikrobiologii, Epidemiologii i Immunologii* 11 (1978), 149–150 (in Russian).

100. Pluzhnikov and Shvedov, "Ubiistva iz Probirki," p. 12.

101. Mangold and Goldberg, *Plague Wars*, p. 109.

102. Shevchenko, *Breaking with Moscow*, pp. 173, 179.

103. It is known that the US did run such an operation (Operation Shocker) in connection with CW. Over a period of 23 years, the US government reportedly passed to the Soviets some 4,500 documents on fictitious US research into a new type of nerve agent. See D. Wise, *Cassidy's Run: The Secret Spy War over Nerve Gas* (New York: Random House, 2000); R. L. Garthoff, "Polyakov's Run," *Bulletin of the Atomic Scientists,* 56 (September/October 2000), 37–40. In 2000 Garthoff wrote (ibid., p. 39): "It is not clear when the US disinformation operations on CBW ended—probably in the mid-1970s—but the operations appear not to have been compromised until 1985. In any case, it is evident that their effects continued long after." Subsequent unpublished background research by US researchers indicates that the BW disinformation effort was halted before the BWC's opening for signature in 1972. For primary documents on the ending of the US offensive BW program, see "Biowar: The Nixon Administration's Decision to End US Biological Warfare Programs," ed. R. A. Wampler, vol. 3 of National Security Archive Electronic Briefing Book No. 58, updated 7 December 2001, http://www.gwu.edu/~nsarchiv/NSAEBB/NSAEBB58/.

104. Cooper, "Life in the Pursuit of Death," p. 72; "Interview—Sergei Popov."

105. See, for example, Raginsky, Rozenblit, and Smirnov, *Bakteriologicheskaya Voina.*

106. This view has periodically been expressed, for example, in Chinese news reports and, on occasion, by Russian officials at conferences where the subject of the Soviet BW program is raised.

107. Miller, Engelberg, and Broad, *Germs,* p. 303.

108. See "International Science and Technology Center: Nonproliferation through Science Cooperation," http://www.istc.ru.

109. See http://www.tacis-medt.ru.

110. Levina, "Eksperimental'naya Biologiya," p. 379.

111. See http://www.mpnt.gov.ru.

112. See http://fasie.ru and http://www.rftr.ru.

113. V. I. Evstigneev, "Biological Weapons and Problems of Ensuring Biological Security," lectures presented at the Moscow Institute of Physics and Technology, 25 March and 8 April 2003, unofficial translation. See also Russian Federation, "Working Paper Submitted by the Russian Federation: On the Procedure for the Management of Microorganisms of the Pathogenicity Groups I–IV in the Territory of the Russian Federation," BWC Meeting of States Parties doc. BWC/MSP.2003/WP.7, 11 November 2003.

114. Darmov, Pogorel'sky, and Velikanov, "Nauchno-Issledovatel'skomu Institutu Mikrobiologii Ministerstva," p. 81.

115. For example, in August 2002 US Senator Richard Lugar was denied access to the Biotechnical Complex Kirov-200. Lugar was traveling on a fact-finding mission to Russia partly to discuss implementation of US-funded cooperative threat reduction programs with Russian officials. According to a Russian Munitions Agency official, access was denied because the facility was no longer under MOD jurisdiction but was part of Kirov University, which is under the authority of the Ministry of Education. The official said that the US should have applied to the Ministry of Education and the Ministry of Science for permission, which, according to him, it did not; "My ne Sobiralis' Primenyat' Biologicheskoe Oruzhie" (We did not plan to use biological weapons), interview of Nikolai Poroskovy (Russian Munitions Agency) by Valery Spirande, *Vremya Novosti,* 74 (24 April 2003), http://www.vremya.ru/2003/74/6/56655.html.

116. DOD, Chemical and Biological Defense Program, *Annual Report to Congress and Performance Plan,* July 2001, p. 8, http://www.defenselink.mil.

117. Alibek, *Biohazard,* p. 189.

118. "Interview—Sergei Popov."

119. US Department of State, "Cooperative Threat Reduction," 4 March 2003, http://www.state.gov/t/vc/rls/rm/18736.htm.

120. "Cooperative Threat Reduction Program," Testimony of Lisa Bronson, Deputy Under Secretary of Defense for Technology Security Policy and Counterproliferation before the Senate Committee on Armed Services Sub-

committee on Emerging Threats and Capabilities, 10 March 2004, available at http://armed_services/senate.gov.

121. For an example of an ongoing effort to clarify BW-related concerns see *Russian BW Monitor,* http://www.russianbwmonitor.co.uk/index.asp.

7. Biological Weapons in Non-Soviet Warsaw Pact Countries

1. Gábor Faludi, "Historical Data on Biological Weapons Project in Hungary" (in Hungarian), *Honvédorvos,* 50 (1998), 189–195; idem, "Challenges of BW Control and Defence during Arms Reduction," Paper presented at the NATO Advanced Research Workshop, Budapest, 5–9 November 1997.

2. Erhard Geissler, *Biologische Waffen—nicht in Hitlers Arsenalen: Biologische und Toxin-Kampfmittel in Deutschland von 1915 bis 1945* (Münster: LIT-Verlag, 1998).

3. Alsos Mission, "Report on the Interrogation of Professor H. Kliewe," Report No. A-D-C-H-H/149, MIS (Military Intelligence Section), War Department, 7–11 May 1945, RG 319, NARA.

4. Dezso Bartos, "A bakteriológiai háború megvilágitása katonai és tudományos szempontból" (Elucidation of bacteriological warfare from military and scientific viewpoints), Health Protection Institute, Hungarian Defense Force, 1955. This document contains technical details and is therefore classified.

5. Gábor Faludi, "The Changing Importance of BW" (in Hungarian), *Honvédorvos,* 50 (1998), 37–69.

6. Lajos Rózsa, interview with Colonel Gábor Faludi, Budapest, 17 September 2001.

7. John Hemsley, *The Soviet Biochemical Threat to NATO: The Neglected Issue* (London: Macmillan, 1988), p. 23.

8. Ken Alibek with Stephen Handelman, *Biohazard: The Chilling True Story of the Largest Covert Biological Weapons Program in the World—Told from the Inside by the Man Who Ran It* (New York: Random House, 1999).

9. J. D. Douglass, "Beyond Nuclear War," *Journal of Social, Political and Economic Studies,* 15 (1990), 141–156.

10. B. C. Garrett, "Czech BW/CW Stocks?" *CBIAC Newsletter,* 8, nos. 1–2 (1994), 1–9; "The Czech Republic Recently Destroyed Two Sets of Viral Strains," *Arms Control Reporter,* 94, no. 4 (18 March 1994).

11. Defense Minister Antonin Baudys, quoted in ibid.

12. J. Mindszenty, *Memoirs* (New York: Macmillan, 1974); L. H. Carlson, *Remembered Prisoners of a Forgotten War: An Oral History of Korean War POWs* (New York: St. Martin's, 2003), pp. 177–194.

13. J. D. Douglass, "Influencing Behavior and Mental Processing in Covert Operations," *Medical Sentinel,* 6, no. 4 (2001), 130–136.

14. Idem, *Red Cocaine: The Drugging of America and the West* (London: Edward Harle, 1999).

15. As described in ibid.
16. Alibek, *Biohazard*.
17. J. Svankmajer, "Out of My Head," *The Guardian*, 19 October 2001.
18. M. Semjakin, "A KGB betege" (The KGB's patient), interview, (in Hungarian), *168 óra*, 15 (2003), 28–29.
19. S. Rump and M. Kowalczyk, "The Military Institute of Hygiene and Epidemiology in Poland Warsaw and Pulawy," *ASA Newsletter*, 27 August 1999, 3–4; S. Jankowska, "Biological Threat," *Przeglad Obrony Cywilnej*, no. 9 (1987), 9–12; B. Sibley, "Combat with Biological Means," ibid., no. 6 (1987), 6–7; M. Bartoszcze, M. Niemcewicz, and M. Malinsky, "The Way to a Polish Biodefense System," *ASA Newsletter*, 30 August 2002, 20–21.
20. Alibek, *Biohazard*, p. 273.
21. Erhard Geissler, "Pockenwäsche und Milzbrandbriefe. Realität und Fiktion der Bio- und Toxin-Waffen," Paper delivered at the Arbeiter Samariter Bund Meeting on Biological and Chemical Threats, Berlin, 10 September 2003, pp. 7–12.
22. K. Lohs, "Toxine als Kampfstoffe?" *Wissenschaft und Fortschritt*, 36 (1986), 307–308; Erhard Geissler, "Biologische und Chemische Waffen: Neue Impulse durch Gentechnik," *Informationsdienst*, 1 (1986), 13–15.
23. J. Anderson and D. van Atta, "Assassination in London: On the Cold Trail of Bulgarian Hit Men," *Washington Post*, 8 September 1991, p. C7.
24. R. Crompton and D. Gall, "Georgi Markov Death in a Pellet," *Medico—Legal Journal*, 48 (1980), 51–62.
25. C. R. Whitney, "Ex-General Cites KGB Murder Link: Dissident Says He Sent Agents to Bulgaria with Poison," *International Herald Tribune*, 14 June 1991, p. 5.
26. R. T. van Keuren, *Chemical and Biological Warfare: An Investigative Guide* (Washington, D.C.: Office of Enforcement, Strategic Investigations Division, US Customs Service, October 1990), p. 90.
27. C. Sokoloff, "Markov Murder Plot Blown Open," *Sunday Times*, 2 June 1991, p. 16.
28. "Suesse Kugel. Der Regenschirmmord wird aufgeklaert—Moskau hat mitgewirkt. Um das Tatmotiv ranken sich die tollsten Agentengeschichten," *Der Spiegel*, 7 July 1992, pp. 168–170.
29. R. Boyes, "Evidence Points to Zhivkov," *The Times*, 7 December 1991, p. 8.
30. I. Black, "Move to Solve 'Umbrella' Murder," *The Guardian*, 27 January 1997, p. 4.

8. The Iraqi Biological Weapons Program

1. www.sipri.org/contents/cbwarfare/cbw_research_doc/cbw_historical/Hist-geneva-res/cbw-hist-geneva-res.html.
2. United Nations Monitoring, Verification and Inspection Commission, "Un-

resolved Issues," UNMOVIC Working Paper, 6 March 2003, p. 153; available at www.unmovic.org.

3. A useful summary appreciation of the Iraqi BW program is provided in United Nations, "Letter dated 27 January 1999 from the Permanent Representatives of the Netherlands and Slovenia to the President of the Security Council," 27 January 1999, S/1999/94, available at http://www.un.org/Depts/unscom/unscmdoc.htm. The UNMOVIC appreciation is summarized in the UNMOVIC Working Paper "Unresolved Issues."

4. CIA, "Comprehensive Report of the Special Advisor to the DCI on Iraq's WMD," 3 vols., 30 September 2004, available at www.cia.gov/cia/reports/iraq_wmd_2004/index.html.

5. UNMOVIC, "Unresolved Issues," p. 153.

6. Ibid., p. 9.

7. Ibid., p. 153.

8. David Kay, "Statement by David Kay on the Interim Progress Report on the Activities of the Iraq Survey Group (ISG) before the House Permanent Select Committee on Intelligence, The House Committee on Appropriations, Subcommittee on Defense, and the Senate Select Committee on Intelligence," 2 October 2003, available at www.cia.gov/cia/public_affairs/speeches/2003/david_kay_10022003.html.

9. See Graham S. Pearson, "The Biological Weapons Convention New Process," Review No. 20, *CBW Conventions Bulletin*, no. 62, December 2001, available at www.sussex.ac.uk/spru/hsp.

10. Arms Control Association, "Searching for the Truth about Iraq's WMD: An Interview with David Kay," *Arms Control Today*, 5 March 2004, available at www.armscontrol.org/aca/midmonth/2004/March/Kay.asp; "David Kay: Il faut reconnaître nos erreurs pour restaurer notre crédibilité," *Le Figaro*, 19 March 2004, available at www.lefigaro.fr.

11. Arms Control Association, "An Interview with Hans Blix," *Arms Control Today*, 16 June 2003, available at www.armscontrol.org/events/blixinterview_june03.asp.

12. CIA, "Comprehensive Report," vol. 3: "Biological Warfare," p. 135.

13. Ibid., vol. 1: "Transmittal Message," p. 135.

14. Ibid., vol. 3, p. 1.

15. US Senate, Select Committee on Intelligence, "Report of the U.S. Intelligence Community's Prewar Intelligence Assessments on Iraq," 7 July 2004, available at http://intelligence.senate.gov/iraqreport2.pdf; *Review of Intelligence on Weapons of Mass Destruction: Report of a Committee of Privy Councillors* (London: Her Majesty's Stationery Office, 14 July 2004), available at www.butlerreview.org.uk/report/intex.asp.

16. Senate Committee on Intelligence, "Report of Prewar Intelligence Assessments," p. 14.

17. *Review of Intelligence on Weapons of Mass Destruction*, para. 49, p. 14.

18. Ibid., para. 209, p. 53.

19. Ibid., para. 41, p. 11.

20. United Nations Security Council, "Letter dated 8 April 1998 from the Executive Chairman of the Special Commission established by the Secretary-General pursuant to paragraph 9 (b) (I) of Resolution 687 (1991) Addressed to the President of the Security Council," S/1998/308, 8 April 1998, p. 11, available at http://www.un.org/Depts/unscom/unscmdoc.htm.

21. Arms Control Association, "Searching for the Truth about Iraq's WMD: An Interview with David Kay."

9. The South African Biological Weapons Program

1. Signe Landgren, *Embargo Disimplemented: South Africa's Military Industry* (Oxford: SIPRI and Oxford University Press, 1989), p. 151.

2. Basson was charged on multiple counts of complicity to murder, dealing in drugs, and fraud. He was acquitted on all charges; however, at the time of writing the state was contesting the court findings.

3. D. J. C Wiseman, "The Second World War 1939–1949 Army: Special Weapons and Types of Warfare," in *Gas Warfare*, vol. 1 (London: War Office, 1951).

4. Chandré Gould and Peter Folb, "The South African Chemical and Biological Warfare Program: An Overview," *Nonproliferation Review*, fall/winter 2000, 12.

5. Chandré Gould, interview with General Constand Viljoen (former chief of the South African Defence Force), Cape Town, 18 May 2000.

6. J. P. de Villiers, G. E. McLouglin, V. P. Joynt, and C. C. Van Der Westhuizen, "Chemical and Biological Warfare in a South African Context in the Seventies," 12 February 1971, Mechem Archives.

7. Testimony of General D. P. Knobel in The State v. Wouter Basson, South African High Court, Transvaal Division, as reported by Marlene Burger in a daily trial report prepared for the Centre for Conflict Resolution's CBW Research Project, 15 November 1999.

8. Ibid.

9. Testimony of Dr. Daan Goosen in The State v. Wouter Basson, 22 May 2000.

10. Chandré Gould and Peter Folb, *Project Coast: Apartheid's Chemical and Biological Warfare Programme* (Geneva: United Nations Institute for Disarmament Research, 2003), p. 8.

11. Roodeplaat Navorsingslaboratoriums, H-Kode Reeks: 1986 [Roodeplaat Research Laboratories, H-code Series: 1986], List of research projects at RRL for the year 1986.

12. Chandré Gould, interview with Dr. Hennie Jordaan, organic chemist formerly employed at Delta G, Pretoria, 18 January 2001.

13. Wouter Basson, "Briefing of the State President," 26 March 1990, South African Defence Force Document, South African History Archives.

14. De Klerk succeeded P. W. Botha as president in 1989. In January 1991 he

unbanned the African National Congress (ANC) and other formerly banned political organizations. Negotiations toward a democratic transition began with the ANC behind the scenes.

15. Basson, "Briefing of the State President," 26 March 1990.

16. Ibid.

17. Wouter Basson, "Presentation to the Reduced Defence Command Council: Proposed Philosophy for Secret Chemical Warfare for the South African Defence Force—Principles and Feedback on Current Status in the SA Defence Force," 3 October 1990, South African History Archives.

18. "Findings and Conclusions," in *Final Report of the Truth and Reconciliation Commission,* vol. 5, pp. 198–238.

19. H. J. Bruwer, "Project Coast Forensic Audit: Supplementary Report of H. J. Bruwer," p. 6, presented by the state in The State v. Wouter Basson, 10 August 2000.

20. Testimony of Goosen in The State v. Wouter Basson, 22 May 1999.

21. Chandré Gould and Peter Folb, "The Role of Professionals in the South African Chemical and Biological Warfare Programme," *Minerva,* no. 40 (2002), 80.

22. Chandré Gould and Mafole Mokolobe conducted structured interviews with 12 scientists from the chemical and biological warfare program during 2001. The survey was aimed at determining, *inter alia,* why scientists joined the companies of the chemical and biological warfare program.

23. Gould and Folb, *Project Coast,* p. 72.

24. Testimony of Goosen in The State v. Wouter Basson, 22 May 2000.

25. Roodeplaat Navorsingslaboratoriums: H-Kode Reeks, 1986.

26. Testimony of Dr. André Immelman in The State v. Wouter Basson, 19 May 2000.

27. Chandré Gould and Marlene Burger, *Secrets and Lies: Wouter Basson and South Africa's Chemical and Biological Warfare Programme* (Cape Town: Zebra, 2002), p. 29.

28. "Minutes of the Seventh Meeting of the Directors held on 14 August 1984," document provided to Chandré Gould by Dr. Schalk Van Rensburg.

29. Basson claimed that he had been using the venom in peptide research.

30. Gould and Folb, *Project Coast,* p. 70.

31. Wouter Basson, "Authorisation for the Sale of Assets: Project Coast," HSF/UG/302/6/C123, SADF Top Secret Document, 19 August 1991, written to secure authorization from the minister of defense, Magnus Malan, for privatization of RRL.

32. Ibid.

33. Affidavit of Dr. André Immelman for the Truth and Reconciliation Commission, Pretoria, 4 June 1988.

34. Ibid.

35. Testimony of Immelman in The State v. Wouter Basson, 29–30 May 2000.

36. Gould and Folb, *Project Coast*, p. 88.
37. Testimony of Immelman, 29 May 2000.
38. Gould and Burger, *Secrets and Lies*, p. 36.
39. Ibid., pp. 36–37.
40. All RRL research proposals had to be authorized by the departmental head and head of research (in this case Mike Odendaal and Schalk Van Rensburg).
41. Chandré Gould, interview with Mike Odendaal, Daan Goosen, and Adriaan Botha, Pretoria, 1 December 1999.
42. Chandré Gould, interview with Adriaan Botha, Onderstepoort, 22 May 2000.
43. Testimony of Dr. Daan Goosen during the TRC hearings into chemical and biological warfare, Cape Town, 10 June 1998.
44. Roodeplaat Navorsingslaboratoriums: H-Kode Reeks, 1986.
45. "Conceptual Design of New Virulent Strain Centre and Fermentation/Mycology Research Laboratories," prepared for RRL by Foster Wheeler South Africa, Pty Ltd, July 1988, Contract No. 1-1600-25112. Document held by the authors.
46. Gould and Folb, *Project Coast*, p. 99.
47. Chandré Gould, discussion with Mike Odendaal, Onderstepoort, 6 October 1999.
48. Basson, "Authorisation for the Sale of Assets: Project Coast," 19 August 1991.
49. Ibid.
50. Gould and Folb, *Project Coast*, p. 151.
51. Personal electronic communication between Chandré Gould and Dr. W. S. Augerson, 15 February 2001.
52. Chandré Gould, interview with General Chris Thirion, Pretoria, 4 September 2001.
53. Gould interview with Odendaal, Goosen, and Botha, 1 December 1999.
54. Chandré Gould, telephone interview with former US ambassador Priceman Lyman, 14 June 2001.
55. Gen. D. P. Knobel, GG/UG/302/6/J1282/5, 18 August 1994.
56. Gould interview with Lyman, 14 June 2001.
57. Ibid.
58. CIA, "Western Platinum Dependence: A Risk Assessment," January 1985, National Security Archive, George Washington University; quoted in Kenneth Mokoena, ed., *South Africa and the United States: The Declassified History, A National Security Archive Documents Reader* (New York: New Press, 1993), p. 34. Most of the documents quoted in Mokoena are found in the National Security Archive's microfiche collection "South Africa: The Making of US Policy, 1962–1989."
59. "Sale of Uranium Made in US to South Africa," 16 April 1975; "South Afri-

can Uranium Enrichment Plant," 17 April 1975; both quoted in Mokoena, *South Africa and the United States,* p. 19.

60. "Supply of Highly Enriched Uranium to South Africa," 15 April 1975; quoted in ibid.

61. "OAU Denounces Western UNSC Veto on Namibia," 11 June 1975; ibid.

62. US State Department, Memorandum for the Senate Foreign Relations Committee, US Policy toward Angola, 16 December 1975; ibid.

63. *Department of State Bulletin,* February 1989, 17; ibid.

64. "US Position on Angola," 24 December 1975; ibid., p. 20.

65. "Possible Contingency Action Regarding Africa," 10 January 1976; ibid.

66. Harold Nelson, *South Africa: A Country Study* (Washington, D.C.: USGPO, 1981); ibid., p. 21.

67. "Angola and the Clark Amendment," 20 November 1982; ibid.

68. Klare, 1981; ibid., p. 22.

69. *Washington Post,* 23 August 1977; ibid., p. 23.

70. "West Vetoes African Resolutions on South Africa in Security Council October," 1 November 1977; ibid., p. 24.

71. "Press Panel Review of South Atlantic Event," 7 February 1978; ibid., p. 27.

72. *Department of State Bulletin,* February 1989, 18; ibid., p. 28.

73. Foreign Broadcast Information Service, Daily Report, Middle East and Africa, Department of Commerce, National Technical Information Service, Springfield, Va.; ibid., p. 29.

74. *New York Times,* 20 September 1982; ibid., p. 31.

75. "G-6 HMSP Gun/Howitzer," 13 September 1982; ibid.

76. Brenda Branaman, Congressional Research Service, "South Africa: Issues for US Policy," IB80032, 1 July 1980, p. 25; quoted in Mokoena, *South Africa and the United States,* p. 32.

77. *New York Times,* 19 November 1986; ibid., p. 37.

78. *New York Times,* 21 September 1989; ibid., p. 44.

79. Letter from the Chief of the Defence Force (Financial Division) to the Minister of Defence, South African Defence Force Document HSF/UG/302/6/C123, "Approval for the Running of Front Companies: Project Coast: April 1991 to March 1992," 15 February 1991.

80. Hendrick Bruwer, "Project Coast, Forensic Audit: Report of HJ Bruwer," 10 November 1999, p. 291, presented in The State v. Wouter Basson.

81. Chandré Gould, interview with Dr. James Davies, Biocon, Roodeplaat, 18 March 1998.

82. Dr. Schalk Van Rensburg and Dr. André Immelman.

83. For a detailed discussion, see Gould and Folb, *Project Coast,* pp. 209–222.

84. Gould interview with Odendaal, Goosen, and Botha, 1 December 1999.

85. Quoted in Gould and Folb, *Project Coast,* p. 214.

86. Alastair Hay, interview with Dr. Daan Goosen, Pretoria, 27 January 2004.

87. Chandré Gould, interview with Dr. Daan Goosen, Johannesburg, 17 February 2003.

88. Ibid.

89. Hay interview with Goosen, 27 January 2004.

90. Facsimile from Don Mayes to Bob Zlockie, 13 March 2002.

91. Facsimile from Tai Minnaar to Don G. Mayes, 4 March 2002, author's collection.

92. Electronic message from Zlockie to Mayes, 20 March 2002.

93. Communication from Minnaar, "Brief Response to the Questions Asked," 22 March 2002.

94. M. W. Odendaal et al., "The Anti-Biotic Sensitivity Patterns of Bacillus Anthracis Isolated from the Kruger National Park," p. 58, 1991, Onderstepoort J Vet Res.

95. Chandré Gould and Alastair Hay, interview with Mike Odendaal, Pretoria, 26 January 2004.

96. Ibid. According to later electronic communication between Minnaar and Zlockie (2 April 2002), 37 people would be involved.

97. Electronic communication between Minnaar and Zlockie, 1 and 2 April 2002.

98. Letter from Donald Mayes, on the letterhead of the ICT Aviations Programs Group, to Rea Bliss (Federal Bureau of Investigation, 301 Simonton Street, Key West, FL 33040), 9 April 2002.

99. Attachment to ibid., headed "SA," 8 April 2002.

100. Ibid.

101. Ibid., p. 7.

102. Warrick, 20 April 2003.

103. US Department of Justice, FBI, Receipt for Property Received: "One toothpaste tube containing one ampoule of E. coli genetically coded with epsilon toxin," 9 May 2002.

104. Warrick, 20 April 2003.

105. Gould interview with Goosen, 17 February 2003.

106. Hay interview with Goosen, 27 January 2004.

107. Sam Sole and Stefaans Brümmer, "Bid to Hijack SA Bio-stocks," *Mail & Guardian,* 25 April 2003.

108. Hay interview with Goosen, 27 January 2004.

109. Sole and Brümmer, "Bid to Hijack SA Bio-stocks."

110. "Notice under Section 13 of the Non-Proliferation of Weapons of Mass Destruction Act, 1993 (Act No. 87 of 1993): Declaration of Certain Goods and Technologies to Be Controlled and Control Measures Applicable to Such Goods," *Government Gazette,* no. 23308, Notice no. 428, 10 April 2002.

111. The Genetically Modified Organisms Act, 1997 (Act No. 15 of 1997).

112. Hay interview with Goosen, 27 January 2004.

113. Gould interview with Goosen, 17 February 2003.

10. Anticrop Biological Weapons Programs

1. The US Environmental Protection Agency offers the following definition of plant growth regulators: "any substance or mixture of substances intended, through physiological action, to accelerate or retard the rate of growth or maturation, or otherwise alter the behavior of plants or their produce . . . regulators are characterized by their low rates of application; high application rates of the same compounds often are considered herbicidal." See epa.gov.

2. See the six-volume set, SIPRI, *The Problem of Chemical and Biological Warfare,* beginning with *The Rise of CB Weapons* (Stockholm: Almqvist & Wiksell, 1971).

3. See, e.g., Julian Perry Robinson, *War and the Environment* (Stockholm: Environmental Advisory Council, 1981), p. 37; Paul Rogers, Simon Whitby, and Malcolm Dando, "Biological Warfare against Crops," *Scientific American,* June 1999, 70–75; and Simon Whitby, *Biological Warfare against Crops* (London: Palgrave, 2001).

4. Erhard Geissler and J. E. van Courtland Moon, eds., *Biological and Toxin Weapons Research, Development and Use from the Middle Ages to 1945: A Critical Comparative Analysis,* SIPRI Chemical and Biological Warfare Studies No. 18 (Oxford: Oxford University Press, 1999).

5. Gradon Carter and G. S. Pearson, "North Atlantic Chemical and Biological Research Collaboration: 1916–1995," *Journal of Strategic Studies,* 19 (March 1996), 74–103.

6. A Note on Herbicide Research, Crop Committee, 1949, AC10031, PRO, WO 195/10027.

7. A Note on the Water Content of Cereals, Crop Committee, 1948, AC10268, PRO, WO 195/10264.

8. A Note on Spore Suspension Research, Crop Committee, 1949, AC10372, PRO, WO 195/10368.

9. A Note on the Use of Aircraft in Agriculture, Crop Committee, 1949, AC10546, PRO, WO 195/10542.

10. Investigations into Anti-Crop Chemical Warfare, Crop Committee, 1949, AC10276, PRO, WO 195/10272.

11. Countermeasures to Anti-Crop and Anti-Animal Warfare, Crop Committee, 1949, AC10742, PRO, WO 195/10738.

12. Report on 2,4-D, Crop Committee, 1950, AC10855, PRO, WO 195/10851.

13. Investigation into Phytotoxicity, Crop Committee, 1950, AC11091, PRO, WO 195/11087.

14. Effects of Insecticides and Herbicides on Animals, Crop Committee, 1951, AC11568, PRO, WO 195/11564.

15. Anti Crop Detection and Destruction of Destructive Agents on Crop Leaf Samples, Crop Committee, 1952, 11728, PRO, WO 195/11725.

16. Anti-Crop Aerial Spray Trials, Crop Committee, 1954, AC12929, PRO, WO 195/12925.

17. Analysis of Rice Blast Epidemic Caused by *Piricularia oryzae*, ADAC, 1954, AC13199, PRO, WO 195/13195.

18. Report on Cereal Rusts, ADAC, 1955, AC13378, PRO, WO 195/13374.

19. Screening of *Piricularia* Isolates for Pathogenicity to Rice, ADAC, 1955, AC13381, PRO, WO 195/13376.

20. Report on Clandestine Attacks on Crops and Livestock of the Commonwealth, ADAC, 1950, AC13154a, PRO, WO 195/13150.

21. Annual Review, 1955, pp. 1–4, ADAC, Advisory Council on Scientific Research and Technical Development, Ministry of Supply, 14 November 1955, PRO, WO 195/13473.

22. P. F. Cecil, *Herbicidal Warfare: The Ranch Hand Project in Vietnam* (New York: Praeger, 1986), p. 17.

23. Brian Balmer, "The Drift of Biological Weapons Policy in the UK, 1945–1965," *Journal of Strategic Studies*, 20, no. 4 (1997), 117.

24. G. D. Heath, "A Note on the Recent Work of the Crops Division at Camp Detrick," December 1955, Ministry of Supply, Advisory Council on Scientific Research and Technical Development, ADAC, AC13538, PRO.

25. D. L. Miller, "History of Air Force Participation in the Biological Warfare Program, 1944–1951," p. 78, Historical Study No. 194, Wright-Patterson Air Force Base, September 1952.

26. Kent Irish, "Anti-Crop Agent Munitions Systems," BWL Technical Study 12, Annex B to Appendix VI, July 1958, of Operations Research Study Group, Study No. 21, CmlC Operations Research Group, Army Chemical Center, Md., vol. 14, p. 4, ORG-0511 (58), AD 347148, 1 August 1958.

27. Whitby, *Biological Warfare against Crops*, p. 169.

28. Anthony Rimmington, "Invisible Weapons of Mass Destruction: The Soviet Union's BW Program and Its Implications for Contemporary Arms Control," *Journal of Slavic Military Studies*, 13 (September 2000), 6.

29. Ibid., p. 16.

30. Ibid.

31. Kenneth Alibek, "The Soviet Union's Anti-Agricultural Biological Weapons," in *Food and Agricultural Security: Guarding against Natural Threats and Terrorist Attacks Affecting Health, National Food Supplies, and Agricultural Economics*, ed. Thomas W. Frazier and Drew C. Richardson (New York: New York Academy of Sciences, 1999), pp. 18–19.

32. Rimmington, "Invisible Weapons of Mass Destruction," p. 12.

33. UNSCOM, United Nations, S/1995/864, para. 75 (0).

34. Letter from Steve Black, historian, UNSCOM, 28 November 1995.

35. R. A. Darrow, Kent. R. Irish, and Charles E. Minarik, "Herbicides Used in South East Asia," Technical Report, US Army, Plant Sciences Laboratories, Fort Detrick, Frederick, Md., August 1969, AD 864362.

36. Cecil, *Herbicidal Warfare*, p. 19.

37. SIPRI, *The Rise of CB Weapons*, p. 171.

38. SIPRI, *The Problem of Chemical and Biological Warfare*, vol. 2: *CB Weapons Today* (Stockholm: Almqvist & Wiksell, 1973), p. 41.

39. J. M. Stellman et al., "The Extent and Patterns of Usage of Agent Orange and Other Herbicides in Vietnam," *Nature*, 422 (2003), 681–687.

40. "Ranch Hand May Ride Again: Opium Poppies the Target?" *Armed Forces Journal*, 108 (5 July 1971), 10. This article warned six months after the end of the program that Ranch Hand "could be reinstituted against opium poppy fields if President Nixon's requests to poppy growing countries to cease production are not successful."

41. Malcolm Dando, *The New Biological Weapons: Threat, Proliferation and Control* (Boulder: Lynne Rienner, 2001), p. 29.

11. Antianimal Biological Weapons Programs

1. Erhard Geissler and J. E. van Courtland Moon, eds., *Biological and Toxin Weapons Research, Development and Use from the Middle Ages to 1945: A Critical Comparative Analysis*, SIPRI Chemical and Biological Warfare Studies No. 18 (Oxford: Oxford University Press, 1999).

2. Charles Loucks, "Fort Terry Historical Report: 1 January 1951–30 June 1952," US Army Chemical Corps.

3. Ibid.; Robert Hurt, "Fort Terry Historical Report: 1 July–30 September 1953," 29 October 1953, Report CMLHO-194, US Army Chemical Corps; idem, "Fort Terry Historical Report: 1 October–31 December 1953," 8 March 1954, Report CMLHO-194, US Army Chemical Corps.

4. Loucks, "Historical Report"; Robert Hurt, "Fort Terry Historical Report: 1 January–30 June 1954," 29 October 1954, Report CMLHO-194, US Army Chemical Corps.

5. Loucks, "Historical Report"; Robert Hurt, "Fort Terry Historical Report: 1 January–31 March 1953," 13 May 1953, Report CMLHO-194, US Army Chemical Corps; idem, "Fort Terry Historical Report: 1 April–30 June 1953," 31 July 1953, Report CMLHO-194, US Army Chemical Corps.

6. NBL—UC—NAMRU Operational Divisions, War Office, London, 1 September 1950, WO 188/703, PRO.

7. D. L. Miller, "History of Air Force Participation in the Biological Warfare Program, 1951–1954," p. 126, January 1957, Historical Study No. 313, Historical Division, Office of Information Services, Air Material Command, Wright-Patterson Air Force Base; JSPC, Chemical (Toxic) and Biological Warfare Readiness, 13 August 1953, JSPC 954/29, CCS385.2 (12-17-43), sec. 18, RG 218, NARA, College Park, Md.

8. D. L. Miller, "History of Air Force Participation in the BW Program, 1944–1951," September 1952, Historical Study No. 194, Historical Office, Office of the Executive, Air Material Command, Wright-Patterson Air Force Base,

pp. 151–152; idem, "BW Program, 1951–1954," pp. 112–116; Colonel Creasy, Presentation to the Secretary of Defense's Ad Hoc Committee on Cebar, p. 12, 24 February 1950, Joint Chiefs of Staff, R&RA Section, The Pentagon.

9. Ibid., p. 9.

10. BRAB, Minutes of the First Meeting of the Biological Research Advisory Board, p. 3, 1946, WO 195/9107, PRO; BRAB, Research on Foot and Mouth Disease, 8 May 1951, WO 195/11426, PRO.

11. BRAB, Foot and Mouth Disease.

12. Agricultural Defence Advisory Committee, Clandestine Attack on the Crops and Livestock of the Commonwealth, 3 October 1955, WO 195/13150, PRO.

13. War Office, Camp Detrick Meeting, January 1945, WO 188/699, PRO.

14. BRAB, Foot and Mouth Disease.

15. Kenneth Alibek, "The Soviet Union's Anti-Agricultural Biological Weapons," *Annals of the New York Academy of Sciences,* 894 (1999), 18–19.

16. Ken Alibek with Stephen Handelman, *Biohazard: The Chilling True Story of the Largest Covert Biological Weapons Program in the World—Told from the Inside by the Man Who Ran It* (London: Arrow Books, 2000), pp. 37–38.

17. Bozheyeva, Kunakbayev, and Yeleukenov, "Former Soviet Biological Weapons Facilities in Kazakhstan: Past, Present, and Future," pp. 11–12, Occasional Paper No. 1, Chemical and Biological Weapons Nonproliferation Project, Center for Nonproliferation Studies, Monterey Institute of International Studies, Monterey, Calif.; Alibek, *Biohazard,* p. 301.

18. Joby Warrick, "Russia's Poorly Guarded Past," *Washington Post,* 17 June 2002, p. 1.

19. Alibek, *Biohazard,* p. 268; Martin Hugh-Jones, personal communication, 25 April 2004.

20. Meryl Nass, "Anthrax Epizootic in Zimbabwe, 1978–1980: Due to Deliberate Spread?" *PSR Quarterly,* 2 (1992), 198–209.

21. Ibid.; Ian Martinez, "The History of the Use of Bacteriological and Chemical Agents during Zimbabwe's Liberation War of 1965–80 by Rhodesian Forces," *Third World Quarterly,* 23, no. 6 (2002), 1159–79.

22. Seth Carus, "Unlawful Acquisition and Use of Biological Agents," in *Biological Weapons: Limiting the Threat,* ed. Joseph Lederberg (Cambridge, Mass.: MIT Press, 1999), p. 219.

23. ProMED Mail (Program for Monitoring Emerging Diseases), "Rabbit Hemorrhagic Disease Virus Introduced—New Zealand (06)," 29 August 1997, ProMED Mail Archive No. 19970827.1811.

24. D. MacKenzee, "Brazil Claims Foot-and-Mouth Sabotage," *New Scientist,* 167 (9 September 2000), 5.

25. Land, buildings, and other facilities of Plum Island have since been transferred back into the security realm; they were subsumed by the Department of Homeland Security in June 2003.

26. Mark Wheelis, "Biological Sabotage in World War I," in Geissler and Moon, *Biological and Toxin Weapons.*

12. Midspectrum Incapacitant Programs

1. G. S. Pearson, "The CBW Spectrum," *ASA Newsletter,* no. 90, 1 and 7–8.
2. E. Kagan, "Bioregulators as Instruments of Terror," *Clinics in Laboratory Medicine,* 21, no. 3 (2001), 607–618.
3. Editorial, "'Non-Lethal' Weapons, the CWC and the BWC," *Chemical and Biological Weapons Conventions Bulletin,* no. 61 (2003), 1–2.
4. J. S. Ketchum and F. R. Sidell, "Incapacitating Agents," in *Medical Aspects of Chemical and Biological Warfare,* ed. F. R. Sidell, E. T. Takafuji, and D. R. Franz (Washington, D.C.: Office of the Surgeon General, US Army, 1997), pp. 287–305; S. Bowman, *Iraqi Chemical and Biological Weapons (CBW) Capabilities,* CRS Issue Brief (Washington, D.C.: Congressional Research Service, 1998); M. R. Dando, "The Danger to the Chemical Weapons Convention from Incapacitating Chemicals," First CWC Review Conference Paper No. 4, Bradford University, Department of Peace Studies, March 2003; J. P. Perry Robinson, "Disabling Chemical Weapons: Some Technical and Historical Aspects," Paper presented to the Pugwash Study Group on Implementation of the CBW Conventions, Den Haag/Noordwijk, 27–29 May 1994; R. Evans, *Gassed: British Chemical Warfare Experiments on Humans at Porton Down* (London: House of Stratus, 2000).
5. G. B. Carter and G. S. Pearson, "Past British Chemical Warfare Capabilities," *RUSI Journal,* February 1996, 59–68.
6. Cabinet Defence Committee, Minutes of a Meeting Held on 3 May 1963, PRO, CAB 131/28.
7. Material quoted from the UK Public Record Office (PRO) was obtained from the Harvard Sussex Archive at the University of Sussex or from the PRO under grant number 054732 to Professor Dando from the Wellcome Trust.
8. CmlC, "Summary of Major Events and Problems: Fiscal Year 1959," CmlC Historical Office, Army Chemical Center, Md., January 1960 (hereafter Cmlc Hist 59); idem, "Summary of Major Events and Problems: Fiscal Years 1961–1962," CmlC Historical Office, Army Chemical Center, Md., June 1962 (hereafter CmlC Hist 61–62).
9. War Department, CDEE, Collected Papers 1962, PRO, WO 188/748; idem, Collected Papers 1964, PRO, WO 188/749.
10. D. F. Downing, "The Chemistry of Psychotomimetic Substances," *Quarterly Review of the Chemical Society of London,* 16, no. 2 (1962), 133–162, in PRO, WO 188/748; R. W. Brimblecombe and A. L. Green, "Effect of Monoamine Oxidase Inhibitors on the Behaviour of Rats in Hall's Open Field," *Nature,* 4832 (9 June 1962), 983; R. W. Brimblecombe, D. F. Downing, D. M. Green, and R. R. Hunt, "Some Pharmacological Effects of a Series of

Tryptamine Derivatives," *British Journal of Pharmacology and Chemotherapy,* 23, no. 1 (1964), 43–54, in PRO, WO 188/749.

11. Defence Correspondent, "British Defence against Germ Warfare: Chemicals That Destroy Will to Fight," *The Times,* 25 May 1964, in file A181, Harvard Sussex Archive.

12. "Chemical Warfare: All Peace at Porton," *Nature,* 222 (June 1969), 1019–20.

13. P. B. Bradley and R. W. Brimblecombe, eds., "Biochemical and Pharmacological Mechanisms Underlying Behaviour," *Progress in Brain Research,* 36 (1972), 1–203.

14. R. W. Brimblecombe and R. M. Pinder, *Tremors and Tremorogenic Agents* (Bristol: Scientechnica Publishers, 1972); R. W. Brimblecombe, *Drug Actions on Cholinergic Systems* (London: Macmillan, 1974); R. W. Brimblecombe and R. M. Pinder, *Hallucinogenic Agents* (Bristol: Wright-Scientechnica, 1975).

15. Advisory Council on Scientific Research and Technical Development, Terms of Reference and Membership of Council, 1960, PRO, WO 195/14926; idem, Constitution, Terms of Reference and Membership of Council, Its Boards and Committees, 1961, PRO, WO 195/14988; idem, The Committee Structure of the Chemical Defence Advisory Board: Note by Professor R. B. Fisher, Chairman, CDAB, 1964, PRO, WO 195/15891.

16. Ketchum and Sidell, "Incapacitating Agents."

17. *Joint CB Technical Data Source Book,* vol. 2: *Riot Control and Incapacitating Agents,* Part 3: *Agent BZ* (Fort Douglas, Utah: Deseret Test Center, 1972).

18. Chemical Casualty Care Division, *Incapacitating Agents* (Fort Detrick, Md.: USAMRID, 2002).

19. C. McLeish, "The Governance of Dual-Use Technologies in Chemical Warfare" (M.Sc. thesis, University of Sussex, 1997); K. McLaughlin, "Technology-Driven Breakout: A Case Study of LSD and the Chemical Weapons Governance Regime" (M.Sc. thesis, University of Sussex, 2001).

20. Directorate of Chemical Defence Research and Development, Progress Report for the Half-Year Ended 30 June 1956, PRO, WO 188/710.

21. Chemistry Committee, Minutes of the 32nd Meeting, 5 March 1959, PRO, WO 195/14637.

22. Advisory Council on Scientific Research and Technical Development, Report for the Year 1959 (January to December), 1959, PRO, WO 195/14876.

23. Idem, Minutes of the 3rd Meeting, 24 November 1960, PRO, WO 195/15078.

24. Chemistry Committee, Minutes of the 34th Meeting, 19 November 1959, PRO, WO 195/14868.

25. Idem, Minutes of the 36th Meeting, 8 June 1960, PRO, WO 195/14987.

26. CDAB, Annual Review of the Work of the Board for the Year Ending September 1961, 1961, PRO, WO 195/15228.

27. Offensive Evaluation Committee, Sixth Meeting of the Committee, 15 July 1960, PRO, WO 195/15014.

28. Biology Committee, Minutes of the 28th Meeting, 23 November 1961, PRO, WO 195/15289.

29. Offensive Evaluation Committee, Minutes of the 5th Meeting, 15 January 1960, PRO, WO 195/14905.

30. Biology Committee, The UK Approach to New Agents, 1960, PRO, WO 195/15031.

31. R. W. Brimblecombe and J. W. Blackburn, "The Biological Testing of Incapacitating Agents, Part I: Review of Testing Methods," Porton Technical Paper No. 765, 1961, PRO, WO 195/15169.

32. Idem, "The Biological Testing of Incapacitating Agents, Part II: Results of Tests Using Drugs with Known Effects on Man," Porton Technical Paper No. 766, 1961, PRO, WO 195/15170.

33. R. W. Brimblecombe, "The Biological Testing of Incapacitating Agents, Part III: Results of Screening Tests on New Compounds," Porton Technical Paper No. 793, 1961, PRO, WO 195/15273.

34. Idem, "The Use of Conditioned Response Tests to Screen Psychotrophic Drugs," Porton Technical Paper No. 805, 1962, PRO, WO 189/327.

35. C. Strafford, "The Synthesis of Some Simple Peptides," Porton Technical Paper No. 769, 1961, PRO, WO 195/15181; D. F. Downing and R. R. Hunt, "The Preparation of Simple Indoles, 1-Substituted Indoles and Tryptamines," Porton Technical Paper No. 770, 1961, PRO, WO 189/1082; idem, "The Preparation of N-N-Diallkyltryptamines," Porton Technical Paper No. 771, 1961, PRO, WO 189/1083.

36. R. W. Brimblecombe, D. F. Downing, D. M. Green, and R. R. Hunt, "The Synthesis and Biological Testing of a Series of Tryptamine Derivatives," Porton Technical Paper No. 822, 1962, PRO, WO 189/344.

37. McLeish, "Dual-Use Technologies in Chemical Warfare."

38. S. Callaway, W. M. Hollyhock, and W. S. S. Ladell, "An Oripavine Derivative (TL2636) as a Potential Incapacitating Agent," Porton Technical Paper No. 835, 1963, PRO, WO 189/357.

39. Chemistry Committee, 42nd Meeting, Item 7(a): Current Investigations in the Search for New Agents, 1962, PRO, WO 195/15388.

40. Offensive Evaluation Committee, Minutes of the 9th Meeting, 3 January 1962, PRO, WO 195/15329.

41. CDAB, Minutes of the 50th Meeting, 24 May 1962, PRO, WO 195/15391.

42. Idem, Minutes of the 53rd Meeting, 6 June 1963, PRO, WO 195/15609.

43. Idem, Minutes of the 54th Meeting, 2 October 1963, PRO, WO 195/15679.

44. Chemistry Committee, Minutes of the 47th Meeting, 10 December 1963, PRO, WO 195/15735.

45. A. Bebbington and D. Shakeshaft, "Acetylenic Amines Related to Tremorine and Tremoram," Porton Technical Paper No. 868, 1963, PRO, WO 189/388; A. Bebbington, "Structural Requirements for Muscarinic Activity in Relation to the Central Action of Tremoram," Porton Technical Pa-

per No. 869, 1963, PRO, WO 189/389; R. W. Brimblecombe and D. C. Parkes, "Pharmacological Studies of Tremorine, Tremoram and Allied Compounds," 1964, Porton Technical Paper No. 871, PRO, WO 189/391.

46. Advisory Council on Scientific Research and Technical Development, Minutes of the 26th Meeting, 3 November 1964, PRO, WO 195/15917.

47. CDAB, Minutes of the 56th Meeting, 4 June 1964, PRO, WO 195/15854.

48. G. J. Bennett, "A Search for Pharmacologically Active Benzimidazoles," 1964, PRO, WO 195/15785.

49. CDAB, Minutes of the 57th Meeting, 8 October 1964, PRO, WO 195/15925.

50. R. W. Brimblecombe, "The Biological Testing of Incapacitating Agents, Part IV: Further Results of Screening Tests on New Compounds," Porton Technical Paper No. 909, 1964, PRO, WO 189/422.

51. D. F. Downing, "Psychotomimetic Compounds," in *Psychopharmacological Agents*, ed. M. Gordon (London: Academic, 1964), pp. 555–618; Brimblecombe et al., "Some Pharmacological Effects of Tryptamine Derivatives."

52. Biological Research Advisory Board / CDAB, Notes on a Visit to CDEE and MRE, Porton, 12 July 1963, PRO, WO 195/15631.

53. Advisory Council on Scientific Research and Technical Development, Report for the Year 1965, 1966, PRO, WO 195/16166.

54. CDAB, Minutes of the 60th Meeting, 8 October 1965, PRO, WO 195/16154.

55. Idem, Minutes of the 59th Meeting, 28 May 1965, PRO, WO 195/16054.

56. Chemistry Committee, Minutes of the 50th Meeting, 18 February 1965, PRO, WO 195/15999.

57. A. Bebbington, "Review of New Agent Program," 1964, PRO, WO 195/15934.

58. R. R. Hunt, "The Synthesis and Pharmacological Activity of Indole Derivatives: A Review of Investigations at CDEE, Porton, 1959–64," 1964, PRO, WO 195/15933.

59. R. W. Brimblecombe and D. G. Rowsell, "The Interaction of Muscarinic Drugs with the Post Ganglionic Cholinergic Receptor," 1964, PRO, WO 195/15941.

60. A. Bebbington and R. W. Brimblecombe, "Muscarinic Receptors in Peripheral and Central Nervous Systems," *Advances in Drug Research*, 2 (1965), 143–172.

61. CDAB, Summary: CDEE Annual Report 1964–65, 1965, PRO, WO 195/16065.

62. Dione J. Berry, Mary Cheetam, W. M. Hollyhock, Frances Lovell, and K. H. Kemp, "A Field Experiment Using LSD25 on Trained Troops," Porton Technical Paper No. 936, 1965, PRO, WO 195/16137; Applied Biology Committee, Minutes of the 1st Meeting, 24 November 1965, PRO, WO 195/16161.

63. CDAB, Annual Review for Period 1.7.65 to 30.6.66, 1966, PRO, WO 195/16310.

64. Idem, Minutes of the 63rd Meeting, 10 October 1966, PRO, WO 195/16381.

65. Idem, Notes on a Visit to CDEE, Porton, 16 June 1966, PRO, WO 195/16281.

66. Applied Biology Committee, Minutes of the 2nd Meeting, 20th April 1966, PRO, WO 195/16273; F. W. Beswick, K. Kemp, and R. J. Moylan-Jones, "Field Experiments—'Exercise Recount'—Preliminary Report of a Field Experiment by CDEE," 1966, PRO, WO 195/16312.

67. R. R. Hunt, "Chemical Research on Toxic Compounds: 6th Progress Report," 1966, PRO, WO 195/16254; Chemistry Committee, Minutes of the 53rd Meeting, 30 June 1966, PRO, WO 195/16277.

68. A. Bebbington, R. W. Brimblecombe, and D. Shakeshaft, "The Central and Peripheral Activity of Acetylenic Amines Related to Oxotremorine," *British Journal of Pharmacology,* 26 (1965), 56–67; A. Bebbington, R. W. Brimblecombe, and D. G. Rowsell, "The Interaction of Muscarinic Drugs with the Postganglionic Acetylcholine Receptor," ibid., 26 (1965), 68–78.

69. R. B. Fisher, "Comments on the CDEE Annual Report," 1966, PRO, WO 195/16309.

70. L. Leadbeater and P. Watts, "The Prediction of Biological Activity," Porton Technical Paper No. 951, 1966, PRO, WO 189/457; R. W. Brimblecombe, D. M. Green, D. C. Parkes, F. A. B. Aldous, and June M. Stratton, "The Pharmacology of Some Anticholinergic Drugs," Porton Technical Paper No. 959, 1966, PRO, WO 189/464.

71. Advisory Council on Scientific Research and Technical Development, Minutes of the 42nd Meeting, 4 April 1967, PRO, WO 195/16451; CDAB, Minutes of the 65th Meeting, 16 June 1967, PRO, WO 195/16496.

72. CDAB, Minutes of the 66th Meeting, 9 October 1967, PRO, WO 195/16553.

73. Biology Committee/Chemistry Committee, Minutes of the 2nd Joint Meeting, 14 February 1967, PRO, WO 195/16461.

74. Advisory Council on Scientific Research and Technical Development, Minutes of the 44th Meeting, 7 November 1967, PRO, WO 195/16573.

75. Applied Biology Committee, Minutes of the 4th Meeting, 26 April 1967, PRO, WO 195/16462.

76. R. J. Moylan-Jones, "U.S. Experience with BZ and Other Benzilates and Glycollates," 1967, PRO, WO 195/16432.

77. K. H. Kemp, "Future Plans for Work in the UK," 1967, PRO, WO 195/16430.

78. Brimblecombe et al., "The Pharmacology of Some Anticholinergic Drugs."

79. R. W. Brimblecombe, F. W. Beswick, and D. F. Downing, "A Review of Some Concepts of Incapacitation," 1967, PRO, WO 195/16429.

80. R. W. Brimblecombe and T. D. Inch, "The Anticholinergic Properties of Enantiomeric Glycollates: A Progress Report," 1967, PRO, WO 195/16558; T. D. Inch, R. V. Ley, and P. Rich, "Stereospecific Synthesis of 2-Alkyl-2-Hydroxy-2-Phenylacetic Acid Esters (Glycollates)," Porton Technical Paper No. 973, 1967, PRO, WO 189/477.

81. L. Leadbeater, "The Interaction of Orvinols with Two Biological Sites: The Active Centre of the N-Dealkylating Enzymes of Rat Liver Microsomes and the Analgesic Receptor in the Rat Central Nervous System," Porton Technical Paper No. 960, 1967, PRO, WO 189/465.

82. R. W. Brimblecombe and Joan V. Sutton, "The Ganglion-Stimulating Effects of Some Amino Acid Esters," Porton Technical Paper No. 978, 1967, PRO, WO 189/481; idem, The Ganglion-Stimulating Effects of Some Amino-Acid Esters," *British Journal of Pharmacology*, 34 (1968), 358–369.

83. CDAB, CDEE Annual Report, 1967–1968, 1968, PRO, WO 195/16822.

84. Advisory Council on Scientific Research and Technical Development, Minutes of the 50th Meeting, 5 November 1968, PRO, WO 195/16796.

85. Idem, Report for the Year 1968, 1968, PRO, WO 195/16832.

86. Chemistry Committee, Minutes of the 48th Meeting, 31 October 1968, PRO, WO 195/16825.

87. Applied Biology Committee, Minutes of the 7th Meeting, 4 December 1968, PRO, WO 195/16855.

88. CDAB, Minutes of the 69th Meeting, 10 January 1969, PRO, WO 195/16867.

89. Applied Biology Committee/Biology Committee, Joint Meeting on "Behavioural Studies," 4 December 1968, PRO, WO 195/16887.

90. P. Holland, "The Behavioural Effects of Drugs on Man as Illustrated by the Glycollates," 1968, PRO, WO 195/16804.

91. R. B. Fisher, "An Outline of CW and Its Problems," 1968, PRO, WO 195/16641.

92. DERA, "An Overview of Research Carried Out on Glycollates and Related Compounds at CBD Porton Down," DERA/CBD/CR990418, September 1999, Ministry of Defence, London.

93. CmlC, "Summary of Major Events and Problems, FY55," CmlC Historical Center, Army Chemical Center, Md., December 1955, pp. 48–49 (hereafter CmlC Hist FY55); Marks John, *Search for the Manchurian Candidate: The CIA and Mind Control* (New York: Times Books, 1979); National Research Council, *Possible Long-Term Health Effects of Short-Term Exposure to Chemical Agents*, 3 vols. (Washington, D.C.: National Academy Press, 1982–1985), vol. 2; L. Wilson Green, "Psychochemical Warfare: A New Concept of War," CmlC, Edgewood Arsenal, 1949.

94. CmlC, "Summary of Major Events and Problems, FY57," CmlC Historical Center, Army Chemical Center, Md., October 1957, pp. 97–99; National Research Council, *Long-Term Health Effects*, vol. 2, p. 47.

95. CmlC, "Summary of Major Events and Problems, FY56," CmlC Historical Center, Army Chemical Center, Md., November 1956), pp. 128–130 (hereafter CmlC Hist FY56).

96. CmlC Hist FY56; CmlC Hist FY57; Allen M. Hornblum, *Acres of Skin: Human Experiments at Holmsburg Prison* (New York: Routledge, 1998), pp. 124–130; National Research Council, *Long-Term Health Effects*, vol. 2, pp. 90–94.

97. CmlC, "Summary of Major Events and Problems, FY58," CmlC Historical Center, Army Chemical Center, Md., March 1959, pp. 100–101; Cmlc Hist 59, p. 32; US House of Representatives, Committee of Science and Astronautics, *Hearings on Chemical, Biological and Radiological Warfare Agents, 86th Congress, June 16 and 22, 1959* (Washington, D.C.: USGPO, 1959), pp. 1–44 (hereafter House CBR 1959).

98. CmlC Hist 61–62, pp. 124–126.

99. Ibid.; J. S. Ketchum and F. R. Sidell, "Incapacitating Agents," in *Textbook of Military Medicine,* ed. R. Zajtchuk (Washington, D.C.: Borden Institute, 1997), pp. 294–298.

100. CmlC Hist 61–62, pp. 61–62; J. P. Perry Robinson, "Disabling Chemical Weapons: A Documentary Chronology of Events, 1945–2003," p. 40 n. 166, Harvard Sussex Program, University of Sussex, 8 October 2003 (hereafter Robinson Chronology). This chronology updates earlier versions (see Robinson, "Disabling Chemical Weapons," in note 4) and gives by far the most complete account of publicly available information on attempts to develop and use incapacitants between 1945 and 2003.

101. House CBR 1959, p. 16.

102. Department of the Army, *Field Manual FM 23-20: Davy Crockett Weapons System in Infantry and Armor Units* (Washington, D.C.: Headquarters, Department of the Army, 19 December 1961), pp. 3–5, 24, 111–159; CmlC, "Summary of Major Events and Problems, FY54 TOP SECRET ANNEX," CmlC Historical Center, Army Chemical Center, Md., September 1954, pp. 2–3; Department of the Army, *Field Manual FM 30-40: Handbook on Soviet Ground Forces* (Washington, D.C.: Headquarters Department of the Army, 30 June 1975), pp. 5-9 and 5-21 to 5-25.

103. US Army Center for Health Promotion and Preventive Medicine, "Detailed Facts about Psychedelic Agent 3—Qunuclidinyl Benzilate (BZ)," USACHPPM Detailed Fact Sheet 218-16-1096, accessed at http://www.certip.org/resources/chem-agents.html on 12 July 2004; Interdepartmental Political-Military Group, "Annual Review of the US Chemical Warfare and Biological Research Programs as of 1 November 1970," p. 8, NSA Document 246 (hereafter 1970 CW Review); Department of the Army, *Field Manual FM 21-40: Small Unit Procedures in Chemical, Biological and Radiological (CBR) Operations* (Washington, D.C.: Headquarters, Department of the Army, October 1963) idem, *Field Manual FM 21-41: Soldier's Handbook for Chemical and Biological Operations and Nuclear Warfare* (Washington, D.C.: Headquarters, Department of the Army, April 1963); idem, *Field Manual FM 21-41: Soldier's Handbook for Chemical and Biological Operations and Nuclear Warfare* (Washington, D.C.: Headquarters, Department of the Army, February 1967); idem, *Field Manual FM 3-8: Chemical Reference Handbook* (Washington, D.C.: Headquarters, Department of the Army, January 1967), pp. 16, 24, 38.

104. Robinson Chronology, p. 52 n. 207.

105. Ibid., p. 55 nn. 217, 219.

106. Interdepartmental Political-Military Group, "US Policy on Chemical and Biological Warfare and Agents: Report to the National Security Council (in Response to NSSM 59)," 10 November 1969, in "National Security Archive Briefing Book No. 58," vol. 3: "Biowar: The Nixon Administration's Decision to End the US Biological Warfare Programs," ed. R. A. Wampler, docs. 6a and 6b, pp. 11, 31–32, National Security Archive, George Washington University; "HAK [Henry A. Kissinger] Talking Points," 25 November 1969, in ibid., doc. 11, p. 7.

107. 1970 CW Review, p. 14; Hornblum, *Acres of Skin,* pp. 127–130.

108. Robinson Chronology, p. 73 n. 269; Departments of the Army, the Navy and the Air Force, *NATO Handbook on the Medical Aspects of NBC Defensive Operations AmedP-6 (Army Field Manual FM 8-9, NAVMED P-5059, Air Force AFP 161-3)* (Washington, D.C.: Departments of the Army, Navy, and Air Force, August 1973 and Change 1, 1 May 1983, pp. 6-1 to 6-7; Department of the Army, *Field Manual FM 21-40: NBC (Nuclear, Biological and Chemical) Defense* (Washington, D.C.: Headquarters, Department of the Army, 14 October 1977), pp. 1-13 to 1-18; US Army Ordnance and Chemical Center and School, *Introduction to Chemical and Radiological Operations and Biological Defense (Interschool Subcourse 220)* (Washington, D.C.: USGPO, 1983), pp. 39–46.

109. Accidental fire report per personal communication, J. Perry Robinson, June 2004; Pine Bluff Chemical Activity (PBCA), Pine Bluff, Ark., accessed at http://www.globalsecurity.org/wmd/facility/pine-bluff.htm on 12 July 2004.

110. US Army, "US Army Non-Stockpile Chemical Materiel Program," Press release, US Army Edgewood MD PRNewswire, 23 March 2004, accessed at www.cpeo.org/lists/military/2004/msg01508.html on 12 July 2004.

111. B. Van Damme, "Moscow Theatre Siege: A Deadly Gamble That Nearly Paid Off," *Pharmaceutical Journal,* 269 (2002), 723–724.

112. M. R. Dando, *A New Form of Warfare: The Rise of Non-Lethal Weapons* (London: Brassey's, 1996), p. 168.

113. N. Davison, "Weapons Focus: Biochemical Weapons," Bradford Non-Lethal Weapons Research Project, Research Report No. 5, May 2004, 27–34.

114. M. R. Dando, "Future Incapacitating Chemical Weapons: The Impact of Genomics," in *The Future of Non-Lethal Weapons: Technologies, Operations, Ethics and Law,* ed. N. Lewer (London: Frank Cass, 2002), pp. 167–181.

115. Editorial, "'Non-Lethal' Weapons, the CWC and the BWC."

13. Allegations of Biological Weapons Use

Acknowledgments: We thank Milton Leitenberg and Matthew Meselson for helpful comments and for provision of documents.

1. Mark Wheelis, "Biological Warfare before 1914," in *Biological and Toxin*

Weapons Research, Development and Use from the Middle Ages to 1945: A Critical Comparative Analysis, ed. Erhard Geissler and J. E. van Courtland Moon, SIPRI Chemical and Biological Warfare Studies No. 18 (Oxford: Oxford University Press, 1999), pp. 8–34; idem, "Biological Sabotage in World War I," ibid., pp. 35–62; and Sheldon H. Harris, *Factories of Death: Japanese Biological Warfare, 1932–1945, and the American Cover-Up* (New York: Routledge, 1995).

2. Chandré Gould and Peter I. Folb, "The South African Chemical and Biological Warfare Program: An Overview," *Nonproliferation Review,* fall/winter 2000, 10–23.

3. Andrew Selth, "Burma and Exotic Weapons," *Strategic Analysis,* 19 (1996), 413–433.

4. Meryl Nass, "Anthrax Epizootic in Zimbabwe, 1978–1980: Due to Deliberate Spread?" *PSR Quarterly,* 2 (1992), 198–209; Ian Martinez, "The History of the Use of Bacteriological and Chemical Agents during Zimbabwe's Liberation War of 1965–80 by Rhodesian Forces," *Third World Quarterly,* 23, no. 6 (2002), 1159–79.

5. Dean Acheson, "'Germ Warfare' Charges Called Fabrication," press release, 4 March 1952, *Department of State Bulletin,* 17 March 1952, 427–428.

6. Milton Leitenberg, "Allegations of Biological Warfare in China and Korea: 1951–1952," in SIPRI, *The Problem of Chemical and Biological Warfare,* vol. 5: *The Prevention of CBW* (Stockholm: Humanities Press, 1971), app. 4, pp. 238–258; Josef Goldblat, "Allegations of Use of Bacteriological and Chemical Weapons in Korea and China," in ibid., vol. 4: *CB Disarmament Negotiations, 1920–1970* (Stockholm: Humanities Press, 1971), pp. 196–223.

7. John Ellis van Courtland Moon, "Biological Warfare Allegations: The Korean War Case," in *The Microbiologist and Biological Defense Research: Ethics, Politics and International Security,* ed. R. A. Zilinskas (Geneva: International Committee of the Red Cross, 1992); Alastair Hay, "Simulants, Stimulants and Diseases: The Evolution of the United States Biological Warfare Programme, 1945–60," *Medicine, Conflict and Survival,* 15 (1999), 198–214.

8. Milton Leitenberg, "Resolution of the Korean War Biological Warfare Allegations," *Critical Reviews in Microbiology,* 24, no. 3 (1998), 169–194; Kathryn Weathersby, "Deceiving the Deceivers: Moscow, Beijing, Pyongyang and the Allegations of Bacteriological Weapons Use in Korea," *Bulletin,* 11 (1998), 176–185, Cold War International History Project, Woodrow Wilson International Center for Scholars; Milton Leitenberg, "New Russian Evidence on the Korean War Biological Warfare Allegations: Background and Analysis," ibid., pp. 185–199; idem, "The Korean War Biological Weapons Allegations: Additional Information and Disclosures," *Asian Perspective,* 24, no. 3 (2000), 159–172.

9. Harris, *Factories of Death.*

10. S. Endicott and E. Hagerman, *The United States and Biological Warfare: Secrets From the Early Cold War and Korea* (Bloomington: Indiana University Press, 1998). Also see Chen Jian, *Mao's China and the Cold War* (Chapel Hill: Uni-

versity of North Carolina Press, 2001), pp. 109–110; M. A. Ryan, *Chinese Attitudes toward Nuclear Weapons: China and the United States during the Korean War* (Armonk, N.Y.: M. E. Sharpe, 1989), pp. 104–109; Xiaobing Li, A. R. Millet, and Bin Yu, *Mao's Generals Remember Korea* (Lawrence: University Press of Kansas, 2001), pp. 157–160.

11. On the Khabarovsk trials, see the anonymous *Materials on the Trial of Former Servicemen of the Japanese Army Charged with Manufacturing and Employing Bacteriological Weapons* (Moscow: Foreign Languages Publishing House, 1950).

12. "Soviets Ask Answer on Trial of Hirohito," *New York Times,* 16 December 1950; "Soviet Organ Sees Confusion in US," ibid., 13 April 1951.

13. United Nations (hereafter UN) doc. S/2142/Rev 1, 18 May 1951.

14. UN doc. S/2684, 30 June 1952, includes texts of the 23 February 1952 North Korean and the 24 February 1952 and 8 March 1952 Chinese statements.

15. A Chinese scientific commission investigated the BW reports from 12 March to 31 March in Korea and northeastern China. Two reports were issued: "Report of the Northeast China Group of the Commission for Investigating the American Crime of Waging Bacteriological Warfare," published as a supplement to *People's China,* 16 April 1952; and "Report on the Crime of American Imperialists in Spreading Bacteria in Korea," 24 April 1952. International Association of Lawyers, "Report on US Crimes in Korea," 31 March 1952; idem, "Report on the Use of Bacterial Weapons in Chinese Territory by the Armed Forces of the United States," 2 April 1952, both in UN doc. S/2684, 30 June 1952. *Report of the International Scientific Commission for the Investigation of the Facts Concerning Bacterial Warfare in Korea and China* (Peking: ISC, 1952). The summary without appendices is available in UN doc. S/2802, 8 October 1952.

16. Confessions of K. L. Enoch and J. Quinn, *Report of the International Scientific Commission,* apps. KK and LL.

17. "The USA Is Waging Germ Warfare in Korea: Statements of Prisoners of War Colonel Frank H. Schwable and Major Roy H. Bley, US Marine Corps," supplement to *New Times,* no. 10 (4 March 1953).

18. Hay, "Simulants, Stimulants and Diseases"; W. M. Creasy, "Manufacturing Potential of Camp Detrick Emergency Planning," 11 June 1953 and 6 August 1953, CMLRE-BWD-4, entry 1b, box 249, RG 175, NARA.

19. Leitenberg, "Resolution of the Korean War Biological Warfare Allegations"; idem, "The Korean War Biological Weapons Allegations"; Weathersby, "Deceiving the Deceivers."

20. A. M. Halpern, *Bacterial Warfare Accusations in Two Asian Communist Propaganda Campaigns* (Santa Monica: RAND, 1952), p. 36. The most comprehensive published summary is "The Case against the United States Germ Warfare Criminals," in Chinese People's Committee for World Peace, *Stop US Germ Warfare!* Part V (Peking, April 1952), pp. 1–20.

21. *Materials on the Trial of Former Servicemen of the Japanese Army.*

22. UN S/2142/Rev 1.
23. "Kuo Mo-Jo Reports to World Peace Council on US Germ Warfare," in Chinese People's Committee for World Peace, *Stop US Germ Warfare!* Part I, pp. 16–22.
24. Ibid.
25. International Association of Lawyers, "Report on US Crimes in Korea," pp. 18–20.
26. Hsinhua News Agency reports from 29 March 1952, in Chinese People's Committee for World Peace, *Stop US Germ Warfare!* See "US Cannot Evade Responsibility for Germ Warfare," Part II, pp. 23–24; and "[Report from] Korean Front," Part III, pp. 41–42.
27. P. Williams and D. Wallace, *Unit 731: The Japanese Army's Secret of Secrets* (London: Hodder and Stoughton, 1989), p. 292.
28. J. P. Craig, "Epidemic Hemorrhagic Fever," Tokyo, 406th GML, February 1951.
29. Martin Furmanski, interview with John P. Craig, 20 June 2000.
30. HQ EUSAK, Office of the Surgeon, Information Bulletin No. 4, 26 January 1952, p. 3, instituting diphtheria and yellow fever vaccination for troops in Korea; ibid., No. 25, 21 June 1952, p. 3, discontinuing these immunizations; both in EUSAK Medical Reports, entry A1 204, box 1553, RG 338, NARA.
31. "Bubonic Plague Ship," *Newsweek,* 9 April 1951, 13; "Award of the Distinguished-Service Cross [to Brig. Gen. Crawford F. Sams], 20 April 1951, General Orders No. 94, Far East Command, in Crawford Sams Papers, box 4, Hoover Institution Archives, Stanford, Calif.
32. A. E. Cowdrey, "'Germ Warfare' and Public Health in the Korean Conflict," *Journal of the History of Medicine and Allied Sciences,* 39 (April 1984), 153–172.
33. Records of the CED, entry 14, box 12, AFEB subfile 45-62, RG 334, NARA.
34. A. V. Hardy, R. P. Mason, and G. A. Martin, "The Dysenteries in the Armed Forces," *American Journal of Tropical Medicine and Hygiene,* 1, no. 1 (1952).
35. "New Weapons Will Get Tried Out in Korea," *US News & World Report,* 21 September 1951.
36. Address of Major General E. F. Bullene, *Congressional Record,* 98, pt. 8 (18 February 1952), A1365–67.
37. "Germs, Gas Seen as Cheapest War Weapons," *Stars and Stripes* (European edition), 27 January 1952.
38. Frederick Kuh, *Chicago Sun-Times,* 2 February 1952.
39. Euchinic Huang, "Anti-Smallpox Drive," *China Monthly Review,* April 1951, 207–298. See also Cowdrey, "'Germ Warfare' and Public Health in Korean Conflict."
40. John E. van Courtland Moon, "Dubious Allegations," book review, *Bulletin of the Atomic Scientists,* 53, no. 3 (1999), 70–72; Leitenberg, "Resolution of the Korean War Biological Warfare Allegations," p. 191 n. 17.

41. Report of the Biological Department, Chemical Corps, to the Panel on Program of the Committee on Biological Warfare, Research and Development Board, Report Series No. 3, 1 October 1949, p. 2, entry 5, box 6, RG 175, NARA; Committee on Biological Warfare, 1951 Program Guidance Report, 5 December 1950, pp. 2, 7, RG 330, NARA.

42. *Hearings before the Select Committee to Study Governmental Operations with Respect to Intelligence Activities of the US Senate, Volume 1: Unauthorized Storage of Toxic Agents, September 16, 17, 18, 1975* (Washington, D.C.: USGPO, 1976), pp. 22–24 (hereafter Church Committee).

43. *Report of the International Scientific Commission for the Investigation of the Facts Concerning Bacterial Warfare in Korea and China* (Beijing: ISC, 1952).

44. China rejected the World Health Organization, a UN entity, because Chinese forces were in combat against UN forces. China rejected the ICRC because it had inspected the Koje Island camp and failed to condemn the US treatment of Communist POWs there.

45. Endicott and Hagerman, *The United States and Biological Warfare.*

46. R. A. Moore, *A Textbook of Pathology* (Philadelphia: W. B. Saunders, 1947), p. 229; W. Boyd, *A Textbook of Pathology* (Philadelphia: Lea and Febiger, 1947), p. 189.

47. T. V. Inglesby et al., "Anthrax as a Biological Weapon," *JAMA*, 287 (2002), 2236–52.

48. Faina A. Abramova, Lev M. Grinberg, Olga V. Yampolskaya, and David H. Walker, "Pathology of Inhalational Anthrax in 42 Cases from the Sverdlovsk Outbreak of 1979," *Proceedings of the National Academy of Sciences*, 90 (1993), 2291–94.

49. J. E. van Courtland Moon, "US Biological Warfare Planning and Preparedness: The Dilemmas of Policy," in Geissler and Moon, *Biological and Toxin Weapons*, pp. 215–254.

50. Quarterly Technical Reports of Pilot Plants Division, Camp Detrick, 1 July 1950 to 30 September 1950, Summary, pp. 1, 29–31; ibid., 1 October 1950 to 31 December 1950, pp. v, 33, 38–39; both in entry 5, box 13, RG 175, NARA.

51. Special Report No. 138, "Feathers as Carriers of Biological Warfare Agents," 15 December 1950, Biological Department, Chemical Corps, RG 218, NARA.

52. D. L. Miller, "History of Air Force Participation in the Biological Warfare Program, 1951–1954," p. 124, January 1957, Historical Study No. 313, Historical Division, Office of Information Services, Air Material Command, Wright-Patterson Air Force Base.

53. R. D. Housewright, *Chemical and Biological Warfare Defense*, US Navy NAVPERS doc. 10098 (Washington, D.C.: USGPO, 1952), p. 233.

54. T. V. Inglesby et al., "Anthrax as a Biological Weapon."

55. H. G. Stockley, "A Fatal Case of Cerebrospinal Anthrax Infection," *Chinese*

Medical Journal (Chengtu), 61A (January 1943), 69; Ling Chao-chi and Chen Yoeh-shu, "Bacillus Anthracis Meningitis: Report of Two Cases," *Chinese Medical Journal,* 66 (August 1948), 431–434.

56. Martin Hugh-Jones, personal communication, April 2004.

57. White House, Memorandum for the Record: Minutes of Special Group Meeting, 21 July 1960, JFK Assassination Record No. 157-10007-10302, NARA.

58. Louis Sola Vila, "Letter dated 12 August 1981 from the representative of Cuba to the Committee on Disarmament addressed to the chairman of the Committee on Disarmament transmitting part of the statement made on 26 July 1981 by Dr. Fidel Castro Ruz, Chairman of the Councils of State and of Ministers of the Republic of Cuba," CD 211, 13 August 1981; Formal Consultative Meeting of States Parties to the Convention, BWC/CONS/1, 27 August 1997; Working Paper submitted by Cuba to the Ad Hoc Group of States Parties to the Convention on the Prohibition of the Development, Production and Stockpiling of Bacteriological (Biological) and Toxin Weapons and on Their Destruction, 21 June 2000, BWC/AD HOC GROUP/WP.417 (hereafter Working Paper 417).

59. Animal diseases: Newcastle disease of fowl (1962); African swine fever (1971 and 1980); bovine nodular pseudodermatosis (1981); bovine ulcerative mammitis (1989); rabbit hemorrhagic disease (1993); varroasis of bees (1996); and ulcerative disease of fish (1996). Plant diseases: sugarcane rust (1978—the report says 1973, but this is clearly a typographic error); tobacco blue mold (1979); plantain black sigatoka (1990); citrus black louse (1992); coffee borer worm (1995); *Thrips palmi,* multiple hosts (1996); and rice mite (1997). Human diseases: dengue fever (1981); enteroviral conjunctivitis (1981); and bacterial dysentery (1981).

60. Mark Wheelis, "Investigation of Suspicious Outbreaks of Disease," in *Biological Warfare: Modern Offense and Defense,* ed. Raymond A. Zilinskas (Boulder: Lynne Rienner, 2000), pp. 105–118.

61. Raymond A. Zilinskas, "Cuban Allegations of Biological Warfare by the United States: Assessing the Evidence," *Critical Reviews in Microbiology,* 25, no. 3 (1999), 173–227.

62. *Foreign Relations of the United States, 1961–1963,* vol. 10: *Cuba, 1961–1962* (Washington, D.C.: USGPO, 1997) (hereafter *FRUS*), doc. 223, pp. 554–560.

63. *FRUS,* doc. 270, pp. 666–672. Membership of the SGA consisted of Special Assistant to the President for National Security Affairs McGeorge Bundy, Deputy Under Secretary of State for Political Affairs Alexis Johnson, Deputy Secretary of Defense Roswell Gilpatrick, Director of Central Intelligence John McCone, Chairman of the JCS General Lyman Lemnitzer, President Kennedy's personal military representative General Maxwell Taylor, and Attorney General Robert Kennedy. General Taylor was appointed chair.

64. William Harvey (CIA), Robert Hurwitch (State), Brig. Gen. Benjamin Harris (DOD), and Don Wilson (USIA).

65. W. K. Harvey, Memorandum to Lansdale, 24 July 1962, JFK Assassination Record No. 157-10007-10286, NARA.

66. Piero Gleijes, *Conflicting Missions: Havana, Washington, and Africa, 1959–1976* (Chapel Hill: University of North Carolina Press, 2002), p. 223.

67. Church Committee.

68. B. J. Brungs, "Status of Biological Warfare in International Law," *Military Law Review,* 24 (1964), 47–95; Moon, "US Biological Warfare Planning and Preparedness."

69. Moon, "US Biological Warfare Planning and Preparedness."

70. Working Paper 417.

71. Zilinskas, "Cuban Allegations of Biological Warfare." At the time Zilinskas was writing, the available source for this allegation did not specifically identify the vaccine.

72. J. E. Lancaster, "Vaccination in the Control of Newcastle Disease," *Bulletin of the Office International des Epizooties,* 62 (1964), 869–876.

73. FAO, *Animal Health Yearbooks,* 1957–1964 (Paris). The reduction in 1964 presumably reflected two years of culling of diseased and exposed birds, a practice that was first introduced as a supplement to vaccination in 1962.

74. ND can be transmitted by aerosol over considerable distances, in infected material such as eggs or birds, by wild birds, on fomites, and on veterinary personnel and vaccine workers. The virus is stable for weeks to years on fomites and in infected tissue, even at room temperature. See John E. Lancaster, *Newcastle Disease: A Review of Some of the Literature Published between 1926 and 1964,* Canada Department of Agriculture, Health of Animals Branch, Monograph No. 3 (Ottawa, 1966).

75. Drew Featherston and John Cummings, "Canadian Says US Paid Him $5,000 to Infect Cuban Poultry," *Washington Post,* 21 March 1977, p. A18.

76. W. H. Craig, untitled draft memorandum, 24 January 1962, JFK Assassination Record No. 178-10002-10414, NARA.

77. W. Craig, Memorandum for the Secretary of Defense, 8 February 1962, JFK Assassination Record No. 178-10002-10426, NARA.

78. Several other crops, such as coffee and tobacco, were identified as export items whose disruption might cause economic harm, but none was specifically mentioned as a sabotage target. No mention of animal targets is found in any document.

79. CIA, "Operation Mongoose, Main Points to Consider," 26 October 1962, JFK Assassination Record No. 178-10003-10010, NARA.

80. W. K. Harvey, Memorandum to Lansdale, 11 October 1962, JFK Assassination Record No. 178-10003-10013, NARA; M. S. Carter, Memorandum for Special Group (Augmented), 16 October 1962, JFK Assassination Record No. 178-10003-10056, NARA.

81. The DIA has yet to respond to a similar request.

82. CIA telegram 005705, 22 March 1977, JFK Assassination Record No. 104-10103-10232, NARA.

83. Stansfield Turner to Daniel Inouye, 13 June 1977, JFK Assassination Records, doc. ID no. 1993.07.22.11:38:57:460600.

84. E.g., *FRUS*, doc. 375, pp. 937–938.

85. CIA Information Report OO-K3-3,231,815, 30 July 1962, JFK Assassination Records, doc. ID no. 1993.07.22.13:44:27:840600, NARA.

86. Memo of Telephone Call, 15 June 1977, JFK Assassination Record No. 104-10103-10213, NARA.

87. Working Paper 417.

88. National Institute of Veterinary Medicine, "Preliminary Report on the African Swine Fever Epizootic in Cuba," *OIE Bulletin*, 7/8 (1971), 415–437.

89. Intelligence sources in Cuba reported seeing Spanish sausage for sale in markets; CIA Directorate of Science and Technology, Weekly Surveyor, OSI-WS-28/71, 12 July 1971, p. 3, MORI DocID 27577, NARA.

90. Drew Featherston and John Cummings, "Cuban Outbreak of Swine Fever Linked to CIA," *Newsday*, 9 January 1977, 5.

91. Zilinskas, "Cuban Allegations of Biological Warfare"; Milton Leitenberg, "Biological Weapons, International Sanctions and Proliferation," *Asian Perspective*, 21, no. 3 (1997), 7–39.

92. "Disease Epidemic in Havana Province Reported," *Granma*, 25 June 1971, CIA translation, CIA, MORI DocID 27557, NARA.

93. Fidel Castro, *Granma Weekly Review*, 16 March 1980.

94. Idem, ibid., 9 August 1981.

95. Idem, *68th Interparliamentary Conference, Havana, 15 September 1981* (Havana: Editora Politica, 1981).

96. Many of the original cables can be found in boxes 1 and 5, CBW Collection, National Security Archive, George Washington University (hereafter NSA). Also see Jane Hamilton-Merritt, *Tragic Mountains: The Hmong, the Americans, and the Secret Wars for Laos, 1942–1992* (Bloomington: Indiana University Press, 1993).

97. Alexander M. Haig Jr., "Chemical Warfare in Southeast Asia and Afghanistan," US Department of State Special Report No. 98, 22 March 1982; George P. Schultz, "Chemical Warfare in Southeast Asia and Afghanistan: An Update," US Department of State Special Report No. 104, November 1982. These reports to Congress and the UN were based on a three-part Special National Intelligence Estimate 11/50/37-82, "Use of Toxins and Other Lethal Chemicals in Southeast Asia and Afghanistan," vol. 1: "Key Judgments" (27 January 1982); vol. 2: "Supporting Analysis" (26 February 1982); vol. 3: "Memorandum to Holders" (2 March 1983), folders 40 and 41, box 20, RG 263 (Records of the CIA), NARA.

98. UN General Assembly Resolution A/35/144C (December 1980); Final Report of the UN Group of Experts, UN Doc. A/37/259 (1 December 1982).

99. The rapid-acting incapacitant, if such existed, would most likely be a biological agent also (see Chapter 12); however, there is insufficient evidence for us to say anything meaningful about it.

100. Committee on Protection against Mycotoxins, *Protection against Trichothecene Mycotoxins* (Washington, D.C.: National Academy Press, 1983); Kenneth S. K. Chinn, "Trichothecenes: Are They Potential Threat Agents?" US Army Dugway Proving Ground, June 1983, Record No. 56236, box 5, CBW Collection, NSA.

101. Haig, "Chemical Warfare in Southeast Asia and Afghanistan"; Schultz, "Chemical Warfare Update."

102. The details of the critique are elaborated in Julian Robinson, Jeanne Guillemin, and Matthew Meselson, "Yellow Rain in Southeast Asia: The Story Collapses," in *Preventing a Biological Arms Race*, ed. Susan Wright (Cambridge, Mass.: MIT Press, 1990), pp. 220–238; Thomas D. Seeley et al., "Yellow Rain," *Scientific American*, 253, no. 3 (September 1985), 128–137; Lois R. Ember, "Yellow Rain," *Chemical and Engineering News*, 9 January 1984, 8–34; Joan W. Nowicke and Matthew Meselson, "Yellow Rain—a Palynological Analysis," *Nature*, 309 (1984), 205–206; Elisa Harris, "Sverdlovsk and Yellow Rain: Two Cases of Soviet Noncompliance?" *International Security*, 11 (spring 1987), 41–95; Thomas Whiteside, "Annals of the Cold War: The Yellow Rain Complex," *New Yorker*, pt. 1 (11 February 1991), 38–67; pt. 2 (18 February 1991), 44–68; Jonathan B. Tucker, "The 'Yellow Rain' Controversy: Lessons for Arms Control Compliance," *Nonproliferation Review*, spring 2001, 25–42.

103. US Embassy (Bangkok) to DIA, Telegram 11615, "CBW sample TH-840209-1DL through TH-840209-11DL," 6 March 1984, NSA.

104. Haig, "Chemical Warfare in Southeast Asia and Afghanistan."

105. US Embassy (Bangkok) to DIA, Telegram TH 0516, 27 January 1981, NSA.

106. T2, diacetoxyscirpenol, nivalenol, and deoxynivalenol.

107. A fifth report of trichothecene detection came from a Rutgers University laboratory that analyzed a sample obtained from Laos by ABC News.

108. US Army, written response to questions from the Senate Committee on Foreign Relations in connection with Senate Hearing 102-719, Chemical Weapons Ban Negotiations, 1 May 1992, unpublished. Thanks to David Hafemeister, former staff member of the Senate Foreign Relations Committee, for providing me with a copy of this testimony.

109. Nowicke and Meselson, "Yellow Rain—a Palynological Analysis"; Seeley et al., "Yellow Rain."

110. S. P. Swanson and R. A. Corley, "The Distribution, Metabolism, and Excretion of Trichothecene Mycotoxins," in *Trichothecene Mycotoxicosis: Pathophysiological Effects*, ed. V. R. Beasley, vol. 1 (Boca Raton: CRC Press, 1989), pp. 37–61; Special National Intelligence Estimate 11/50/37-82, "Memorandum to Holders." Aflatoxins were considered not to be a component of yellow rain, although this possibility was considered.

111. J. J. Norman and J. G. Purdon, "Final Summary Report on the Investigation of 'Yellow Rain' Samples from Southeast Asia," Defence Research Establishment Ottawa Report No. 912, Ottawa, 1986; R. M. Black, R. J. Clarks, and

R. W. Read, "Detection of Trace Levels of Trichothecene Mycotoxins in Environmental Residues and Foodstuffs Using Gas Chromatography with Mass Spectrometric or Electron-Capture Detection," *Journal of Chromatography,* 388 (1987), 365–378.

112. Stuart Johnson et al., "Enteritis necroticans among Khmer Children at an Evacuation Site in Thailand," *Lancet,* no. 2 (1987), 496–500.

113. Rebecca Katz and Burton Singer, "Revisiting the Investigation into Chemical or Toxin Use in Southeast Asia and Afghanistan, 1976–1985," forthcoming.

114. DIA, Foreign Technology Weapons and Systems: Trends and Developments, pp. 9–14, 3 March 1980, NSA; CIA, National Foreign Assessment Center, Soviet Biological Warfare Agent: Probable Cause of the Anthrax Epidemic in Sverdlovsk, 1981, NSA.

115. S. Bezdenezhnykh and V. N. Nikiforov, "An Epidemiological Analysis of Incidences of Anthrax in Sverdlovsk," *Journal of Microbiology, Immunology and Epidemiology,* no. 5 (May 1980), 111–113. See also Matthew S. Meselson, "The Biological Weapons Convention and the Sverdlovsk Anthrax Outbreak of 1979," *Federation of American Scientists Public Interest Report,* 41, no. 7 (1988), 1–6.

116. Nicholas Sims, *The Diplomacy of Biological Disarmament: Vicissitudes of a Treaty in Force, 1975–85* (New York: St. Martin's, 1988).

117. Jeanne Guillemin, *Anthrax: The Investigation of a Deadly Outbreak* (Berkeley: University of California Press, 1999).

118. Faina A. Abramova, Lev M. Grinberg, Olga V. Yampolskaya, and David Walker, "Pathology of Inhalational Anthrax in 42 Cases from the Sverdlovsk Outbreak of 1979," *Proceedings of the National Academy of Sciences USA,* 90 (1993), 2291–94.

119. Matthew Meselson et al., "The Sverdlovsk Anthrax Outbreak of 1979," *Science,* 266 (1994), 1202–08. The wind was also from the NW on Sunday April 1, but many fewer victims would have been in the area then.

120. Ken Alibek with Stephen Handelman, *Biohazard: The Chilling True Story of the Largest Covert Biological Weapons Program in the World—Told from the Inside by the Man Who Ran It* (New York: Random House, 1999).

121. Matthew Meselson, "Note regarding Source Strength," *ASA Newsletter,* no. 48 (8 June 1995), 1, 20–21.

122. Boris Yeltsin, interview, *Komsomolskaya Pravda* 27 May 1992; David Hoffman, "A Puzzle of Epidemic Proportions: Source of 1979 Anthrax Outbreak in Russia Still Clouded," *Washington Post,* 16 December 1998, p. A1. Continuing denials of responsibility for the outbreak are commonly heard from Russian officials speaking informally to Western colleagues.

123. Mark L. Wheelis, "Strengthening the Biological Weapons Convention through Global Epidemiological Surveillance," *Politics and the Life Sciences,* 11 (1992), 179–189.

124. UN General Assembly Resolutions 42/37 C (1987) and 43/74 A (1988).

125. Peter Barss, "Epidemic Field Investigation as Applied to Allegations of Chemical, Biological, or Toxin Warfare," *Politics and the Life Sciences,* 11, no. 1 (1992), 5–22.

126. Article V has been invoked only once, by Cuba, to consider its allegation that the US introduced *Thrips palmi* into its territory. See Nicholas A. Sims, *The Evolution of Biological Disarmament* (Oxford: Oxford University Press, 2001), pp. 37–50.

127. Furmanski, work in preparation.

14. Terrorist Use of Biological Weapons

Acknowledgments: We thank Milton Leitenberg and Julian Robinson for helpful comments.

1. See Seth Carus, "Bioterrorism and Biocrimes: The Illicit Use of Biological Agents in the 20th Century," Working Paper from the Center for Counterproliferation Research, National Defense University, March 1999; available from the website of the Center for Nonproliferation Studies of the Monterey Institute of International Studies.

2. The closest to primary sources are "Germ Warfare against Indians Is Charged in Brazil," (New York) *Medical Tribune and Medical News,* 10, no. 98 (8 December 1969); and Norman Lewis, "From Fire and Sword to Arsenic and Bullets: Civilisation Has Sent Six Million Indians to Extinction," (London) *Sunday Times Magazine,* 23 February 1969, pp. 36–59. Useful secondary sources include John Hemming, *Die If You Must: Brazilian Indians in the Twentieth Century* (New York: Macmillan, 2003), pp. 226–234; S. H. Davis, *Victims of the Miracle: Development and the Indians of Brazil* (New York: Cambridge University Press, 1977), p. 138; Seth Garfield, "Brazil and the Charges of Genocide against Its Indigenous Population," *Columbia International Affairs Online,* Columbia University Press, www.ciaonet.org/wps/gas01, accessed in January 1995; Susanna Hecht and Alexander Cockburn, *The Fate of the Forest: Developers, Destroyers, and Defenders of the Amazon* (New York: Verso, 1989).

3. Lewis, "From Fire and Sword," p. 43.

4. Mark Wheelis, "Biological Warfare before 1914," in *Biological and Toxin Weapons Research, Development and Use from the Middle Ages to 1945: A Critical Comparative Analysis,* ed. Erhard Geissler and J. E. van Courtland Moon, SIPRI Chemical and Biological Warfare Studies No. 18 (Oxford: Oxford University Press, 1999), pp. 8–34.

5. Unless otherwise noted, details of the outbreak are from Thomas J. Török et al., "A Large Community Outbreak of Salmonellosis Caused by Intentional Contamination of Restaurant Salad Bars," *JAMA,* 278 (1997), 389–395; and CDC, "Restaurant-Associated *Salmonella typhimurium* Gastroenteritis, The

Dalles, Oregon, 1984," limited-distribution report, EPI-84-93-2, 1 November 1988.

6. B. Aserkoff, S. A. Schroeder, and P. S. Brechman, "Salmonellosis in the United States: A Five-Year Review," *American Journal of Epidemiology*, 92 (1970), 13.

7. Adjacent Wasco and Sherman Counties share a single hospital and public health department.

8. The CDC considered 10 restaurants, with at least three associated cases each, to be definite sources, and an additional 12, with one or two associated cases each, to be possible sources.

9. Laurence R. Foster, "Preliminary Report—Salmonellosis Outbreak, The Dalles, Oregon, September 1984," Memo for the record, 7 November 1984.

10. The account of the Rajneesh is drawn from Lewis F. Carter, *Charisma and Control in Rajneeshpuram: The Role of Shared Values in the Creation of a Community* (Cambridge: Cambridge University Press, 1990).

11. Sheela Patel Silverman, whose Rajneesh name was Ma Anand Sheela.

12. Diane Onang, aka Ma Anand Puja.

13. W. Seth Carus, "The Rajneeshees" (1984), in *Toxic Terror: Assessing Terrorist Use of Chemical and Biological Weapons*, ed. Jonathan B. Tucker (Cambridge, Mass.: MIT Press, 2000), pp. 115–137.

14. Carter, *Charisma and Control in Rajneeshpuram*, p. 237.

15. See, e.g., US Code, title 18, pt. I, chap. 10 (Biological Weapons).

16. For profiles of Asahara and various crimes committed by the Aum, see *Global Proliferation of Weapons of Mass Destruction (Part One): Hearings before the Permanent Subcommittee on Investigations of the Committee on Governmental Affairs, United States Senate, One Hundred Fourth Congress, First Session, October 31 and November 1, 1995* (Washington, D.C.: USGPO, 1996); David Kaplan, "Aum Shinrikyo," in Tucker, *Toxic Terror*, pp. 207–226. The following works by Japanese scholars are also informative: Hiromi Shimada, *Aum* (Tokyo: Tarnsview, 2001) (in Japanese); Toshiki Takeoka, *Aum Shinrikyo Jiken Kanzen Kaidoku* (Perfect decryption of the Aum incident) (Tokyo: Ben-Sei Shuppan, 1999).

17. Speech by Shoko Asahara, 7 April 1989. Here, "transform his soul" is a euphemism for murder.

18. Tokyo District Court, Judgment on Murder and Other Cases of Tomomitsu Niimi, 26 June 2002, *Hanrei Jiho*, no. 1795 (11 November 2001), pp. 45–96, esp. p. 61 (in Japanese); Masaaki Sugishima, "Biocrimes in Japan," in *A Comprehensive Study on Bioterrorism*, ed. Sugishima (Legal Research Institute of Asahi University, 2003), pp. 86–121.

19. "Police Newly Identified Aum's Microbiological Laboratory," *Yomiuri Shinbun*, 5 May 1995 (in Japanese); "Aum's Two Scientists Were Addicted to Scientific Experiments," *Mainichi Shinbun*, 30 May 1995 (in Japanese).

20. There are about a dozen types of *C. botulinum*, only a few of which produce a toxin lethal for humans.

21. "Wearing Protective Suits to Test Germs: Ex-Senior Follower Talked about His Secret Mission," *Yomiuri Shinbun*, 15 April 1995 (in Japanese).

22. Robert C. Mikesh, *Japan's World War II Balloon Bomb Attacks on North America* (Washington, D.C.: Smithsonian Institute Press, 1973).

23. Testimony of Endo, 10 January 2002, in Ken-ichi Furihata, *Aum Hotei* (Court trials of the Aum), vol. 12 (Tokyo: Asahi Shinbun, 2003), p. 292.

24. Hiroshi Takahashi et al., "*Bacillus anthracis* Incident, Kameido, Tokyo, 1993," *Emerging Infectious Diseases*, 10, no. 1 (January 2004), 117–120; http://www.cdc.gov/ncidod/EID/vol10no1/03-0238.htm; Paul Keim et al., "Molecular Investigation of the Aum Shinrikyo Anthrax Release in Kameido, Japan," *Journal of Clinical Microbiology*, 39 (December 2001), 4566–67; Iku Aso, *Gokuhi Sosa* (Secret investigation) (Tokyo: Bungei Shunjyu, 1997), pp. 11–15.

25. Prosecutors' Opening Statement at Asahara's Court Trial, 25 April 1996, in Kyodo Tsushin Shakaibu (Social Affairs Bureau of the Kyodo News Agency), *Sabakareru Kyoso* (Guru under the court trial) (Tokyo: Kyodo Tsushin, 1997), pp. 254–255.

26. Takahashi et al., "*Bacillus anthracis* Incident, Kameido, Tokyo, 1993."

27. Raymond Zilinskas, "Biotechnology and Its Possible Risks Including Criminal Applications," in Sugishima, *Comprehensive Study on Bioterrorism*, p. 23.

28. Kaplan, "Aum Shinrikyo."

29. Gary Matsumoto, "Anthrax Powder: State of the Art?" *Science*, 302 (2003), 1492–97.

30. Although the US ended its offensive BW program in 1969, it has continued to make anthrax spore preparations for use in its biodefense program.

31. Milton Leitenberg, *The Problem of Biological Weapons* (Stockholm: Swedish National Defense College, 2004), p. 146; Leonard A. Cole, *The Anthrax Letters: A Medical Detective Story* (Washington, D.C.: Joseph Henry, 2003), pp. 160–184.

32. Allan Lengel, "Suspicious Powders, Packages Keep FBI Unit on Edge," *Washington Post*, 13 April 2004, p. A1.

33. Cole, *The Anthrax Letters*.

34. Tucker, *Toxic Terror*; Carus, "Bioterrorism and Biocrimes."

35. Jonathan B. Tucker and Jason Pate, "The Minnesota Patriots Council" (1991), in Tucker, *Toxic Terror*, pp. 159–184.

36. Dan Eggen, "FBI Releases Details of Letter with Ricin Sent to White House," *Washington Post*, 24 February 2004, p. A3.

37. Leitenberg, *The Problem of Biological Weapons*, pp. 119–136.

38. Joby Warrick, "An al Qaeda 'Chemist' and the Quest for Ricin," *Washington Post*, 5 May 2004, p. A1.

15. The Politics of Biological Disarmament

1. SIPRI, *The Problem of Chemical and Biological Warfare*, vol. 4: *CB Disarmament Negotiations, 1920–1970* (Stockholm: Almqvist & Wiksell, 1971).

2. Forrest Russell Frank, "U.S. Arms Control Policymaking: The 1972 Biological Weapons Convention Case" (Ph.D. diss., Stanford University, 1974).

3. Elisa D. Harris, "The Biological and Toxin Weapons Convention," in *Superpower Arms Control*, ed. Albert Carnesale and Richard N. Haass (Cambridge, Mass.: Ballinger, 1987), pp. 191–219.

4. See Brian Balmer, *Britain and Biological Warfare: Expert Advice and Science Policy, 1930–65* (Basingstoke: Palgrave, 2001); Jonathan Tucker, "A Farewell to Germs: The U.S. Renunciation of Biological and Toxin Weapons, 1969–1970," *International Security*, 27 (summer 2002), 107–148; and Susan Wright, "Geopolitical Origins," in *Biological Warfare and Disarmament: New Problems/New Perspectives*, ed. Wright (Lanham, Md.: Rowman and Littlefield, 2002), pp. 313–342.

5. SIPRI, *CB Disarmament Negotiations*.

6. In the UN context, "'General' means that every member State would be committed to disarmament. 'Complete Disarmament' means that weapons and forces would be scaled down to a minimum, defined as the point at which States retain just enough military capability to maintain order, but sufficient to assist the UN-sanctioned international operations"; "Research Note 5 1997–98: United Nations—General and Complete Disarmament," Parliamentary Library, Parliament of Australia, Canberra, accessed at www.aph.gov.au on 29 June 2004.

7. See Hedley Bull, *The Control of the Arms Race: Disarmament and Arms Control in the Missile Age*, 2d ed. (New York: Praeger, 1965), especially "Disarmament and Chemical, Biological, and Radiological Warfare," pp. 123–136; and Thomas C. Schelling and Morton H. Halperin, *Strategy and Arms Control* (New York: Twentieth Century Fund, 1961).

8. Wright questions whether Western insistence on on-site verification was sincere; "Geopolitical Origins," pp. 324–336.

9. Western European Union, *Text of the Brussels Treaty and of the Protocols Modifying and Completing That Treaty, Signed in Paris, 23rd October, 1954*, Article I, Protocol No. III.

10. Confidential communication, E. J. W. Barnes to Lord Hood, "Western European Union: Control of Biological Weapons," 23 November 1967, FCO 41/275, PRO.

11. P. Wilkinson to R. McC. Andrew, Esq., 2 April 1963, FO 371/171134, PRO.

12. J. H. Lambert to R. McC. Andrew, Esq., 18 March 1963, ibid.

13. Wilkinson to Andrew, 31 January 1964, FO 371/176376, PRO.

14. R. S. Faber to H. B. Shepherd, Esq., 22 September 1964, ibid.

15. Minutes, S. J. Barrett, "Reactions to Idea of Western Initiative in C.B. War-fare," 31 August 1964, IAD 1039/5/G, ibid.

16. Jack Stephens to H. B. Shepherd, Esq., 13 August 1964, ibid.

17. Minutes, S. J. Barrett, 31 August 1964.

18. Frank, "U.S. Arms Control Policymaking," pp. 43–73.

19. See "Statement by ACDA Director Foster to the First Committee of the General Assembly, November 14, 1966," in ACDA, *Documents on Disarmament, 1966* (Washington, D.C.: USGPO, 1967), p. 740.

20. The Conference on Disarmament is the principal multilateral disarmament forum in the world. The CD was established in 1978; its predecessor organizations were the Conference of the Committee on Disarmament (CCD), 1969–1978; the Eighteen Nation Committee on Disarmament (ENDC), 1962–1969; and the Ten Nation Disarmament Committee, 1960–1962; John Allphin Moore Jr. and Jerry Pubantz, eds., *Encyclopedia of the United Nations* (New York: Facts of File, 2002), p. 66. The Soviet Union and its allies preferred to introduce resolutions to the First Committee. Because the General Assembly included all UN members, the Eastern-bloc states were more likely to achieve success in the First Committee than in the ENDC, whose membership was more carefully constructed to balance East and West.

21. "Hungarian Draft Resolution Submitted to the First Committee of the General Assembly: Use of Chemical and Bacteriological Weapons," 7 November 1966, A/C.1/L.374, in ACDA, *Documents on Disarmament, 1966*, pp. 694–695.

22. The use of chemical defoliants in Vietnam could have been considered a use of chemicals to destroy the means of existence for human beings.

23. The proposed amendments, for example, replaced several paragraphs condemning CBW use and declaring that such use was an international crime by a single paragraph deploring such use; "Hungarian Draft," pp. 694–697 n. 3.

24. Foster statement, in ACDA, *Documents on Disarmament, 1966*, p. 740.

25. "Revised Western Amendments to the Revised Hungarian Draft Resolution on the Use of Chemical and Bacteriological Weapons," 23 November 1966, in ibid., pp. 762–763.

26. E. J. Richardson to A. C. Stuart, Esq., 3 November 1965, FO 371/181378, PRO. The British apparently did not consider the Vietnam conflict a "war."

27. Ibid.

28. "Maltese Draft Resolution Introduced in the First Committee of the General Assembly: Use of Chemical, Biological, and Radiological Weapons," 7 December 1967, A/C.1/L.411, in ACDA, *Documents on Disarmament, 1967* (Washington, D.C.: USGPO, 1968), pp. 625–626.

29. "Statement by the Maltese Representative (Pardo) to the First Committee of the General Assembly: Chemical and Bacteriological Weapons," 12 December 1967, A/C.1/PV.1547, 2-33, in ibid., pp. 634–647.

30. Ibid., p. 646.

31. "Statement by the Hungarian Representative (Csatorday) to the First Committee of the General Assembly: Chemical and Biological Weapons," 12 December 1967, A/C.1/PV.1547, in ibid., pp. 657–662.

32. "Statement by the Soviet Representative (Shevchenko) to the First Committee of the General Assembly: Chemical and Bacteriological Weapons," 13 December 1967, A/C.1/PV.1548, in ibid., p. 667.

33. "Statement by the Netherlands Representative (Eschauzier) to the First Committee of the General Assembly: Chemical and Biological Weapons," 13 December 1967, A/C.1/PV.1549, in ibid., p. 668.

34. Department of State Telegram, from US Mission United Nations, New York, to Secretary of State, Washington, D.C., 16 December 1967, General Records of the Department of State, Central Foreign Policy Files, Political and Defense, POL 27-10, NARA.

35. United Nations Report No. E 69 I 24, *Chemical and Bacteriological (Biological) Weapons and the Effects of Their Possible Use* (New York: Ballantine Books, 1970).

36. Department of State Telegram, from US Mission Geneva to Secretary of State, Washington, D.C., 31 October 1969, pp. 2–3, POL 27-10, NARA.

37. See Balmer, *Britain and Biological Warfare,* pp. 176–178.

38. Chemical and Biological Warfare Summary, 3 May 1962, WO 32/18931, PRO.

39. I. F. Porter, United Kingdom Delegation to the 18 Nation Disarmament Conference, to Mulley, 31 March 1969, FCO 73/114, PRO.

40. Talking Points on Biological Warfare for ENDC Informal Meeting on 14 May 1969, ibid.

41. R. D. Holmes AD/BD to DS22, Revised Draft FCO Note on Biological Warfare, 8 May 1969, DEFE 19/97, PRO.

42. Talking Points on Biological Warfare, FCO 73/114, PRO.

43. Ibid.

44. Ibid.

45. Comment from J. E. Quinlan, D of DP (D), to DS22 Mr. Heyhoe BW, Reference DefDP (D)36/69/38, 7 May 1969, DEFE 19/97, PRO.

46. D. C. R. Heyhoe, DS22, to D. L. Benest, Esq., Disarmament Department, FCO, 16 June 1969, FCO 66/136, PRO.

47. R. Holmes AB/BD to DS22, "Verification Aspects of BW," 8 May 1969, DEFE 19/97, PRO; repeated in D. C. R. Heyhoe to D. K. Timms, Disarmament Department, 12 May 1969, ibid.

48. I. F. Porter, United Kingdom Delegation to the 18 Nation Disarmament Conference, to R. C. Hope-Jones, Esq., Disarmament Department, 31 January 1969, DEFE A/97, PRO.

49. Biological Warfare Convention, R. C. Hope-Jones, Disarmament Depart-

ment, to Sir T. Brimelow and Mr. Williams, 9 June 1969, FCO 66/136, PRO.

50. W. N. Hillier-Fry, United Kingdom Delegation to the 18 Nation Disarmament Conference, to R. C. Hope-Jones, Esq., Disarmament Department, 16 June 1969, ibid.

51. Hope-Jones to Hillier-Fry, 18 June 1969, ibid.

52. United Kingdom Delegation to the 18 Nation Disarmament Conference to Sir Thomas Brimelow, Foreign and Commonwealth Office, 19 May 1969, ibid.

53. ACDA, Memorandum of Conversation, 21 June 1967, p. 1, Secret, POL 27-10, NARA.

54. CIA, Intelligence Report, "Disarmament: Chemical-Biological Warfare Controls and Prospects for Improvement," 18 August 1969, pp., 14, 19, declassified 3/19/02; NSC Files: Subject Files, Chemical, Biological Warfare (Toxins, etc.), box 310, Richard M. Nixon Presidential Materials Staff, NARA.

55. Report to the NSC, US Policy on Chemical and Biological Warfare and Agents, Submitted by the Interdepartmental Political-Military Group in response to NSSM 59, 10 November 1969. An earlier draft of the report, dated 15 October 1969, contains somewhat different language. NSC Files, box 310, NARA.

56. Confidential Saving Telegram, Foreign and Commonwealth Office to Abidjan, 29 September 1969, DEFE 24/551, PRO.

57. Department of State Telegram, from US Mission Geneva to Secretary of State, Washington, D.C., 23 September 1969, Confidential, p. 1, POL 27-10, NARA.

58. Tucker, "A Farewell to Germs."

59. The JCS argued that retention of BW would keep military options available and that ratification of the Geneva Protocol would set an undesirable precedent for nuclear weapons; ibid., pp. 126–127.

60. Ibid., pp. 127, 128.

61. Frank, "U.S. Arms Control Policymaking," pp. 182–186.

62. The original British draft did not include toxins in its ban, but was amended in 1970.

63. Confidential Saving Telegram, Foreign and Commonwealth Office to Abidjan, 29 September 1969, DEFE 24/551, PRO.

64. Ibid., p. 4.

65. Memorandum for the President from Henry A. Kissinger, Negotiation of a Convention Banning Biological Weapons, 23 April 1971, NSC Files, Chemical, Biological Warfare (Toxins, Etc.), vol. 4, pt. 1, box 312, NARA.

66. Department of State Telegram, ACDA/IR:RLMCCORMACK, April 1971, ibid.

67. Department of State Telegram, from Secretary of State to US Mission in

Geneva, Subj: CCD: Consultations with UK on BW Convention, 6 May 1971, ibid.

68. Memorandum for Kissinger from Guhin, Convention Banning Biological Weapons and Toxins, 17 September 1971, ibid.

69. BW Convention: Proposed Changes, 8 September 1971, ibid.

70. Nicholas A. Sims, *The Diplomacy of Biological Disarmament: Vicissitudes of a Treaty in Force, 1975–1985* (New York: St. Martin's, 1988), pp. 174–179.

71. Julian Perry Robinson, "East-West Fencing at Geneva," *Nature,* 284 (3 April 1980), 393.

72. See Marie Isabelle Chevrier and Iris Hunger, "Confidence Building Measures for the BTWC: Performance and Potential," *Nonproliferation Review,* fall–winter 2000.

73. United Nations, "Summary Report, Ad Hoc Group of Governmental Experts to Identify and Examine Potential Verification Measures from a Scientific and Technical Standpoint," Fourth Session, Geneva, BWC/CONF.III/VEREX/8, 13–24 September 1993, pp. 1–20.

74. See Malcolm R. Dando, *Preventing Biological Warfare: The Failure of American Leadership* (Basingstoke: Palgrave, 2002); and Marie Isabelle Chevrier, "Chemical and Biological Weapons," in *Arms Control: Cooperative Security in a Changing Environment,* ed. Jeffrey A. Larsen (Boulder: Lynne Rienner, 2002), pp. 152–155.

75. Marie Isabelle Chevrier, "Waiting for Godot or Saving the Show? The BWC Review Conference Reaches Modest Agreement," *Disarmament Diplomacy,* no. 68 (December 2002/January 2003).

76. Conversation with a member of US delegation to the Fifth Review Conference of the BWC, 14 November 2002.

16. Legal Constraints on Biological Weapons

1. *World Armaments and Disarmament: SIPRI Yearbook 1976* (Stockholm: Almqvist & Wiksell; Cambridge, Mass.: MIT Press, 1976), pp. 468, 474. The Irish declaration of 7 February 1972 was a Note received by the Depositary for the Geneva Protocol on 10 February 1972 and in slightly adapted form was also attached to the Irish signature to the BWC on 10 April 1972. It thereby links the two treaties and forms part of the legal documentation of both.

2. BWC/CONF.IV/9, 6 December 1996.

3. SIPRI, *The Problem of Chemical and Biological Warfare,* vol. 3: *CBW and the Law of War* (Stockholm: Almqvist & Wiksell, 1973), p. 150.

4. SIPRI, *The Problem of Chemical and Biological Warfare,* vol. 5: *The Prevention of CBW* (Stockholm: Almqvist & Wiksell, 1971), p. 214.

5. Ibid., pp. 200–215.

6. Adam Roberts and Richard Guelff, *Documents on the Laws of War,* 3d ed. (Oxford: Oxford University Press, 2000), pp. 160–167.

7. BWC/CONF.IV/9, 6 December 1996, p. 22.

8. Roberts and Guelff, *Documents on the Laws of War,* p. 157.

9. Adapted from Nicholas A. Sims, *The Evolution of Biological Disarmament,* SIPRI Chemical and Biological Warfare Studies No. 19 (Oxford: Oxford University Press, 2001), p. 6.

10. Some, mainly US, lists add Taiwan as an original State Party because the US, as a depositary for the BWC, allowed the Taipei authorities to sign in 1972 and ratify in 1973 as "the Republic of China." These actions were denounced by the People's Republic of China as illegal, null, and void when it acceded to the BWC in 1984. The rosters kept by the British and Soviet/Russian depositaries have never included Taiwan or "the Republic of China"; neither have the lists used by the UN. The lists that do include Taiwan count 47 original parties and therefore show a number higher by one for each year (for example, 155 for 2005).

11. "The term 'bacteriological' as used in the Protocol is not sufficiently comprehensive to include the whole range of microbiological agents that might be used in hostilities"; UK, Working Paper on Microbiological Warfare, ENDC/231, 6 August 1968.

12. R. R. Baxter and Thomas Buergenthal, "Legal Aspects of the Geneva Protocol of 1925," in *The Control of Chemical and Biological Weapons* (New York: Carnegie Endowment for International Peace, 1971), p. 16, quoting from the UN General Assembly resolution of 16 December 1969, GA Res.2603A (XXIV).

13. George Bunn, "Banning Poison Gas and Germ Warfare: Should the US Agree?" *Wisconsin Law Review,* 375, no. 2 (1969), 413 n. 180; reproduced in *The Geneva Protocol of 1925: Hearings before the Committee on Foreign Relations, US Senate, 92nd Congress, March 1971* (Washington, D.C.: USGPO, 1972), pp. 72–113.

14. SIPRI, *CBW and the Law of War,* p. 47 n. 22.

15. Ibid., p. 43. Emphasis in original.

16. BWC/CONF.I/10, 21 March 1980.

17. BWC/CONF.II/13/II, 26 September 1986.

18. BWC/CONF.III/23, Geneva, 1992. The declaration had been adopted on 27 September 1991.

19. Ibid.

20. BWC/CONF.IV/9, 6 December 1996.

21. BWC/CONF.III/23, 1992.

22. UK, Draft Microbiological Warfare Convention, ENDC/255, 10 July 1969.

23. Sims, *The Evolution of Biological Disarmament,* p. 181.

24. Judith Miller, Stephen Engelberg, and William Broad, *Germs: The Ultimate*

Weapon (London: Simon and Schuster UK, 2001), p. 288; and, on the legal issues surrounding the CIA's Clear Vision project of 1999–2001, pp. 293–296.

25. Malcolm R. Dando and Simon M. Whitby, "Article I—Scope," in *Strengthening the BWC: Key Points for the Fifth Review Conference*, ed. Graham S. Pearson, Malcolm R. Dando, and Nicholas A. Sims (Bradford, UK: Department of Peace Studies, University of Bradford, 2001), p. 23, para. 34.

26. BWC Formal Consultative Meeting of States Parties, BWC/CONS/1, 27 August 1997.

27. Report of the Chairman of the BWC Formal Consultative Meeting, contained in a letter to States Parties from the UK Permanent Representation to the Conference on Disarmament, 15 December 1997.

28. ENDC/PV.418, 10 July 1969.

29. ENDC/255, 10 July 1969.

30. BWC/CONF.III/23, 1992.

31. Sims, *The Evolution of Biological Disarmament*, p. 46.

32. Ibid., pp. 36–50.

17. Analysis and Implications

1. Fourth Review Conference of the Parties to the Convention on the Prohibition of the Development, Production and Stockpiling of Bacteriological (Biological) and Toxin Weapons and on Their Destruction, BWC.CONF.IV/9 Part II, p. 15; Convention on the Prohibition of the Development, Production, Stockpiling and Use of Chemical Weapons and on Their Destruction, Article II, available at http://www.opcw.org/html/db/cwc/eng/cwc_frameset.html..

2. "The Changing Status of Chemical and Biological Warfare: Recent Technical, Military and Political Developments," in *SIPRI Yearbook 1982* (London: Taylor and Francis, 1982), pp.317–336.

3. Editorial, "Preventing the Hostile Use of Biotechnology: The Way Forward Now," *CBW Conventions Bulletin*, 57 (2002), 1–2.

4. International Committee of the Red Cross, *Biotechnology, Weapons and Humanity: Summary Report of an Informal Meeting of Government and Independent Experts, Montreux, Switzerland, 23–24 September* (Geneva, 2002).

5. J. B. Petro, T. R. Plasse, and J. A. McNulty, "Biotechnology: Impact on Biological Warfare and Biodefense," *Biosecurity and Bioterrorism: Biodefense Strategy, Practice and Science*, 1 (2003), 161–168; Mark Wheelis, "Will the New Biology Lead to New Weapons?" *Arms Control Today*, 34 (July/August 2004), http://www.armscontrol.org/act/2004_07–08/Wheelis.asp.

6. G. S. Pearson, "Prospects for Chemical and Biological Arms Control," *Washington Quarterly*, 16 (1993), 145–162.

7. M. R. Dando, "The Danger to the Chemical Weapons Convention from In-

capacitating Chemicals," First CWC Review Conference Paper No. 4, Bradford University, Department of Peace Studies, March 2003; available at <www.brad.ac.uk/acad/scwc>.

8. Idem, *Preventing Biological Warfare: The Failure of American Leadership* (Basingstoke: Palgrave, 2002).

9. Jez Littlewood, "Preparing for a Successful Outcome to the BWC Sixth Review Conference in 2006," Briefing Paper No. 11 (second series), Bradford University, Department of Peace Studies, December 2003; available at <www.brad.ac.uk/acad/sbtwc>.

10. Ad Hoc Group of Governmental Experts, Final Report, BWC/CONF.III/VEREX/8, Geneva, 1993.

11. Milton Leitenberg, "Distinguishing Offensive from Defensive Biological Weapons Research," *Critical Reviews in Microbiology*, 29 (2003), 223–257; Mark Wheelis and Malcolm Dando, "Back to Bioweapons?" *Bulletin of the Atomic Scientists*, 59 (January/February 2003), 40–46.

12. World Health Organization, *Public Health Response to Biological and Chemical Weapons: WHO Guidance*, 2d ed. (Geneva, 2004); ICRC, "Initiative on Biotechnology, Weapons and Humanity," available at http://www.icrc.org/Web/eng/siteeng0.nsf/html/bwh/OpenDocument.

Contributors

DONALD HOWARD AVERY is Professor of History at the University of Western Ontario, Canada.

BRIAN BALMER is Senior Lecturer in Science Policy Studies, University College London, UK.

MARIE ISABELLE CHEVRIER is Associate Professor of Public Policy and Political Economy, University of Texas, Dallas.

MALCOLM DANDO is Professor of International Security, University of Bradford, UK.

MARTIN FURMANSKI is a Fellow of the College of American Pathologists and is board certified in Anatomic and Clinical Pathology in the US.

CHANDRÉ GOULD is the Global Network Coordinator for the BioWeapons Prevention Project and lives in South Africa.

JOHN HART is a researcher at the Stockholm International Peace Research Institute (SIPRI), Sweden.

ALASTAIR HAY is Professor of Environmental Toxicology in the School of Medicine, University of Leeds, UK.

OLIVIER LEPICK is a senior researcher with the Fondation pour la Recherche Stratégique, France.

PIERS MILLET is a Research Fellow in the Department of Peace Studies, University of Bradford, UK.

JOHN ELLIS VAN COURTLAND MOON is Professor of History, Emeritus, Fitchburg State College, Massachusetts.

KATHRYN NIXDORFF is Professor of Microbiology at the Technical University of Darmstadt, Germany.

GRAHAM S. PEARSON is Visiting Professor of International Security, University of Bradford, UK (previously Director General and Chief Executive, Chemical and Biological Defence Establishment, Porton Down, Salisbury, UK).

JULIAN PERRY ROBINSON is a professorial fellow of SPRU–Science and Technology Research, University of Sussex, UK.

LAJOS RÓZSA is Senior Scientist in the Animal Ecology Research Group of the Hungarian Academy of Sciences and the Hungarian Natural History Museum.

NICHOLAS A. SIMS is Reader in International Relations at the London School of Economics and Political Science (University of London), UK.

MASAAKI SUGISHIMA is Professor of International Law at the Asahi University, Japan.

MARK WHEELIS is Senior Lecturer in Microbiology at the University of California, Davis.

SIMON M. WHITBY is a Research Fellow in the Department of Peace Studies, University of Bradford, UK

Index